ANIMAL DIVERSITY

third edition

Milton Fingerman

Tulane University

SAUNDERS COLLEGE PUBLISHING

Philadelphia □ New York □ Chicago
San Francisco □ Montreal □ Toronto
London □ Sydney □ Tokyo □ Mexico City
Rio de Janeiro □ Madrid

To Stephen and David Clay.

I hope they know why.

Address orders to:
383 Madison Avenue
New York, NY 10017
Address editorial correspondence to:
West Washington Square
Philadelphia, PA 19105

This book was set in Bodoni by Centennial Graphics.
The editors were Mike Brown, Carol Field and Sarah Fitz-Hugh.
The art director was Nancy E. J. Grossman.
The text and cover design was done by Phoenix Studio, Inc.
The production manager was Tom O'Connor.
New artwork was drawn by Eileen Rudnick.
The printer was Vail Ballou.

LIBRARY OF CONGRESS
CATALOG CARD NO. : 80-53918

Fingerman, Milton
Animal diversity.
Philadelphia, Pa. : Saunders College
256 p.
8101 801010

ANIMAL DIVERSITY ISBN 0-03-049611-X

1234 090 987654321

CBS COLLEGE PUBLISHING
Saunders College Publishing
Holt, Rinehart and Winston
The Dryden Press

Preface

The objective of this third edition of *Animal Diversity* remains the same as it was in the first two — to integrate a survey of the animal kingdom with a discussion of what I consider are the main pathways along which animals most likely evolved. I hope that the readers of this book will gain insight into the order that exists among the diversity in the animal kingdom. During the planning of the first edition of this book, I decided that an evolutionary approach to the subject of animal diversity would be the one most meaningful to the student. This approach still seems the best one, and has been retained. My views concerning the relationships of the various animal groups have been shaped to a large extent by the writings of Dr. Libbie H. Hyman. It is an obvious impossibility to present in a text of this length more than the principal routes of animal evolution. Details must, of necessity, be left to lengthy, more specialized texts. The reader should realize at the outset that the evolutionary sequences favored in this book represent an interpretation of data gathered from the work of numerous investigators and refer to a series of events that occurred during millions of years in the past. As more information is accumulated, the interpretation of the data may have to be modified. Such is the nature of science.

Numerous changes were made in the preparation of this third edition to assure that its use will continue to be a profitable experience for the student. Many portions of the text were rewritten, and expanded slightly, in an attempt to increase their clarity. Where the reasoning behind certain conclusions may not have been obvious to the student in the second edition, supporting evidence has been included in this third edition that explains why these conclusions were drawn. New tables and illustrations have been added and some of the old illustrations have been modified in an attempt to improve their usefulness. The sequence of several topics in Chapter 1 was changed to provide more clarity. The new order should make the transition from one topic to another flow more easily, thus enabling the reader to grasp the information more readily. The long chapter on chordates in the second edition has been divided into two chapters, one dealing with the lower chordates and the second with the vertebrates. This change substituted two smaller, more uniform chapters for one large one; in the second edition, the reader constantly had to jump mentally back and forth from lower to higher chordates, which was not as pleasant an experience as, it is hoped, reading two smaller, more coherent chapters will be. Breaking the material down into two chapters should provide for better comprehension of it. Important new information has been added, as in the discussion of human evolution in the last chapter.

This book is particularly suitable for use in General Biology and General Zoology courses. In addition, it can be helpful as a supplementary text to both students and instructors in more advanced courses such as Comparative Animal Physiology and Comparative Endocrinology. Students in the more advanced courses can become acquainted with or refresh their memories of the characteristics of the animals being discussed. Their instructors will gain the opportunity to devote their lectures fully to the appropriate subject matter and will not have to digress and explain what sorts of animals are being referred to when, for example, mention in class is made of polychaetes or cephalopods.

I am deeply indebted to all who have contributed to the preparation of this edition, including my students, who, after using the second edition, recommended some of the changes that were incorporated into this edition, those colleagues (especially Dr. Robert C. Cashner of the University of New Orleans, Dr. Willaim C. Grant, Jr., of Williams College, Dr.

Richard L. Turner of the Florida Institute of Technology, and Dr. Carol Turpen of Pennsylvania State University) who have constructively reviewed the manuscript, and last, but far from least, my wife, Dr. Sue W. Fingerman.

M.F.
New Orleans, Louisiana

Modern Biology Series | **Published Titles**

Delevoryas	**Plant Diversification,** *second edition*
Ebert-Sussex	**Interacting Systems in Development,** *second edition*
Fingerman	**Animal Diversity,** *third edition*
Griffin-Novick	**Animal Structure and Function,** *second edition*
Levine	**Genetics,** *second edition*
Novikoff-Holtzman	**Cells and Organelles,** *second edition*
Odum	**Ecology,** *second edition*
Ray	**The Living Plant,** *second edition*
Savage	**Evolution,** *third edition*

Contents

chapter **1**

The Concept of Animal Diversity

The diversity in kind among animals is immense. Approximately 1,250,000 kinds of living animals have been described already, and at least another million will probably be found as biologists continue their investigations, particularly when little-known parts of the earth are explored further. In spite of this prolific diversity in kind, all animals must be able to perform the same basic functions necessary for survival, such as feeding and respiration. However, because of this diversity, different animals will often perform the same function in different ways. This book will elucidate the diversity in form that is encountered among animals, for we cannot fully appreciate how an animal performs its various functions unless we comprehend its structure.

The diversity among animals has always been of keen interest, not only to biologists dealing with structure, but also to those concerned with function. Biologists concerned with the functional approach make use of this diversity by selecting for their experiments animals that possess a structure particularly advantageous for the study they wish to undertake. For example, much of our basic information concerning conduction of nerve impulses was gained through using giant axons from squids. With a microelectrode, an investigator is able to penetrate these giant axons, which are sometimes as large as 1 millimeter in diameter, with minimum difficulty.

Although the diversity among the members of the animal kingdom is great, at the cellular level we find many shared similarities, as in the structure of the *organelles* (the several specialized structures within or on a cell), the chemical nature of the genetic material (all cells use the same "four-letter alphabet" in their genetic codes), and the steps in carbohydrate metabolism. It is our concern here to examine the diversity among animals without losing sight of the basic similarities among their individual cells.

ANIMAL TRAITS The use of the term "animal" evokes a very basic question: How do we define "animal"? We can ask essentially the same question in other ways, such as "Why do we call a sponge an animal?" or "What are the differences between animals and plants?" The simplest classification scheme merely divides all living things into two kingdoms, one for plants and another for animals (Scheme I in Table 1-1). Everyone can tell that a dandelion is a plant and a tiger is an animal; the higher plants and animals present no

Table 1–1 ***Some Schemes Used to Divide the Living World into Kingdoms***

I	*II*	*III*	*IV*	*V*
Plantae	*Monera*	*Protista*	*Monera*	*Monera*
Bacteria	Bacteria	Bacteria	Bacteria	Bacteria
Algae	Blue-green algae	Blue-green algae	Blue-green algae	Blue-green algae
Slime molds	*Plantae*	Protozoans	*Protista*	*Protista*
True fungi	Chrysophytes	Slime molds	Chrysophytes	Chrysophytes
Bryophytes	Green algae	*Plantae*	Green algae	Protozoans
Tracheophytes	Brown algae	Chrysophytes	Brown algae	Slime molds
Animalia	Red algae	Green algae	Red algae	*Fungi*
Protozoans	Slime molds	Brown algae	Protozoans	True fungi
Metazoans	True fungi	Red algae	Slime molds	*Plantae*
	Bryophytes	True fungi	*Plantae*	Green algae
	Tracheophytes	Bryophytes	Bryophytes	Brown algae
	Animalia	Tracheophytes	Tracheophytes	Red algae
	Protozoans	*Animalia*	*Animalia*	Bryophytes
	Metazoans	Metazoans	Metazoans	Tracheophytes
				Animalia
				Metazoans

problem. However, when we deal with some of the less highly evolved organisms, particularly the unicellular organisms, even the simplest classification is often difficult to decide, because these organisms exhibit not only characteristics that have traditionally been attributed to plants alone, but also some that have been thought to belong only to animals. A number of these organisms have been classified as plants by some biologists, as animals by others, while yet another group of biologists, in an effort to solve the dilemma, has established new kingdoms into which these borderline organisms have been placed. The traditional method to determine whether an organism might fit into the animal kingdom or the plant kingdom has been to base the decision on whether animal or plant characteristics predominate. It must here be emphasized that there is no one decisive trait; but there are some general criteria that have been used. Animals ordinarily store carbohydrates in the form of glycogen; plants, in the form of starch. Animals usually require complex, synthesized foods that are used for growth or metabolized for energy, whereas plants usually use simple molecules and the energy of sunlight to synthesize complex molecules. Animal cells usually do not have cell walls, whereas plant cells usually have rigid walls. Animals are usually able to move about, whereas plants usually cannot, and animals show more rapid responses to stimuli than do plants.

The traditional approach of treating the unicellular animals, called *protozoans*, as one of the major groups in the animal kingdom has been adopted herein. The rest of the animals are multicellular organisms, called *metazoans*. The number of kingdoms into which living organisms should be divided has been debated for quite a while; from two to five kingdoms have been proposed. Table 1-1 presents the more commonly encountered kingdom classifications. The simplest opinion, that all living things belong to either of two kingdoms, plants, Plantae, or animals, Animalia (Scheme I in Table 1-1), also seems to be the oldest. However, it is generally accepted now that organisms should be divided into at least three kingdoms. The bacteria and blue-green algae are indeed fundamentally different from all other organisms because they are *prokaryotic* organisms (that is, they lack a membrane-bound nucleus), whereas all other organisms are *eukaryotic*. (They possess a membrane-bound nucleus.) The prokaryotes themselves constitute a kingdom that most biologists call the Monera.

As Table 1-1 shows, all the unicellular organisms that appear to be animals, such as the amoeba, have often been combined with other unicellular and various problematic multicellular organisms into a kingdom called the Protista. However, as Schemes III, IV, and V of Table 1-1 demonstrate, there is not much agreement among authors as to what organisms belong in the kingdom Protista with the protozoans. Setting up the kingdom Protista seems to create more problems than it solves. Organisms such as the brown algae are sometimes lumped (as in

Scheme IV) with the protozoans; brown algae seem less closely related to the protozoans, particularly the more highly evolved protozoans, such as the ciliates (Chapter 2), than do protozoans and sponges. These schemes of dividing organisms into kingdoms are attempts by biologists to help us better understand the relationships and diversity among living organisms. There will probably always be debate about the best way to classify organisms into kingdoms, but such debate is welcome because it stimulates us to study the organisms and to think about their relationships to each other. The tradition of including the protozoans in the animal kingdom has been upheld in this book, not only because of their "animal-like" characteristics (Chapter 2) but also because we wish to emphasize the evolutionary origin of the multicellular animals from them. Considering the protozoans animals seems reasonable at this time; it does not seem reasonable to conclude that an organism, such as an amoeba, is not an animal merely because it consists of a single cell.

SPECIATION

Each kind of organism belongs to a single species, and the individuals of one species are related more closely to one another than to the individuals of any other species. *Speciation* is the process by which new species evolve. Species are not artificial groupings established by humans, but are a definite reality in nature. By definition, *a species is a series of populations that are capable, in nature, of interbreeding with one another to yield fertile offspring, but are ordinarily unable to interbreed with other populations.* Different species usually cannot mate and produce fertile offspring, and consequently they exchange genes with other species rarely or not at all. This definition of a species is, of course, applicable only to organisms that reproduce sexually; for organisms that reproduce only asexually, the determination of whether an investigator has perhaps discovered a previously unknown, hence unnamed, species can be made only on the basis of anatomical, physiological, behavioral, and other comparisons.

Geographic or *allopatric speciation* appears to have been the major pathway for the multiplication of species (Fig. 1-1). This type of speciation can get underway when a single population becomes divided into two or more populations that are geographically isolated from each other by some external barrier, such as a river or mountain range. This barrier would prevent interbreeding among the populations. These separated habitats would almost certainly be different from each other in one way or another. Absence of the opportunity to interbreed, or *hybridize,* is an essential element of speciation. With the passage of sufficient time, the characteristics of the populations will most likely diverge from each other because of their different environments, each population taking a different evolutionary path. Ultimately the popula-

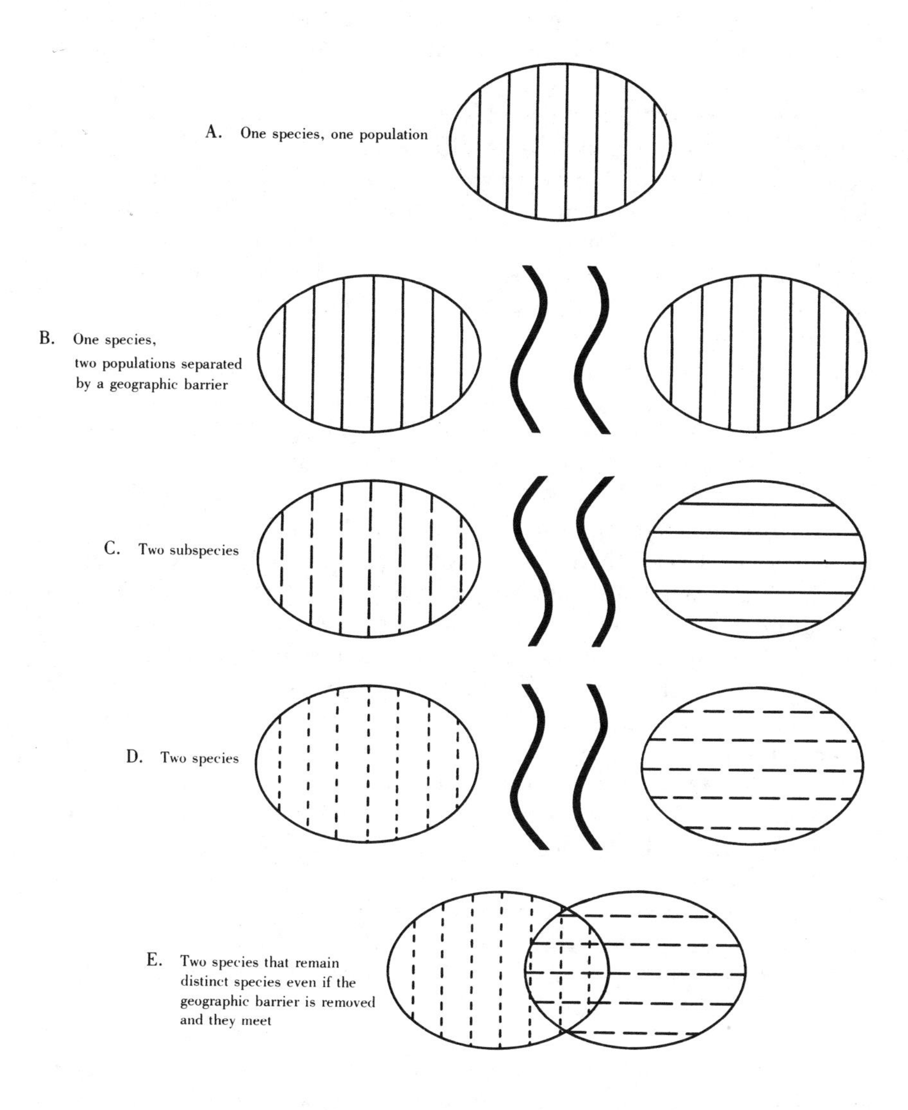

Fig. 1-1 *Speciation as a result of geographic isolation. (A), single parent population; (B), a geographic barrier is formed that divides the population into two portions that are isolated from each other in different environments; (C), with the passage of time the two populations accumulate enough differences to become distinct subspecies; (D), with the passage of still more time the subspecies evolve further into two distinct species; (E), no interbreeding will occur even if the geographic barrier disappears and the species meet.*

tions will become incapable of interbreeding if they should happen to come back together. When the populations are no longer capable of interbreeding, they are no longer a single species. Isolated populations of a single species that have accumulated different heritable characteristics but have not yet evolved into different species are known as *subspecies*. If the geographic barrier were removed before these subspecies could become separate species, when the members of the two populations met they would be free to interbreed with each other; thus they would still be a single species. Two populations of tuft-eared squirrels that inhabit the rims of the Grand Canyon are frequently cited as illustrations of allopatric speciation. The Abert squirrel on the southern rim and the Kaibab squirrel on the northern rim are considered separate species by some biologists, whereas others consider them only well-differentiated subspecies, because some hybridization is possible in the laboratory. No matter which view is correct, these two populations doubtless evolved from a single species that became split as the deep canyon formed, creating a geographic barrier that prevented interbreeding between them; as a result, they evolved divergently until they have at least approached being two distinct species. If the two populations are not now separate species, they are very close to becoming so. Most biologists agree that geographic separation was the initiating factor in the evolution of the vast majority of animal species.

It is also possible for speciation to occur within a single population without any geographic barrier; this is *sympatric speciation*. Although it seems to have occurred among some plants and insects, its occurrence in insects has not been demonstrated with certainty. If it has occurred in the animal kingdom, it has probably been a rare event at best. Sympatric speciation could occur, for example, if (a) mutations (changes in the hereditary material) caused a portion of the population to become unable to interbreed with the rest or (b) there were a sudden doubling of the chromosome number (polyploidy) in some members of the population, immediately isolating them reproductively from the rest of the population.

Two or more species that occupy the same area or have overlapping ranges (that is, are *sympatric*) will ordinarily be prevented from interbreeding (thereby their gene pools are kept separate and they are maintained as distinct species) because of *intrinsic isolating mechanisms*, biological characteristics that prevent, or at least minimize, successful interbreeding among species that coexist in the same place. They are called "intrinsic" to distinguish them from extrinsic or geographic barriers that physically isolate populations. When two populations of the same species are geographically isolated, this extrinsic isolation provides the opportunity for speciation to occur. But after speciation has occurred, the intrinsic isolating mechanisms, which were evolving and accumulating in the isolated populations while speciation was going on, will help to maintain the newly formed, closely related species distinct from one another, should they happen to come to occupy the same territory.

Intrinsic isolating mechanisms can be divided into two categories: premating and postmating. Premating intrinsic isolating mechanisms are ecological, behavioral, seasonal, structural, and physiological barriers to reproduction. When there is ecological isolation, closely related species inhabiting the same geographic area remain distinct from one another as separate species because each species occupies, or at least breeds, in a different habitat within its common range. Behavioral isolation, for example, can occur if each species has its own courtship ritual. Seasonal isolation occurs when species have different breeding seasons. Structural isolation occurs when the copulatory organs of each species have evolved in such a way that these organs are physically incompatible, thereby precluding successful mating. Physiological isolation can be due to chemical differences in gametes or seminal fluid that hinder fertilization.

By contrast, postmating intrinsic isolating mechanisms are factors that prevent a hybrid offspring from reproducing. For example, the hybrid may die before reaching sexual maturity and consequently not reproduce *(hybrid inviability);* or, if such a hybrid survives, it may be sterile *(hybrid sterility).* The mule, the result of a cross between a horse and donkey, is a classical example of hybrid sterility.

Even among closely related sympatric species, however, interbreeding is rare, not only because of these isolating mechanisms, but also because they are often reinforced by *character displacement,* the tendency of closely related sympatric species to diverge (grow increasingly unlike) rapidly. In the process, not only are the chances for interbreeding reduced, but also competition for food and habitats is minimized. It is to the mutual advantage of sympatric species to be less alike. Furthermore, when closely related species have overlapping habitats, the tendency is for them to differ most in the overlapping region. That is, they are more dissimilar where they overlap than where they are geographically separate, evidence that character displacement has occurred.

For centuries, humankind had been impressed by the great number of species of organisms living on earth. Until 1858, almost everyone accepted the concept that all the species of organisms then living had been placed on the earth at the time it was formed and had remained unchanged ever since. In 1858, Charles Darwin and Alfred Russel Wallace changed that concept with evidence that all living organisms are modified descendants of previously living forms. They also proposed the theory of natural selection, which provides an explanation of how a species can change with the passage of time, that is, how it evolves. *Evolution* is a change in the characteristics of a species from generation to generation, and how it comes about is best explained by the theory of natural selection.

When the theory of natural selection was first proposed, it was heavily criticized by some individuals because of its implications that humanity has evolved from a lower form. It has grown in acceptance through the years, however, and is now almost universally accepted. Darwin and Wallace conceived the theory of natural selection independently, but Darwin accumulated significantly more evidence in favor of the theory than did Wallace, and is consequently given more credit.

The theory is based on three observations from which two deductions were drawn. The first two observations—(a) that populations have the capability to increase in numbers at an exponential rate, and (b) that in spite of this capability, they tend to maintain a relatively constant number of individuals—allow the deduction that the potential number of individuals in a population is not realized, and therefore there must be a struggle for survival among the members. The third observation was (c) that the members of a population are not all alike because in every generation new, heritable variation arises. The deduction followed that those individuals bearing variations favorable for survival in their particular environments have a competitive advantage, and will survive and pass this combination of favorable traits to their offspring. With successive generations, those traits favorable for survival will increase within the population, while those that are unfavorable will decrease. The theory of natural selection revealed that a species is not a static assemblage, but a dynamic one instead, inexorably changing with time, with natural selection being the guiding force of evolution. It was Darwin who actually gave the name "natural selection" to this process by which individuals interact with the environment in such a way that those having traits favorable to survival will reproduce. In contrast, what he called "artificial selection" is the process that breeders of crops and domestic animals practice, in which the breeder selects for breeding only those specimens that have the desired traits. Another way of expressing the second deduction is that natural selection will favor (for survival and reproduction) those organisms that are well-adapted to their environments. The well-adapted organisms compete more successfully for food and mates, are better able to escape from predators, and can withstand the rigors of a harsh climate.

A heritable characteristic, such as having lungs for air-breathing, which better enables an organism to survive and reproduce in its habitat, is called an *adaptation*, and any such characteristic is said to have *adaptive significance*. Adaptations may be physiological, behavioral, or structural. In the case of geographically isolated populations of a species, different adaptations will begin to accumulate in each population because it is highly unlikely that the two populations would be exposed to the same set of environmental factors. Hence, the populations will become better fitted to their respective habitats, and in the process, will diverge more and more as they accumulate these adaptations, natural selection favoring the well-adapted individuals. Eventually, as explained earlier, these populations will diverge from each other so much that they will become separate species.

Darwin and Wallace were unable to explain how variation arises within a population and how this variation is transmitted to the next generation. Gregor Mendel first published the basic laws of inheritance in 1866, although, unfortunately, the significance of his principles was not recognized until 1900. Subsequent study by geneticists revealed the natural, spontaneous occurrence of mutations. These changes in the hereditary material provide the potential for the appearance of variation in future generations. Thus, the theories of natural selection and genetics not only provide the explanation of how a single species can change with the passage of time, as its members continue to become adapted to their changing environment, but they also explain how new species can evolve from populations of an older species that have become geographically isolated from each other. These isolated populations ultimately will, through natural selection, accumulate enough heritable differences to evolve into new species. Although the changes in the favorable heritable characteristics from one generation to the next may be small, given enough time, major evolutionary changes can be produced.

When a stock has split into two, and the two have become more and more dissimilar with the passage of time, the organisms are said to have undergone *divergent evolution*, and we say that *divergence* has occured. In contrast, when two groups of organisms are not closely related, but in becoming adapted to similar habitats have come to resemble each other more and more as they have developed similar adaptations, we refer to them as having undergone *convergent evolution*, and we say that *convergence* has occurred. There is a third evolutionary pattern, called *parallelism*, in which organisms are said to have undergone *parallel evolution* after a single stock became split into two and the two then evolved in much the same way for a long period of time. Presumably the two stocks were exposed to similar environments where natural selection would favor the same traits in both populations, similar traits having been selected to solve similar problems.

The chief rival to the theory of natural selection is the concept of the inheritance of acquired characteristics. This concept was proposed in 1809 by Jean Baptiste de Lamarck. It holds that changes acquired by an organism during its lifetime, whether in the size or structure of a body part or in some ability, are passed on to its offspring. Lamarck used the giraffe as an example that he thought supported his concept. He claimed that giraffes have long necks because for generations they stretched their necks to reach leaves high up in trees. As each generation stretched its neck a bit further, this newly acquired characteristic would be passed on to its offspring. So, each generation had a slightly longer neck. However, there is no evidence to support Lamarck's concept, but many people, many animal breeders in particular, still adhere to it.

The theory of natural selection would explain the long neck of giraffes by proposing that in the past, giraffes had short necks but that there was variation in the length of the necks, some being somewhat longer than others. If the supply of food was limited, those giraffes with the slightly longer necks would have a better chance of feeding on the higher leaves, hence of surviving, reproducing, and passing on the trait of a longer neck to their offspring. If, with the passage of each generation, there was an advantage in being able to feed even higher on the trees, the necks of the population would eventually become longer and longer as a result of this natural selection process. Mutations are reflected in alterations of the organism, such as the elongated neck of the giraffe, but a change in a body part acquired during the lifetime of an individual does not produce a mutation that would result in the appearance of this change in the next generation. The hereditary material directs the formation of the body but does not accept alterations back from the body.

Mutations occur at random; pure chance determines in which individual a mutation will occur. According to the theory of natural selection, individuals bearing such a randomly arising mutation that provides a trait with adaptive significance will survive, reproduce, and pass this newly acquired mutation on to their offspring. The concept of the inheritance of acquired characteristics implies the opposite view, namely that a newly acquired characteristic will somehow cause a mutation that would enable this new characteristic to be handed down to the next generation.

ANIMAL CLASSIFICATION AND THE DENDROGRAM

The main purpose of classifying animals is to show the most probable evolutionary relationships of the different species to one another. Classification also enables us to retrieve information about each species without difficulty and to determine in what groups the number of species is continuing to increase—how and why. The current system of nomenclature is based on a method devised in 1753 by the Swedish biologist, Carolus Linnaeus. Until that time, an organism's common name was generally used, and this practice led to considerable confusion among biologists of different nationalities. In an effort to end the confusion, Linnaeus gave each species known at that time a latinized name consisting of two words: a generic name (the *genus;* plural, *genera*) and a specific name (the *specific epithet*). The specific name is ordinarily adjectival, describing the species to some degree. Latin was chosen because, at that time, it was the international language of science. The same generic name, always capitalized, is given to a group of closely related species; the specific name, not capitalized, is given only to members of a single species within a genus. Both names are italicized. For example, all domestic dogs are *Canis familiaris*, but the gray wolf, also called the timber wolf, is *Canis lupus*. (The generic name is common to both, and the specific names differ.) There is, however, a certain amount of subjectivity in classifying animals above the species level. For example, one expert may consider several species of crayfishes sufficiently similar to warrant combining them into a single genus, whereas another might consider the differences among the species large enough to justify their being split into two or more genera. When the generic name, but not the specific name, is known (which could occur when a biologist has collected a specimen but has not had the opportunity to identify it fully) this lack of information can be indicated by the use of "sp.," as, for example, *Drosophila* sp.

INTERNATIONAL CODE OF ZOOLOGICAL NOMENCLATURE

The rules for the scientific naming of animals are contained in the *International Code of Zoological Nomenclature*. This document aims to ensure uniform practice in the scientific naming of animals. Specifically, these rules assure that only one scientific name is given to a single species and that two species are not given the same scientific name. Additional groupings showing broader relationships, above the generic level, will be considered after the following discussion of homology.

Homology indicates a relationship between features in two or more animals that can be traced back to the same feature in a common ancestor. Homologous structures develop from similar embryonic rudiments. We identify homologous structures in adults according to the similarity of their location on or in the organism, similarity of shape, and number of component parts. For example, the arm of a man, the wing of a bird, and the wing of a bat all develop in a similar manner—they have similar bone structure, musculature, and innervation, and are all homologous (Fig. 1-2). We would conclude from the study of these appendages that the animals bearing them are fairly closely related. On the other hand, if we compare the wing of a bird or a bat with the wing of a butterfly, we find that although these wings are used for the same purpose, namely flight, the wing of a bird or of a bat is structurally and developmentally unrelated to the wing of an insect, and these wings are not at all homologous. Structures that are not homologous but nevertheless have a functional similarity are termed *analogous*, that is, they have a similar function but different embryonic development and dissimilar structural plans. In the classificatory scheme used in biology today, different species are placed in a series of increasingly inclusive categories or *taxa* (sing., *taxon*) as the number of homologous structures they share decreases.

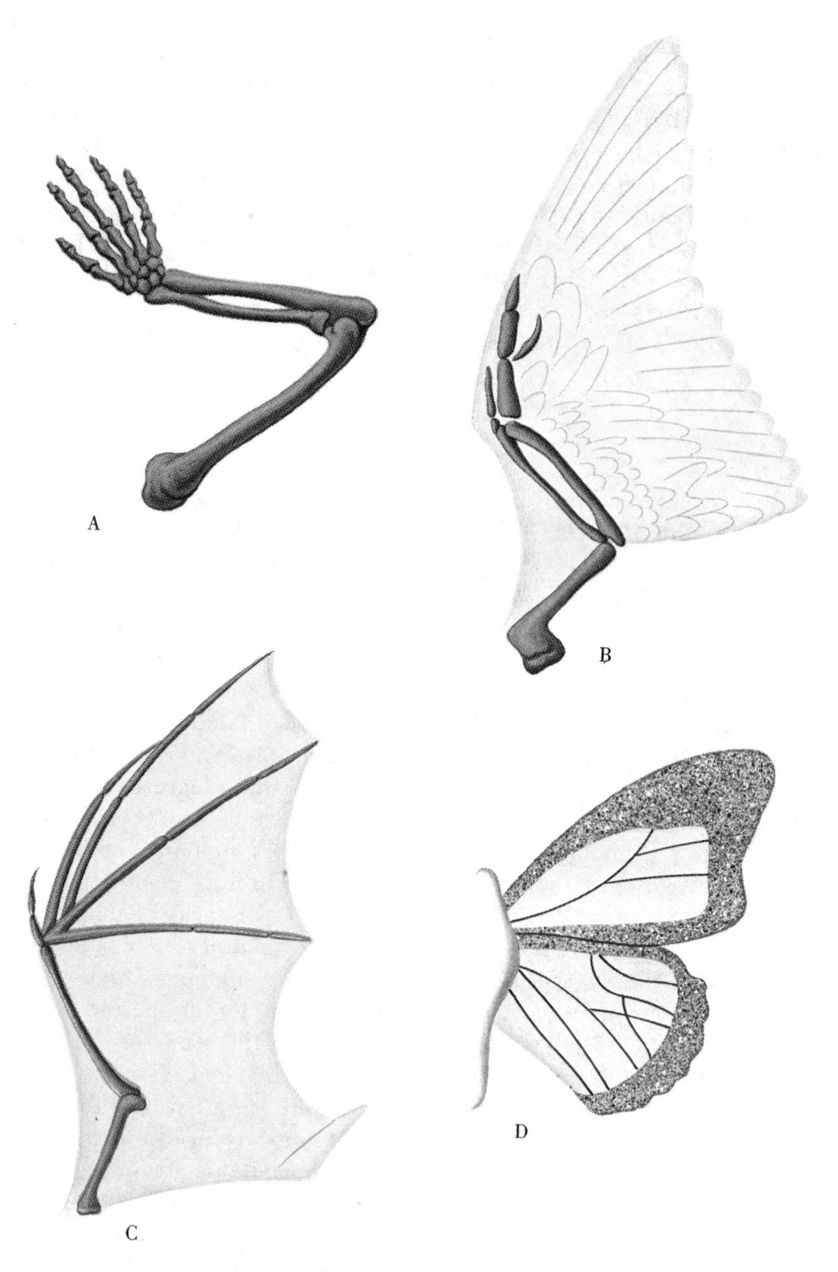

Fig. 1-2 *Homology and analogy. The arm of a man (A), wing of a bird (B), and wing of a bat (C) are homologous structures, having the same basic internal skeletal structure. The insect wing (D), which develops without an internal skeleton, is analogous to the wings of birds and bats but not to the arm of man. (After Cockrum and McCauley,* Zoology, *W. B. Saunders Company, 1965.)*

Let us now turn to the major categories above the generic level. The first one is the *family.* Genera that have a large number of homologous structures in common would be placed in the same family. Different species of any one genus in the family would, however, share more homologous structures than would the genera that compose the family. *Families* are combined into an *order,* orders into a *class*, and classes into a *phylum*. Finally, all the phyla of animals compose the *animal kingdom*. The kingdom is the most inclusive category of all. The number of phyla that the animal kingdom should be divided into is not fully agreed upon; some experts feel there is justification for giving certain groups phylum status, whereas others feel that these same groups deserve only class rank in a phylum. Here we shall consider the animal kingdom divided into 32 phyla (Table 1-2). Included in the 19 phyla discussed in Chapters 2 through 13 are about 99.8 percent of the known living species of animals. Examples of the system of classification, using the major categories, are given in Table 1-3 for the medicinal leech, the American lobster, and the human being.

A phylum represents a broad grouping of related animals that presumably have a common ancestry and are characterized by having similar overall body plans. In phyla with a large number of species, additional categories such as a suborder or a superclass have been created by splitting the major categories in order to show more precisely the relationships of the species. A suborder would consist of families that are more closely related to each other than to the other families in the order. Similarly, a superclass would be a group of classes more closely related to each other than to the rest of the classes in the phylum.

The hypothetical evolutionary relationships among the phyla can be illustrated by means of a diagram called a *dendrogram*, or *phylogenetic tree*. The dendrogram shown here (Fig. 1-3) incorporates the views of L. H. Hyman. Her views on the phylogeny of the animal kingdom seem to be those most widely accepted. Dendrograms that differ strikingly from the one shown in Figure 1-3 have been devised by others, but they have not received much support from biologists. The evidence in favor of the dendrogram shown here will be presented in the following chapters. Also to be discussed in subsequent chapters are many of the ideas embodied in the less widely accepted dendrograms.

Each dendrogram represents the particular author's best guess of what has occurred through millions of years in the past. As such, it is only a working model, which is revised as new discoveries are made. New data concerning the relationships between major animal groups continue to show up with surprising frequency, and with further exploration of the deeper regions of the oceans and more genetic, physiological, behavioral, cytological, and biochemical studies, we have great hope that some of the unanswered questions of the origins and relationships of the phyla will be resolved.

Table 1–2 The Generally Recognized Phyla of Animals

Scientific name	*Common name*	*Description*
Protozoa	Protozoans	See Chapter 2
Mesozoa	Mesozoans	Minute, ciliated, wormlike parasites. Kinship unknown; may be an offshoot from the early multicellular animals or degenerate flatworms; 50 species.
Porifera	Sponges	See Chapter 3
Cnidaria	Cnidarians or coelenterates	See Chapter 4
Ctenophora	Comb jellies or sea walnuts	See Chapter 4
PROTOSTOMIA		
Acoelomates		
Platyhelminthes	Flatworms	See Chapter 5
Gnathostomulida	Gnathostomulids	Minute marine worms. Each cell of the body surface has a single cilium, a unique feature of this phylum. Most recently discovered phylum, first description published in 1956; 90 species.
Nemertinea	Proboscis worms or ribbon worms	See Chapter 5
Pseudocoelomates		
Acanthocephala	Spiny-headed worms	See Chapter 6
Entoprocta	Entoprocts	Stalked, attached animals that are mostly colonial. Have a U-shaped digestive tract with the mouth and anus within a ring of ciliated tentacles; 60 species.
Rotifera	Wheel animalcules	See Chapter 6
Gastrotricha	Gastrotrichs	Microscopic animals that resemble rotifers, but lack the rotifer corona, a ciliated crown. However, gastrotrichs do have cilia elsewhere on their external surface; 400 species.

Table 1–2 (Continued)

Scientific name	Common name	Description
Kinorhyncha	Kinorhynchs	Small (less than 1 mm) spiny marine animals that resemble rotifers except that they have no cilia anywhere on the outer surface; 100 species.
Nematoda	Roundworms	See Chapter 6
Nematomorpha	Horsehair worms	Slender, elongated worms. Juveniles are parasitic but adults are free-living; 250 species.
Schizocoelomates		
Sipuncula	Peanut worms	Cylindrical, marine worms. Anterior portion of the worm narrower than posterior portion and can be retracted into it. Mouth, located at the anterior end of the worm, is surrounded by lobes or tentacles. Digestive tract is U-shaped; 330 species.
Priapulida	Priapulids	Cucumber-shaped, small (about 6 cm long) animals with an anterior proboscis and posterior trunk. The trunk is covered with small spines and tubercles; 8 species.
Echiura	Echiurans	Cylindrical marine worms. Body consists of a nonretractable proboscis and a trunk; 100 species.
Pogonophora	Beard worms	Deep-water, tube-dwelling, marine worms. Anteriorly, have long ciliated tentacles, hence the origin of the common name. No mouth or digestive tract; 100 species.
Annelida	Annelids	See Chapter 7
Mollusca	Mollusks	See Chapter 8
Tardigrada	Water bears	Minute, segmented animals with four pairs of legs that terminate in claws. Probably branched off the annelid-arthropod line; 350 species.
Onychophora	Onychophorans	See Chapter 9

Table 1–2 (Continued)

Scientific name	*Common name*	*Description*
Phoronida	Phoronids	A marine group of animals having a wormlike body with ciliated tentacles encircling the mouth. Posterior portion of animal somewhat wider than anterior part; 15 species.
Pentastomida	Tongue worms	Parasites in the nasal passages and lungs of vertebrates. Closely related to arthropods. Body has five anterior, short projections; 70 species.
Arthropoda	Arthropods	See Chapter 9
Bryozoa	Moss animals	See Chapter 10
Schizocoelomates and Enterocoelomates		
Brachiopoda	Lamp shells	See Chapter 10
DEUTEROSTOMIA		
Enterocoelomates		
Chaetognatha	Arrow worms	See Chapter 10
Echinodermata	Echinoderms	See Chapter 10
Hemichordata	Hemichordates	See Chapter 10
Chordata	Chordates	See Chapters 11, 12, 13

Note: Only the very minor phyla that are not discussed in a more or less detailed manner in the following chapters are described in this table.

Table 1–3 Classification of the Medicinal Leech, the American Lobster, and Man

	Medicinal Leech	*American Lobster*	*Man*
Kingdom	Animalia	Animalia	Animalia
Phylum	Annelida	Arthropoda	Chordata
Class	Hirudinea	Crustacea	Mammalia
Order	Gnathobdellida	Decapoda	Primates
Family	Hirudidae	Homaridae	Hominidae
Genus	*Hirudo*	*Homarus*	*Homo*
Specific epithet	*medicinalis*	*americanus*	*sapiens*

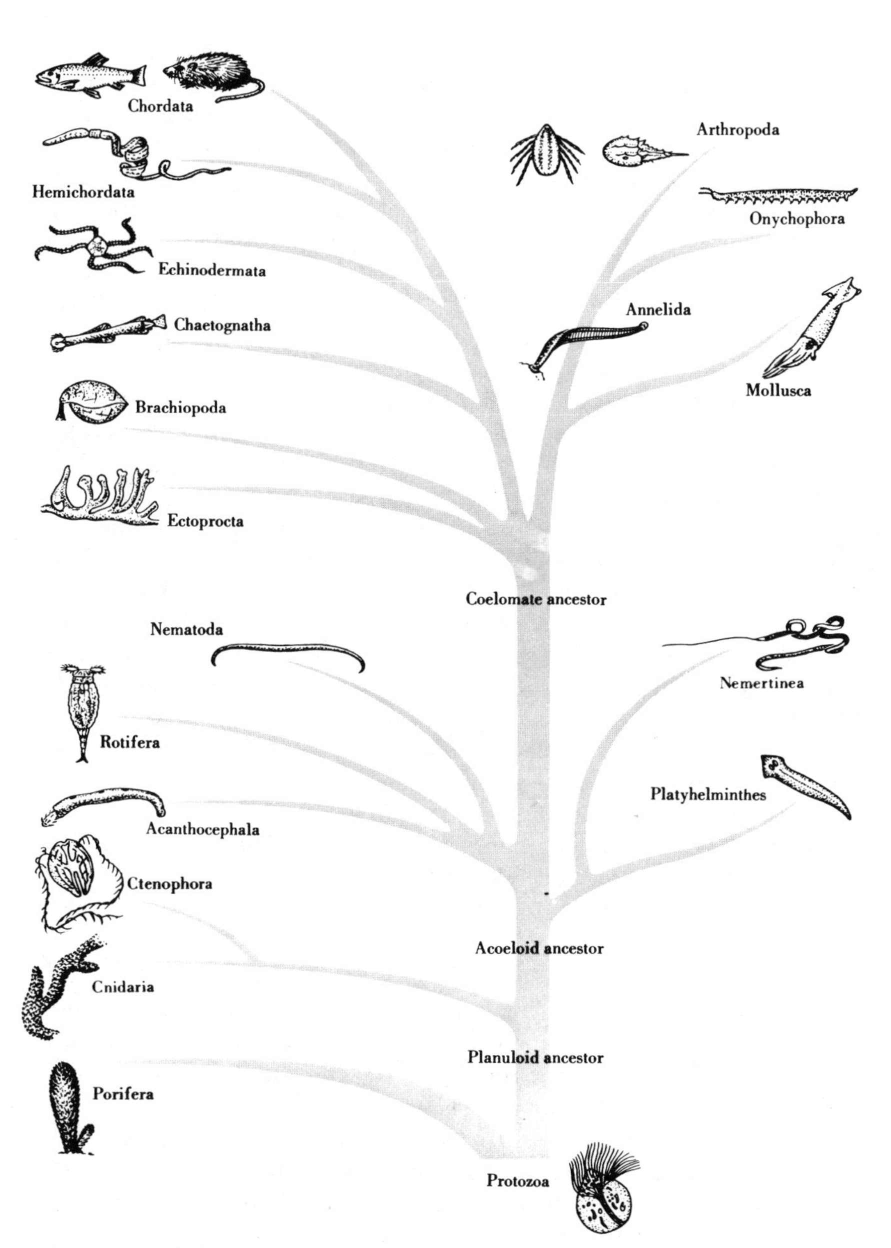

Fig. 1-3 *A dendrogram or phylogenetic tree showing what are considered in this book to be the most likely evolutionary relationships of the phyla discussed in the following chapters.*

In discussing animals, terms such as "lower" or "higher," "primitive" or "advanced," "unspecialized" or "specialized," and "unsuccessful" or "successful" are often used. The "higher" animals, such as the chordates, are those that have advanced farther along the main lines of evolution than have the "lower" ones, such as the sponges. "Primitive" and "advanced" are used in comparing species within a single group of animals. The "primitive" ones retain many of the characteristics of the ancestral stock from which they arose. In contrast, "advanced" species are those that have changed considerably from the ancestral stock, usually as a result of adaptation to a different environment or upon taking on a new mode of existence, such as shifting from being free-living to life as a parasite. "Unspecialized" and "specialized" are usually used to refer to parts of animals rather than to species. An "unspecialized" structure is theoretically capable of further modification and improvement through natural selection. On the other hand, a "specialized" structure has minimal evolutionary flexibility, that is, the structure has become extremely adapted for life in a particular habitat. The decision whether a group of animals is "unsuccessful" or "successful" is usually based on the number of species in the group, the total number of individuals in the group, its geographic range, and the diversity of its *ecological niches* or roles in the ecosystem. The *ecosystem* consists of all the organisms in an area and the physical environment in which they live. The successful groups are, of course, those that exist in large numbers (both in number of species and number of individuals), have a broad geographic distribution, and have a wide diversity of niches. Insects and placental mammals by these criteria are obviously "successful" animals, whereas echinoderms and amphibians are much less "successful." In another sense, however, all living animals are "successful" because in the game of life, survival itself is really the fundamental measure of success.

TEN MAJOR PHYLA There are ten major phyla in the animal kingdom. The following brief introduction to these ten phyla will be augmented by more extensive discussion in succeeding chapters. These major phyla contain about 99.2 percent of the known, living species of animals. In addition, nine of the minor phyla will be briefly discussed in subsequent chapters. These nine minor phyla deserve to be included either because of their phylogenetic significance or their fairly common occurrence in nature in numbers of individuals if not of species. *Phylogeny* is the evolutionary history of a particular group of organisms.

The phylum Protozoa, of which about 50,000 living species have been described, consists of the unicellular organisms that have traditionally been viewed as animals (Fig. 1-4). It will be recalled that the problem of what kingdom to place the protozoans in and the decision to

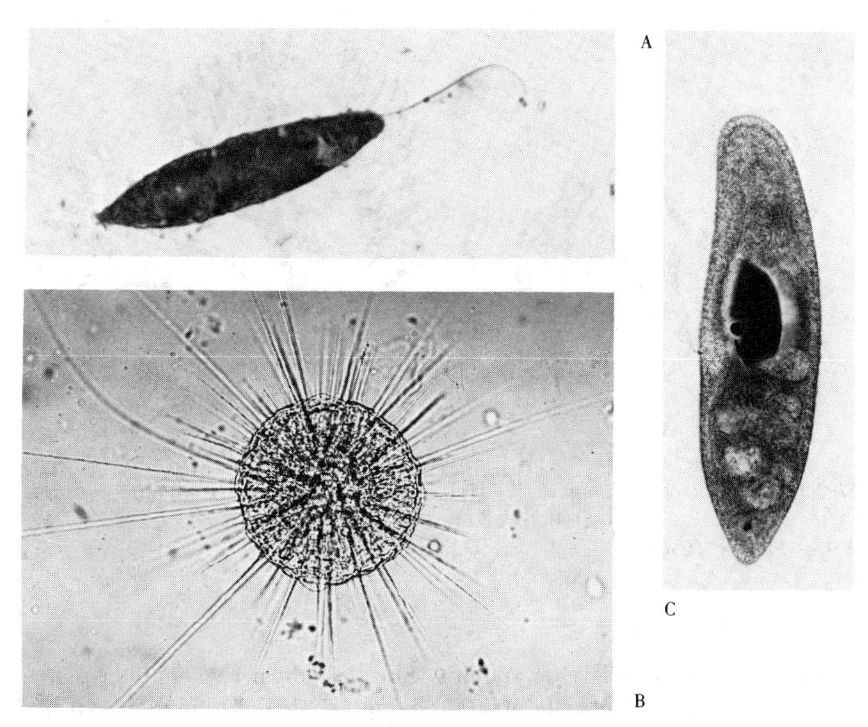

Fig. 1-4 *Selected protozoans. (A)* Euglena gracilis; *(B)* Actinosphaerium nucleofilum, *a heliozoan or sun animalcule; (C)* Paramecium caudatum. *(Courtesy of Carolina Biological Supply Company.)*

include them in the animal kingdom for this text were mentioned earlier in this chapter. The distribution of protozoans is worldwide, as they are found in the sea, in fresh water, and on land. In addition to the free-living species, some are parasitic.

The phylum Porifera, the sponges, is exclusively marine except for one family that inhabits fresh water (Fig. 1-5). These animals have many pores to allow a water current bearing food and oxygen to flow through the body, and about 4,500 living species are known.

The phylum Cnidaria is represented by about 10,000 aquatic (primarily marine) species and includes the hydras, jellyfishes, sea anemones, and corals (Fig. 1-6). One of the basic characteristics of the members of this phylum is that they have stinging cells.

The phylum Platyhelminthes consists of about 7,000 living species; the platyhelminths are commonly known as flatworms (Fig. 1-7), because they are dorsoventrally flattened. Included in the phylum are free-living forms, such as the planarians, and parasitic species, the flukes and tapeworms.

The phylum Nematoda comprises a number of unsegmented, wormlike organisms commonly referred to as roundworms (Fig. 1-8). There are already about 12,000 known species of nematodes, but L. H. Hyman suggested that these 12,000 species represent only a small frac-

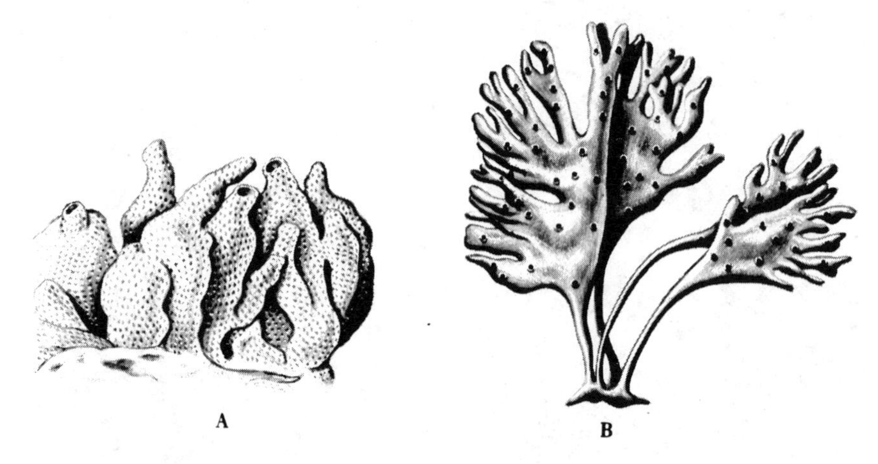

Fig. 1-5 *Selected sponges. (A)* Halichondria panicea; *(B)* Chalina oculata. *(After Burbanck, in* Selected Invertebrate Types, *F. A. Brown, Jr., ed., John Wiley & Sons, 1950).*

tion of the number of living species. She thought it reasonable to suppose that there are actually 500,000 living species of nematodes in the world. It is a ubiquitous phylum, and includes both free-living and parasitic species. Commonly found roundworms include the free-living vinegar eel and the parasitic pinworm, the nematode depicted in Fig. 1-8.

The phylum Annelida consists of about 7,000 species of segmented worms (Fig. 1-9). Included in this phylum are the earthworms, polychaetes, and leeches.

The phylum Mollusca is a diverse one of approximately 100,000 species (Fig. 1-10). Included are such animals as the chiton, snail, clam, squid, and octopus. The vast majority of mollusks are aquatic. Most mollusks have external shells, but in some species the shell may be absent (as in the case of the octopus), or reduced and internal (as in the squid).

The phylum Arthropoda is the largest phylum in number of known living species, comprising about 1,000,000, more than all the other phyla combined (Fig. 1-11). Of this total about 900,000 species are insects. Arthropod characteristics include an external skeleton and jointed appendages. These animals are found in habitats that range from mountain tops to the ocean depths and from the polar regions to the tropics. Among the representatives of this phylum, in addition to the insects, are the crustaceans (crabs, shrimps, crayfishes), centipedes, millipedes, spiders, ticks, and the horseshoe crab.

The phylum Echinodermata is a phylum of approximately 6,500 exclusively marine species of animals (Fig. 1-12). Representatives of this

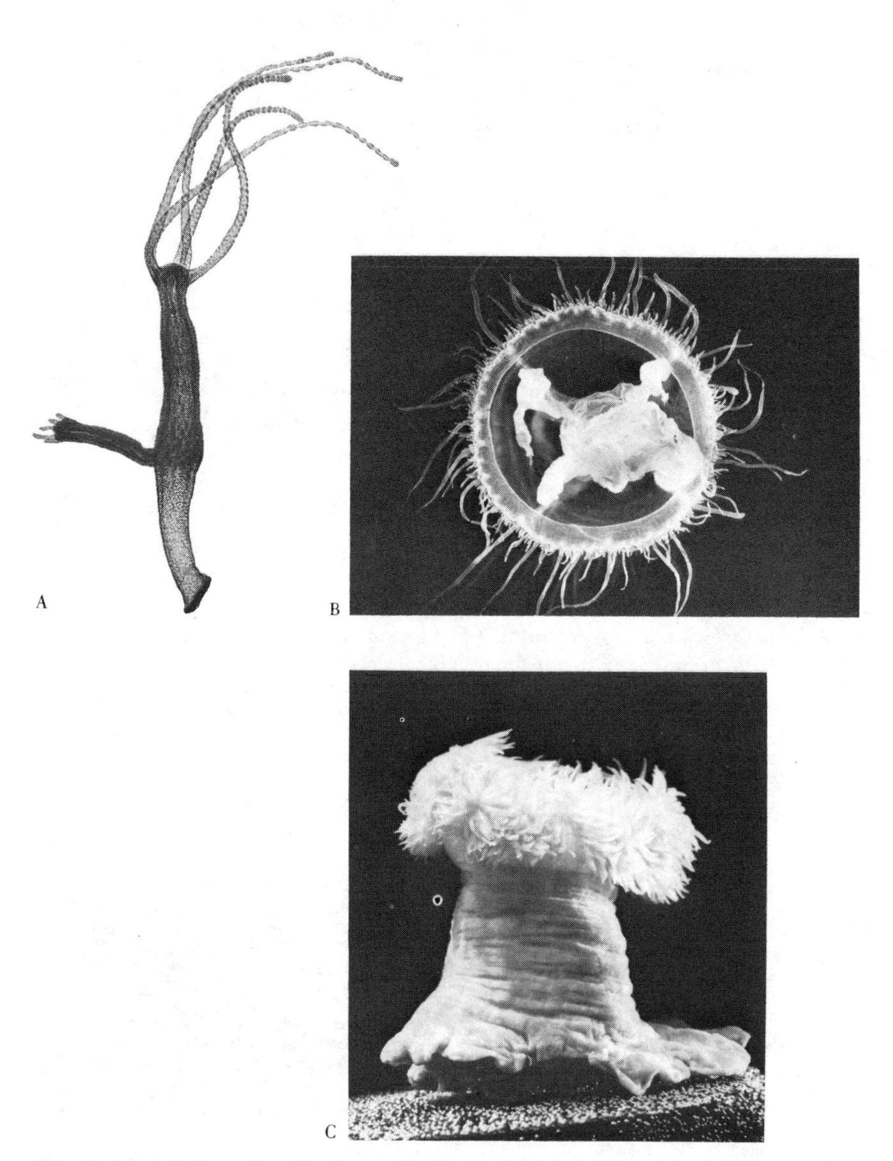

Fig. 1-6 *Selected cnidarians. (A)* Hydra littoralis *with a bud; (B) a jellyfish,* Craspedacusta sowerbyi; *(C) a sea anemone,* Metridium marginatum *(Courtesy of Carolina Biological Supply Company.)*

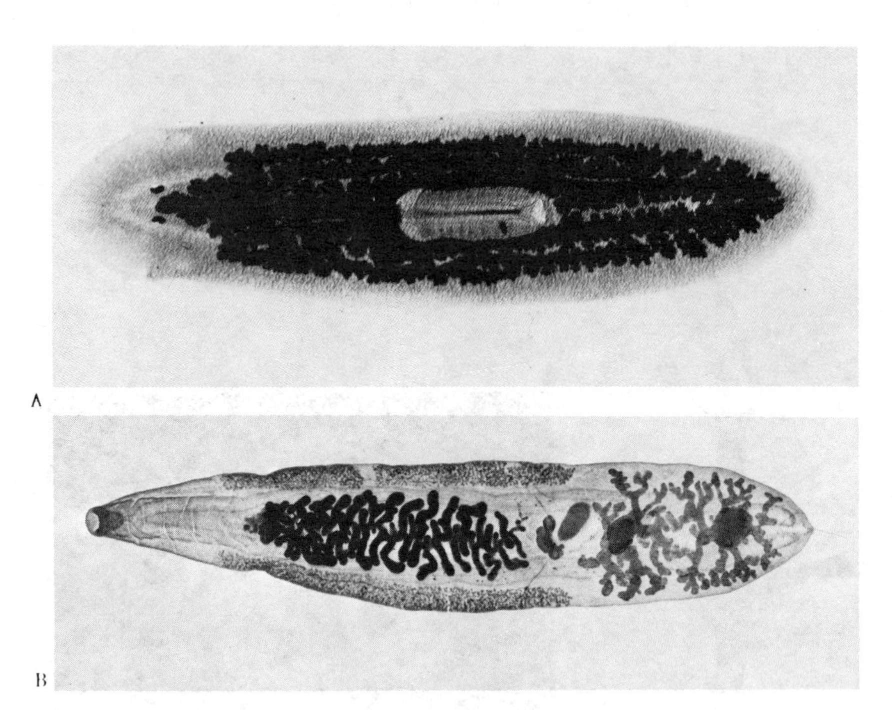

Fig. 1-7 *Selected platyhelminths. (A) A planarian,* Dugesia tigrina *(Courtesy of Ward's Natural Science Establishment, Inc.) (B) Oriental liver fluke,* Opisthorchis sinensis *(Courtesy of Carolina Biological Supply Company.)*

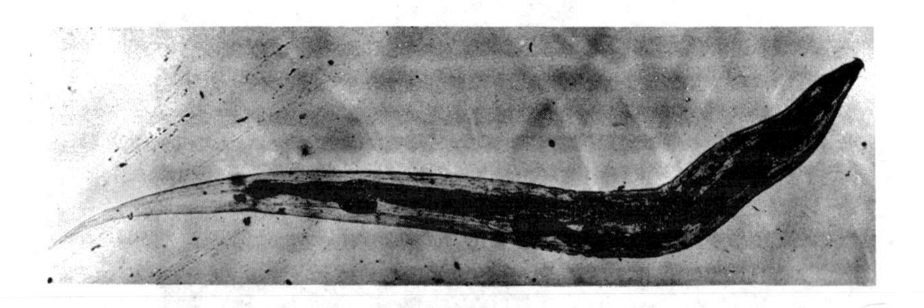

Fig. 1-8 *A nematode,* Enterobius vermicularis, *which is commonly known as the pinworm. It is a parasite in the intestine of man. (From Burnett and Eisner,* Animal Adaptation, *Holt, Rinehart and Winston, 1964.)*

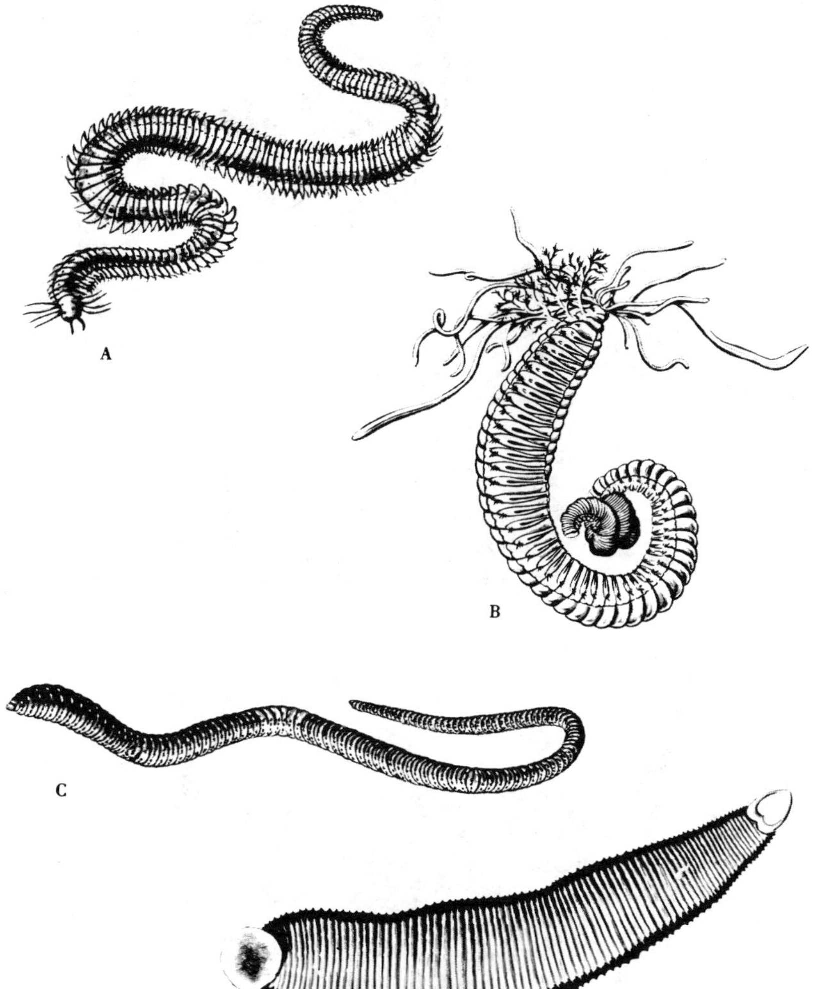

A

B

C

D

Fig. 1-9 *Selected annelids. (A) An errant polychaete,* Phyllodoce *sp.; (B) a sedentary polychaete,* Amphitrite ornata, *removed from the tube in which it dwells; (C) an earthworm,* Lumbricus terrestris; *(D) a leech,* Hirudo medicinalis. *[(A) Reprinted by permission of Dodd, Mead and Company, Inc. from* Biology in Action *by N. J. Berrill. Copyright © 1966 by Dodd, Mead and Company, Inc. (B) and (C) After Brown, in* Selected Invertebrate Types, *F. A. Brown, Jr., ed., John Wiley & Sons, 1950.]*

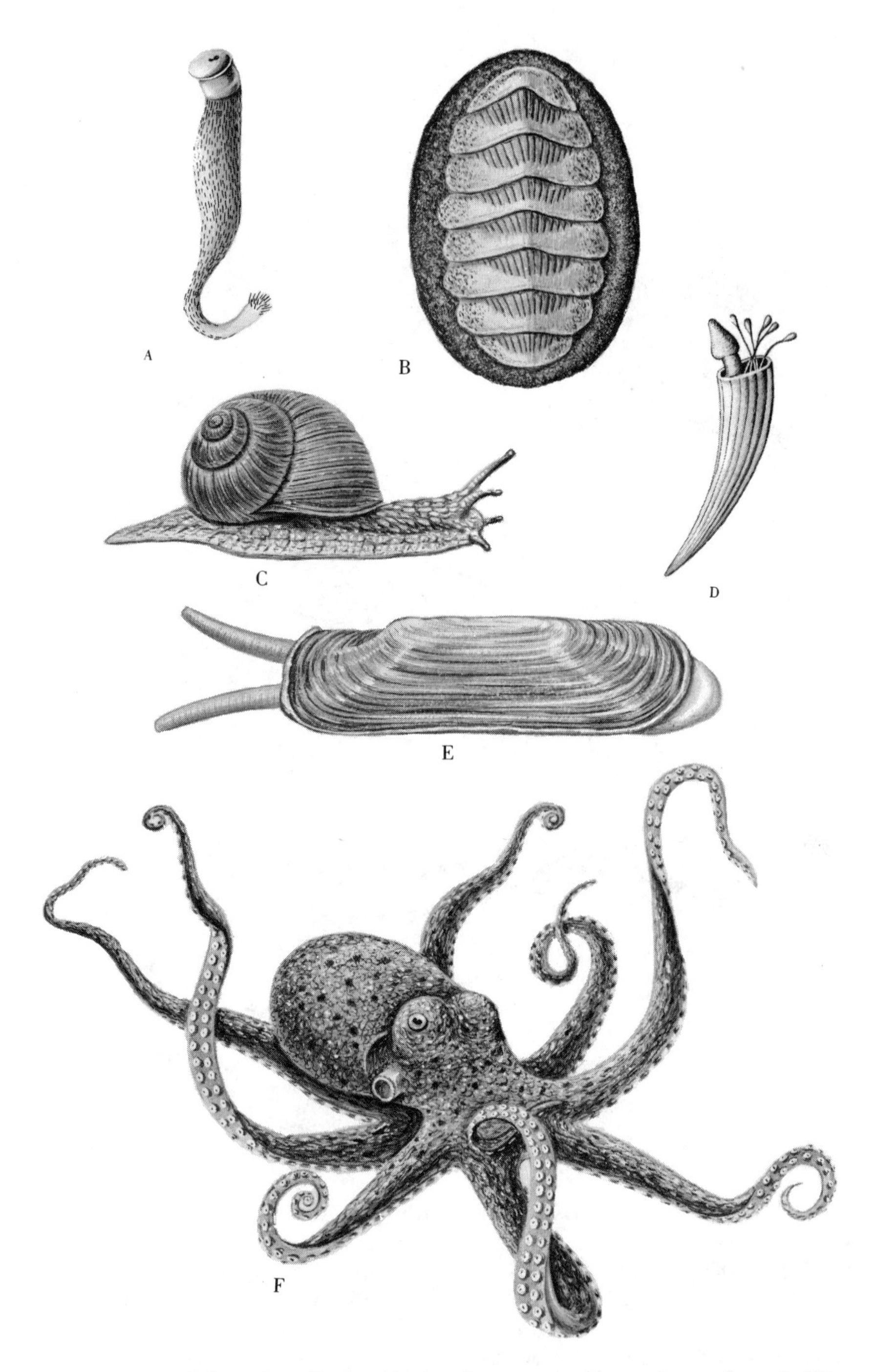

Fig. 1-10 *Selected mollusks. (A) A solenogaster,* Chaetoderma loveni; *(B) a chiton,* Chaetopleura apiculata; *(C) a snail,* Helix pomatia; *(D) a scaphopod,* Dentalium pretiosum; *(E) a clam,* Tagelus gibbus; *(F) an octopus,* Octopus *sp.* [*(A) After Hyman,* The Invertebrates, *Vol. VI. Copyright © 1967, McGraw-Hill Book Company. Used by permission.*]

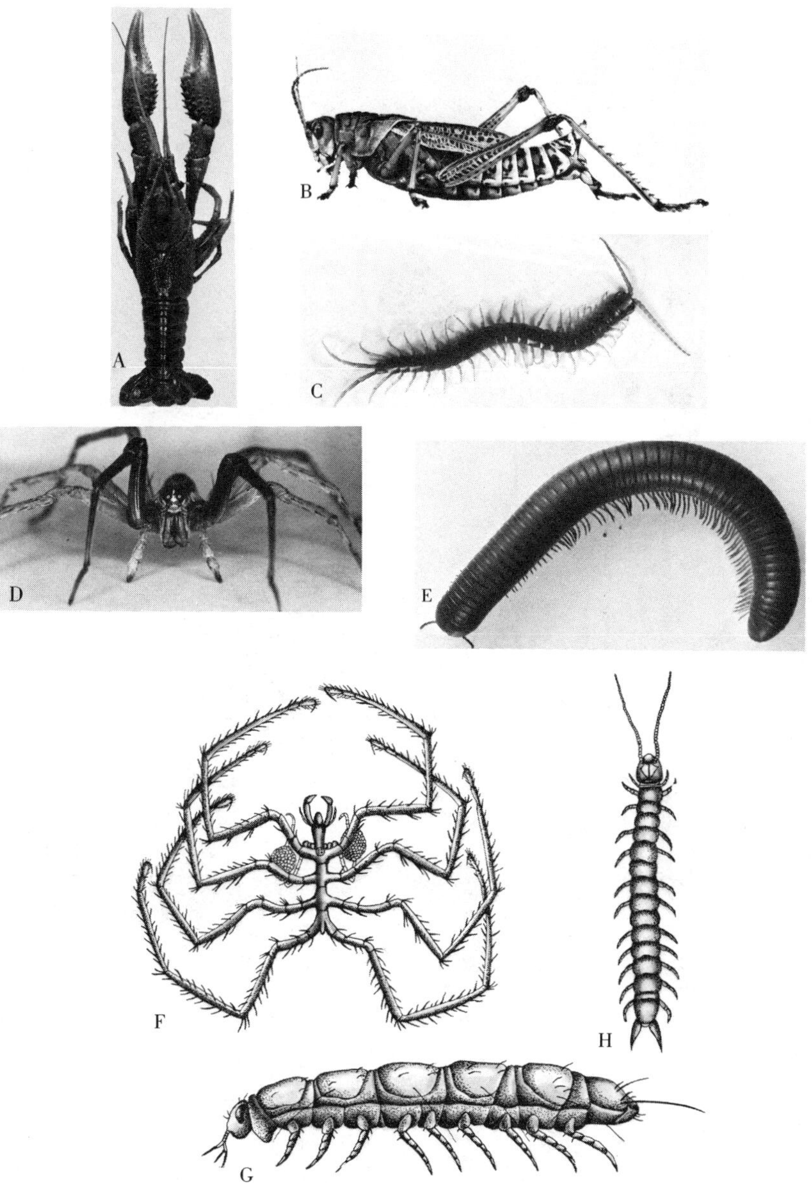

Fig. 1-11 *Selected arthropods. (A) A crayfish,* Procambarus clarki; *(B) a grasshopper,* Romalea microptera; *(C) a centipede,* Scolopendra *sp.; (D) a spider,* Lycosa rabida; *(E) a millipede,* Spirobolus *sp.; (F) a pycnogonid,* Nymphon rubrum; *(G) a pauropod,* Pauropus silvaticus; *(H) a symphylan,* Scutigerella immaculata. [*(A)–(C), (E) Courtesy of Carolina Biological Supply Company. (D) from J. S. Rovner,* Science, *vol. 152. Copyright 1960 by the American Association for the Advancement of Science. (F) After Sars from Fage, in* Traité de Zoologie. *Vol. VI, P.-P. Grassé, ed. Masson et Cie., 1949. (G) After Tiegs. From the* Quarterly Journal of Microscopial Science, *Vol. 88, 1947.* Used by permission. (H) *After Snodgrass.* A Textbook of Arthropod Anatomy. *Copyright © 1952 by Cornell University. Used by permission of the publisher, Cornell University Press.*]

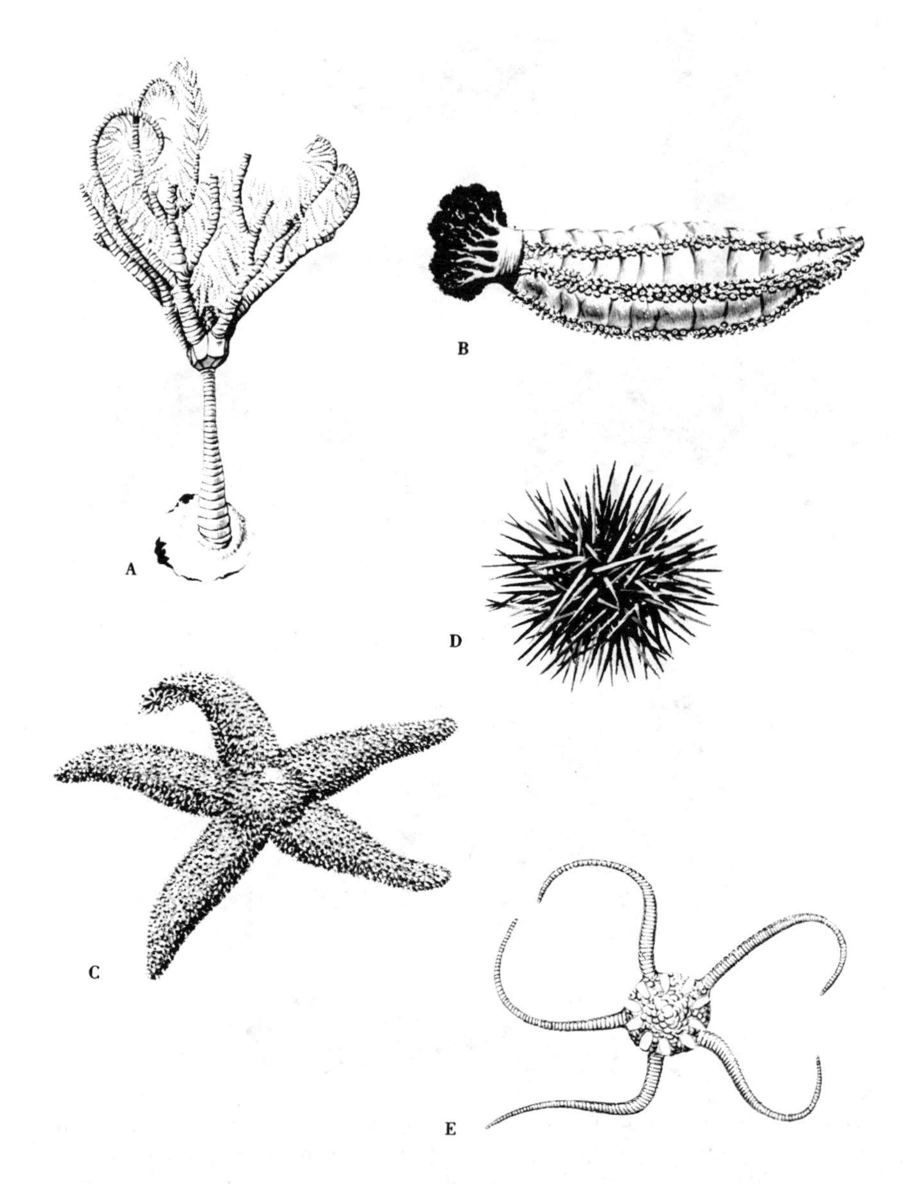

Fig. 1-12 *Selected echinoderms. (A) A sea lily,* Calamocrinus diomedae; *(B) a sea cucumber,* Cucumaria frondosa; *(C) a sea star* (starfish), Asterias forbesi; *(D) a sea urchin,* Arbacia punctulata; *(E) a brittle star,* Ophiolepis elegans. [*(A) After Cuénot after Agassiz, in* Traité de Zoologie. *Vol. XI, P.-P. Grassé, ed. Masson et Cie., 1948. (E) From* The Invertebrates, *Vol. IV by Hyman. Copyright © 1955, McGraw-Hill Book Company. Used by permission.*]

phylum include the sea stars (starfishes), brittle stars, sea urchins, sea cucumbers, and sea lilies. Most adult echinoderms are radially symmetrical, but all echinoderm larvae are bilaterally symmetrical.

The phylum Chordata includes about 43,000 living species, all of which possess (at least during their early embryonic life) three salient structural features: a longitudinal supporting rod or *notochord,* a dorsal hollow nerve cord, and a series of clefts in the body wall that open into the pharynx (Fig. 1-13). Each of these features may be lost or modified in the adult stage. Among the chordates are the fishes, amphibians, reptiles, birds, and mammals, as well as the more primitive chordates, the tunicates, and the lancelet, which is also called amphioxus. The fishes, amphibians, reptiles, birds, and mammals have a vertebral column (consisting of separate, cartilaginous or bony units called vertebrae) for support of the body, and hence are called *vertebrates.* All other animals, regardless of the phylum to which they belong, lack a vertebral column and are known as *invertebrates.*

DETERMINING PHYLOGENETIC RELATIONSHIPS

In determining phylogenetic relationships, biologists rely on several different approaches to obtain sufficient data that enable them to arrive as nearly as possible at a correct phylogeny. As more and more evidence is obtained, the inferences drawn about the phylogenetic relationships of different species to each other are more likely to be correct.

COMPARATIVE MORPHOLOGY

Biologists rely very heavily on the *comparative morphology* of adults to assist them in arriving at correct phylogenetic relationships. That is, they compare the form and structure of adults. These studies have included not only comparisons of gross morphology, both external and internal, but also cytological examination of cell structures such as the chromosomes to determine the number and shape. Another factor of importance in determining an organism's relationships to other organisms is whether the organism is unicellular or multicellular. In addition, living species are compared, when possible, with fossil remains. The type of symmetry an animal exhibits is important morphological information for determining its phylogenetic relationships.

Symmetry (Fig. 1-14) refers to the correspondence of the parts of an animal's body that are on opposite sides of a line or plane. Most of the sponges and some of the protozoans are *asymmetrical*; there is no plane that can be used to divide them into equivalent parts. Most ani-

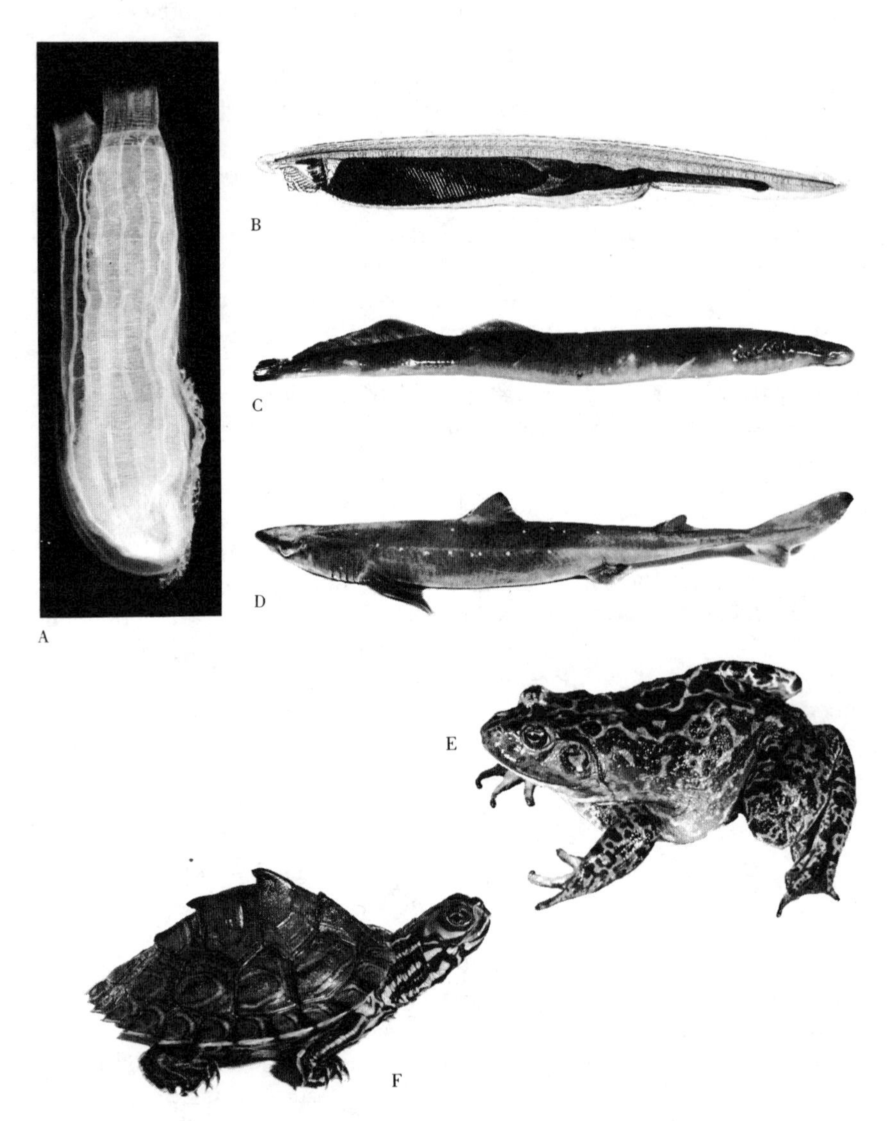

Fig. 1-13 *Selected chordates. (A) An ascidian tunicate,* Ciona intestinalis; *(B) a lancelet or amphioxus,* Branchiostoma caribaeum; *(C) a lamprey,* Petromyzon marinus; *(D) a dogfish shark,* Squalus acanthias; *(E) a frog,* Rana catesbeiana; *(F) a turtle,* Graptemys barbouri. [*(A), (C)–(E) Courtesy of Carolina Biological Supply Company; (B) Courtesy of Ward's Natural Science Establishment, Inc.; (F) Courtesy of F. R. and J. A. Cagle, Tulane University.*]

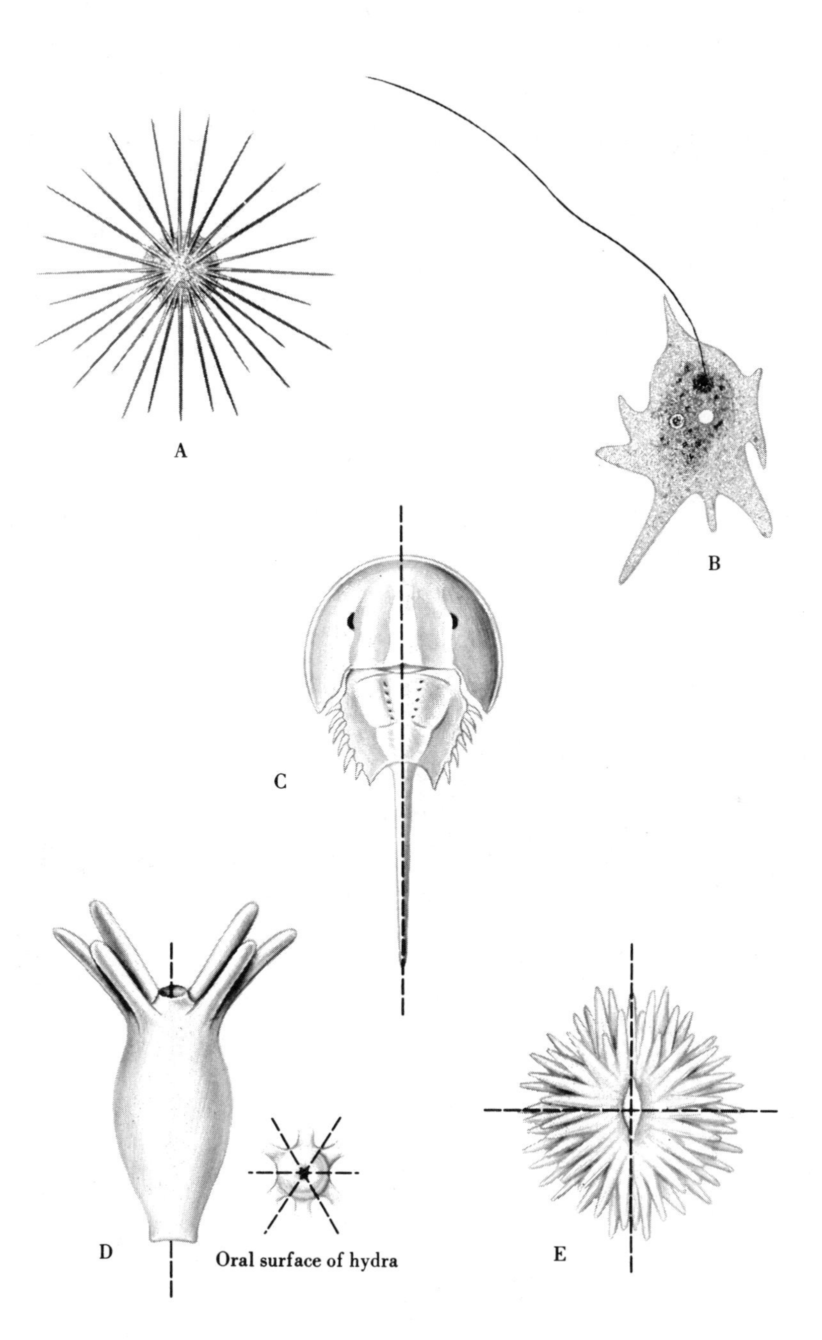

Fig. 1-14 *Types of symmetry. (A) Spherical—a heliozoan or sun animalcule, a protozoan; (B) asymmetrical—a rhizomastigid, a protozoan; (C) bilateral—a horseshoe crab; (D) radial—a hydra; (E) biradial—a sea anemone.*

mals, such as man, are *bilaterally symmetrical*; that is, their bodies can be divided by a single plane into left and right halves that are mirror images of one another. Six terms are used to describe the location of structures in bilaterally symmetrical animals (Fig. 1-15): *anterior*, toward or pertaining to the front or head end; *posterior*, toward or pertaining to the hind or tail end; *dorsal*, toward or pertaining to the upper surface or back; *ventral*, toward or pertaining to the lower surface or belly; *medial*, in, toward, or pertaining to the midline or middle of the body; and *lateral*, toward or pertaining to the side of the body. Some animals, such as the hydra, exhibit *radial symmetry*. Their bodies are built around a medial axis that runs from the mouth to the opposite surface. The surface on which the mouth lies is called the *oral surface*, while the opposite surface is called the *aboral surface*. Any cut through the axis of the cylinder will divide the animal into equal sectors. *Spherical symmetry* is found only among a few protozoans, such as the heliozoans or sun animalcules. The bodies of such creatures are essentially spheres, and any slice through the center will divide the or-

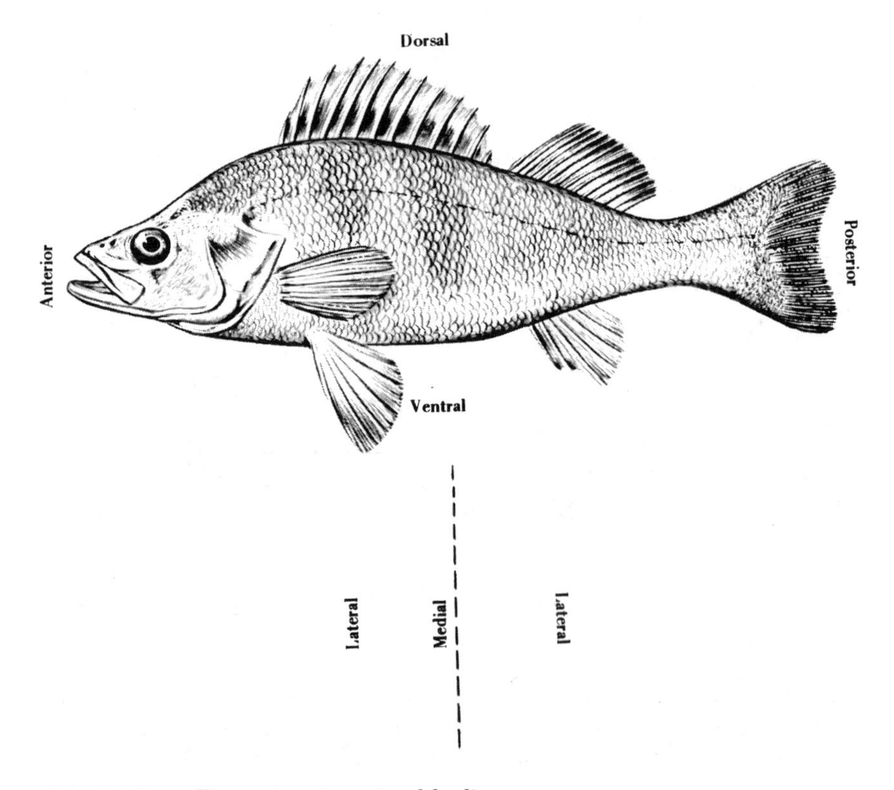

Fig. 1-15 *The regions in animal bodies.*

ganism into equal portions. *Biradial symmetry* is found in those few animals, some of the sea anemones, for example. In animals with biradial symmetry, there are two planes along which the body can be sectioned to produce mirror halves. However, the mirror halves produced by cutting through one plane are different from those produced by a second cut through a plane at an angle 90° to the first.

The presence or absence of segmentation is another important morphological characteristic of a species that aids in determining its phylogenetic relationships. *Segmentation*, the linear repetition of body parts along the anteroposterior axis, is synonymous with *metamerism;* parts which do not extend the length of the animal are serially repeated at regular intervals. This repetition is the essence of segmentation. It is readily apparent both externally and internally in earthworms, for example. Jellyfishes, on the other hand, are an example of animals that show no segmentation, either internally or externally. In most segmented animals, the most anterior segment is the oldest and the most posterior segment is the youngest. But in the segmented tapeworms, the opposite is true: the most posterior segment is the oldest and the most anterior segment is the youngest. As will be seen in subsequent chapters, segmentation appears to have been an adaptation which, depending upon the phylum, provided for either improved locomotor ability or for the production of a greater number of eggs and sperm.

BIOCHEMICAL AND IMMUNOLOGICAL TECHNIQUES

Biochemical techniques are being used more and more to determine phylogenetic relationships. Techniques are available for determining the degree of similarity of (a) the deoxyribonucleic acid and (b) the amino acid sequences in proteins from different organisms. For example, the sequence of amino acids in the chain of these acids that forms the protein portion of cytochrome *c*, a molecule involved in the electron-transport system of mitochondria, has been determined for about 50 animal species. This chain consists of about 104 amino acids. The less closely related one animal is to another, the more unlike are the base sequences in their deoxyribonucleic acids, and the amino acid sequences in their cytochromes *c*. Inasmuch as the base sequence in deoxyribonucleic acid determines (through messenger ribonucleic acid) the amino acid sequence in a protein, one would expect to find differences in the base sequences if there are differences in the amino acid sequences. The implication is that the more similar base and amino acid sequences are descendants of common ancestral base and cytochrome *c* amino acid sequences.

Immunology has also contributed to unraveling relationships. If a protein, such as serum-albumin, from another animal is injected into a rabbit, the rabbit will produce an antibody against that protein. This antibody will subsequently react strongly against serum-albumin from the original donor species, because it was synthesized specifically to combat this particular serum-albumin; it will react less strongly against serum-albumin from yet another species. The less closely the original donor is related to the second donor, the weaker the reaction. The techniques devised by molecular biologists, biochemists, and immunologists have, for example, confirmed the conclusion of anatomists that humans are more closely related to chimpanzees and gorillas than to any other animals.

Investigations at the molecular level are important not only because they confirm conclusions based on older techniques, but also because by expressing the differences between species in precise quantitative terms, they provide a basis for calculating rates of molecular evolution. For example, let us assume that the number of differences per 100 amino acids between the cytochromes *c* of two species is a direct measure of the time that has passed since their two lines diverged. Assuming that the rate of evolution of cytochrome *c* has remained fairly constant, we can plot the number of differences per 100 amino acids between the cytochromes *c* of any two species against the time (determined from the fossil record) since their evolutionary histories diverged. The slope of the line drawn after such data obtained from several groups of animals have been plotted represents the rate of evolution of cytochrome *c*. The radioactive isotope dating methods which have been developed in recent years give a more precise age to rocks and fossils than was previously possible. Isotope dating methods are based upon the fact that each radioactive substance has a characteristic, constant rate of decay. The decay rate is usually expressed as *half-life*, the time required for half of the initial radioactive substance to become disintegrated. It is assumed that all the radioactivity in an object, such as a rock or fossil, was incorporated into it at the time the rock formed or when the organism was alive and that none was added afterwards. Two highly useful "atomic clocks" are based upon the decay rates of uranium 238, which has a half-life of 4.5 billion years, and of carbon 14, which has a half-life of 5,730 years. Uranium 238 decays to lead 206. Knowing the half-life of uranium 238 and determining the proportion of uranium 238 to lead 206 in a sample enables an investigator to calculate its age. In the case of carbon 14, during their lifetime, organisms incorporate the radioactive carbon 14 and the stable carbon 12 into their tissues in the same proportion as carbon 14 and carbon 12 occur in their environment. When an organism dies, its decaying carbon 14 is no longer being replenished, and the proportion of carbon 14 to carbon 12 decreases. The time elapsed since the organism died can then be calculated after determining the proportion of carbon 14 to carbon 12 in its remains. The competing forces of formation and decay of carbon 14 have been going on in nature at a relatively constant rate for so long that the carbon 14 to carbon 12 ratio does indeed provide a valid dating technique. Carbon 14 is formed in the upper atmosphere by the reaction of nitrogen 14 atoms with neutrons produced by cosmic rays. The carbon 14 ultimately combines with oxygen, forming radioactive carbon dioxide which mixes with ordinary carbon dioxide, that containing carbon 12, in the atmosphere. Plants, making use of carbon dioxide in photosynthesis, take up both types of carbon and incorporate them into their tissues in the same proportion as is in the atmosphere. Animals then eat the plants, thereby acquiring the carbon 12 and carbon 14. Before isotope dating was developed, the method most often used to calculate geologic time was based on estimates of the annual rate at which sedimentary deposits accumulate in the earth's crust, thickness of deposits being assumed proportional to elapsed time. However, the great difficulty in assessing the rate of deposition of a layer of sediment led to wide variation in the age estimates.

BEHAVIORAL AND FUNCTIONAL OBSERVATIONS

Phylogenetic relationships are also being determined by analyzing the sounds of birds, insects, and amphibians by electronic techniques. Observations of courtship behavior of birds, crabs, and other animals have also been profitable in this regard. Comparative studies combining functional and structural observations (functional morphology) are also being carried out more and more frequently. For example, elucidation of arthropod phylogeny is being attempted by combining studies of the structure of legs and leg joints with careful analyses of stepping movements.

DEVELOPMENTAL PATTERNS

The developmental patterns of embryos and larvae also frequently provide important information about phylogenetic relationships, as explained earlier in the discussion of the concept of homology. Once established, the basic developmental pattern of each phylum appears to have remained fairly stable during the subsequent course of evolution. Consequently, embyros and larvae have been used to determine some evolutionary relationships that are not apparent from the study of adults. This stability is probably due to the fact that a mutation that expressed itself during early development would likely have such profound effect on the form of the developing animal that there would be little chance of the embryo or larva becoming an adult, or of the adult surviving long enough to transmit the mutation to the next generation. Larvae, all of which lead an existence free of their parents, are much more common among marine species than among freshwater inhabitants. This is true presumably because the sea is a less hostile environment to such free-swimming forms. Freshwater species are more subject to osmotic stress, greater temperature variations, and a less abundant food supply; consequently, freshwater inhabitants tend to brood their young.

The phyla consisting of those metazoans that are bilaterally symmetrical either throughout life or only as larvae have been divided into two groups, the *Deuterostomia* and *Protostomia*. Each group has a characteristic mode of development, including a basic difference in the ultimate destiny of the *blastopore*, the first opening that develops from the exterior cell layer into the digestive cavity of an early embryo. The Deuterostomia include the animals belonging to the phyla Chaetognatha, Echinodermata, Hemichordata, and Chordata. In deuterostomes, the blastopore becomes the anus, or the anus forms very near the closed blastopore. The mouth is a new opening that is a considerable distance away from the blastopore. All other phyla of bilaterally symmetrical metazoans, including the Platyhelminthes, Annelida, Mollusca, and Arthropoda, constitute the Protostomia. Among protostomes, the blastopore becomes the mouth, or the mouth forms near the closed blastopore. If an anus develops in a protostome (an anus is lacking in the phylum Platyhelminthes), it is a new opening, well removed from the site of the blastopore. The terms "protostome" (first mouth) and "deuterostome" (second mouth) were in fact coined to describe the difference between the embryonic origins of the adult mouth of these two groups. Deuterostomes and protostomes differ also in their pattern of *cleavage*, the division of the fertilized egg into many smaller cells, called *blastomeres*, which are much nearer to the size of normal body cells than is the egg. (The embryo does not grow during cleavage.) Deuterostomes show *radial cleavage;* that is, the cleavage plane is either perpendicular to or arranged symmetrically around the *polar axis* of the egg, the axis through the animal (upper) and vegetal (lower) poles of the egg (Fig. 1-16). *Spiral cleavage* is characteristic of protostomes; that is, the cleavage planes during many of the divisions are oblique to the polar axis, thereby resulting in a twisted or spiral arrangement of successive generations of blastomeres (Fig. 1-17). Consequently, because of the spiral cleavage, the blastomeres, for example, of the top tier in the eight-cell stage do not lie directly above the four in the lower tier, but instead lie over the furrows between the blastomeres of the bottom tier.

Still another difference between the modes of development of deuterostomes and protostomes is that the eggs of deuterostomes characteristically undergo *indeterminate cleavage*; that is, the developmental fate of a cell in an early embryo is not irreversibly determined. In contrast, the eggs of protostomes characteristically have *determinate cleavage*; the fate of a given cell is fixed very early in development. The difference between indeterminate cleavage and determinate cleavage can be seen after the cells of a four-cell stage embryo are separated from

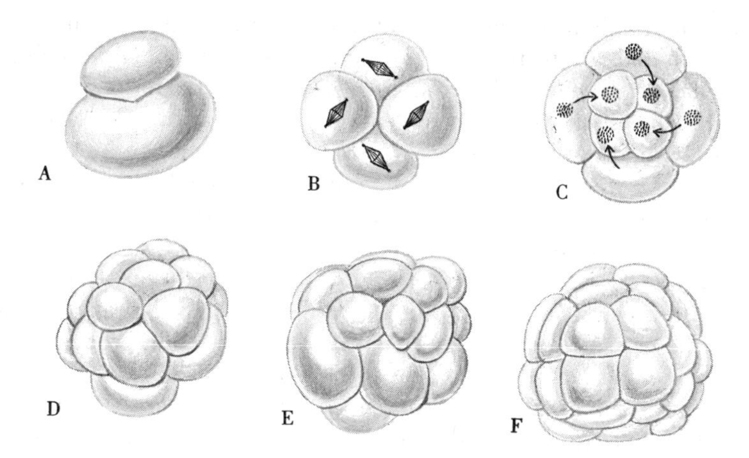

Fig. 1-16 *Cleavage of an egg from the sea cucumber,* Synapta digitata. *(A) 2-cell stage; (B) 4-cell stage viewed from the animal pole; (C) 8-cell stage, lateral view; (D) 16-cell stage; (E) 32-cell stage; (F) blastula, vertical section. (After Ebert and Sussex,* Interacting Systems in Development. *2nd ed. Holt, Rinehart and Winston, 1970; and Selenka, from Korschelt.* Vergleichende Entwicklungsgeschichte der Tiere, *Gustav Fischer Verlag, 1936.)*

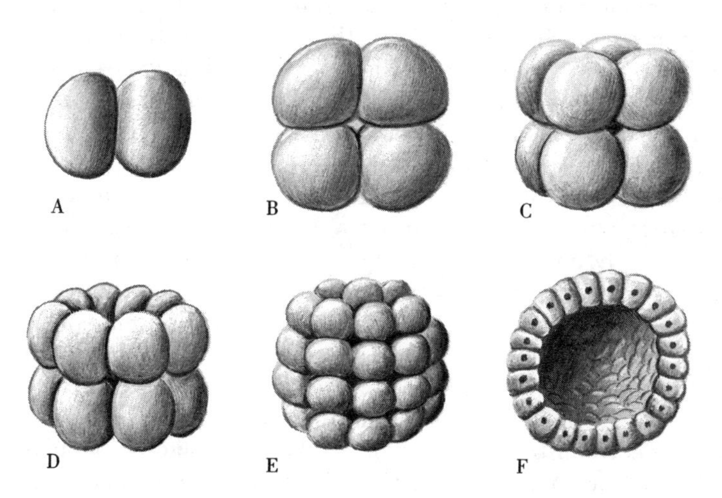

Fig. 1-17 *Cleavage in an egg of the annelid* Nereis *sp. (A) 2-cell stage, viewed from the animal pole; (B) 4-cell stage, viewed from the animal pole; (C) 8-cell stage, viewed from the animal pole; (D) 16-cell stage, viewed from the animal pole; (E) 16-cell stage, viewed from the right side; (F) 32-cell stage, viewed from the animal pole. The orientation of the spindle fibers is indicated in B. The arrows in C indicate the direction of shift of the upper tier in relation to the lower. (From Ebert and Sussex,* Interacting Systems in Development, *2nd ed., 1970, after T. W. Torrey,* Morphogenesis of the Vertebrates, *John Wiley & Sons, 1962.)*

one another. When this experiment is performed with a sea urchin embryo, each cell will form a complete larva. With an annelid, each of the four cells will develop only into a predetermined quarter of a larva; that is, each cell will produce only that part of the larva it would have formed if the cells had not been separated from each other.

In *isolecithal eggs*, those containing a small amount of evenly distributed yolk, all the cells in the embryo at the end of the cleavage process are of approximately the same size, as in the sea cucumber (Fig. 1-16), and cleavage is said to be *equal*. But when there is a larger, but still moderate, amount of yolk and it is concentrated near the vegetal pole of the egg, as in a frog egg, cleavage is *unequal*, and the cells of the blastula are unequal in size. Cleavage of the fertilized egg results in the formation of a blastula. A *blastula* is an early embryo that usually is hollow because the cells that compose it have arranged themselves that way; it is solid only in some of the animals that never develop a body cavity. When a blastula is hollow, its wall consists of a single cell layer. Even the first cleavage may be unequal, as in the annelid *Nereis* (Fig. 1-17).

Some eggs, however, such as those of a chicken or squid, have such massive amounts of yolk that the patterns of development characteristic of their respective phyla are greatly modified because this large amount of yolk prevents the typical cleavage patterns from appearing. When the amount of yolk is still moderate, as in a frog egg, the cell divisions can cut completely through the yolk; but with a large mass of inert yolk, as in a chicken egg, division of the yolk is impossible, and all cell division is restricted to the small disk of cytoplasm on the surface of the undivided yolk. Eggs, such as those of the frog and chicken, that have the yolk concentrated near the vegetal pole are called *telolecithal eggs*. When an egg divides completely during cleavage, the cleavage is called *holoblastic*, or *total* (as in a frog), but when the egg does not divide completely and an undivided yolk remains (as in a chicken), the cleavage is called *meroblastic*, or *partial*.

Still another type of egg is the *centrolecithal egg*, and it undergoes still another type of cleavage, *superficial cleavage*. In centrolecithal eggs, such as those which occur among arthropods, there is a peripheral layer of yolk-free cytoplasm that surrounds a mass of yolk; but in the center of the yolk mass is a yolk-free island of cytoplasm that contains the nucleus. Because of this large amount of yolk, cleavage is *superficial*. The centrally located nucleus divides several times. These nuclei migrate through the yolk to the outer layer of cytoplasm, and cell membranes eventually form, dividing the outer cytoplasmic layer into separate cells. The cells of the cleaved egg thus form an external layer, one cell thick, enclosing the undivided yolk mass. In meroblastic cleavage, there is an undivided yolk at one end of the egg and a cap of cells at the other, whereas in superficial cleavage, the undivided yolk is surrounded by a layer of cells.

A body cavity (a fluid-filled space between the body wall and the digestive tract) is found in most of the bilaterally symmetrical metazoans. The body cavity serves the animal that has one in a variety of ways. Being fluid-filled, it can provide a cushion for protection of the internal organs from impact by external forces. In some soft-bodied animals, such as earthworms, it is an aid in locomotion; it serves as an incompressible hydrostatic skeleton to which pressure is applied by contraction of muscles in the body wall, resulting in movement of the body. Without the fluid-filled body cavity, earthworms would merely telescope when the longitudinal muscles in the body wall contracted, and these animals certainly would not be as efficient at burrowing as they are. Another interesting use of the body cavity is seen in such animals as the sea cucumber, *Thyone rubra*, where it serves as the site of egg fertilization. Presumably, occurrence of fertilization inside the body cavity increases the chances of an egg and sperm meeting, compared with the chances in the open sea.

Those metazoans with primary bilateral symmetry that do not have a body cavity are called *acoelomates*. The term "acoelomate" has traditionally been used to refer only to those animals that lack a body cavity and also have primary bilateral symmetry. The primary symmetry of the animals in a phylum is the type exhibited by its larvae. However, the lower multicellular animals, such as the sponges and cnidarians, that do not have primary bilateral symmetry do not have body cavities, either. Their body plans do not provide for them. A body cavity develops only in an animal that has a primary bilateral symmetry, but as stated above, even in some of the animals that have a primary bilateral symmetry a body cavity never develops. If the body cavity in an adult animal is completely derived from the *blastocoel*, or *primary body cavity* of the blastula, the body cavity of this adult is called a *pseudocoel*, or *false coelom*, and these animals are called *pseudocoelomates*. The cavity inside a hollow blastula is called a blastocoel; a pseudocoel is essentially a persistent blastocoel.

The blastula develops into a *gastrula*. The early gastrula is a two-layered embryo, consisting of an outer cell layer, the *ectoderm*, and an inner cell layer or mass, the *endoderm*. In the gastrulas of most animals a third layer, the *mesoderm*, develops between the ectoderm and endoderm. During gastrulation (the process of gastrula formation) an additional cavity, the *gastrocoel*, usually develops. It is the cavity of the *archenteron*, or primitive gut, and not a body cavity, and it is this cavity that opens to the exterior through the blastopore. However, a solid gastrula, one with no archenteron, and therefore with no gastrocoel or blastopore, is formed by many sponges and cnidarians. Sponges have a unique body plan, as will be explained in Chapter 3, and even when they form an archenteron, it does not serve as a digestive cavity, despite its name. The archenteron is an endodermal structure. In a large majority of the animal species, another type of body cavity, a *true coelom* or *secondary body cavity,* also forms. It is usually, however, simply referred to as a coelom, the word "true" being omitted, and the animals that have a coelom are called *coelomates.* A coelom has two modes of development, both involving the mesoderm. By one method, the coelom arises as a split inside bands, plates, or masses of mesodermal cells, and such a coelom is more specifically called a *schizocoel* (Fig. 1-18). If, however, the coelom originates from the cavities in mesodermal pouches that formed as evaginations from the endodermal wall of the archenteron, it is known as an *enterocoel.* Clearly, a coelom, unlike a pseudocoel, does not develop from the blastocoel. The blastocoel of some species that form a coelom becomes obliterated during development. However, in most species that possess a coelom (especially the arthropods), portions of the blastocoel persist in the adults as blood-filled spaces that are part of the circulatory system, and these spaces are collectively termed a *hemocoel.* Animals that form a coelom are called *coelomates* regardless of whether they also have a hemocoel or whether the coelom is reduced in size because of crowding by the hemocoel.

All deuterostomes are enterocoelous coelomates (also called enterocoelomates). The protostomes, in contrast, do not show such uniformity. Among the members of one phylum of protostomes, the Brachiopoda, both the schizocoelous and enterocoelous methods of coelom formation have been observed; they are the only enterocoelous protostomes in the animal kingdom. The rest of the protostomes are either acoelomates, pseudocoelomates, or schizocoelous coelomates (also called schizocoelomates).

The acoelomate, pseudocoelomate, and coelomate body plans are diagrammatically represented in Figure 1-19. For example, the flatworms are acoelomates; the space between their integument and organs is filled with muscle cells and *mesenchyme.* Mesenchyme is a loosely organized connective tissue consisting of an intracellular gelatinous substance, called *mesogloea,* and amoebocytes (wandering cells). In coelomates, annelids for example, the body cavity (coelom) is completely lined with a thin membrane called the peritoneum. This not only lines the internal surface of the body wall but also covers the organs (for example, the intestine). As a consequence of the two modes of development of a coelom (the schizocoelous and enterocoelous), the peritoneum always consists of mesodermally derived cells. Furthermore, the organs within the coelom, such as the intestine, do not hang free; they are held in place by *mesenteries* (sheets of mesodermally derived tissue that are continuous with the peritoneal covering of the organs and the peritoneal lining of the coelom). In pseudocoelomates, as discussed above, there is also a body cavity (pseudocoel), but in contrast to the situation in coelomates, the organs in pseudocoelomates lie free in the pseudocoel. There is neither peritoneum nor are there mesenteries in pseudocoelomates.

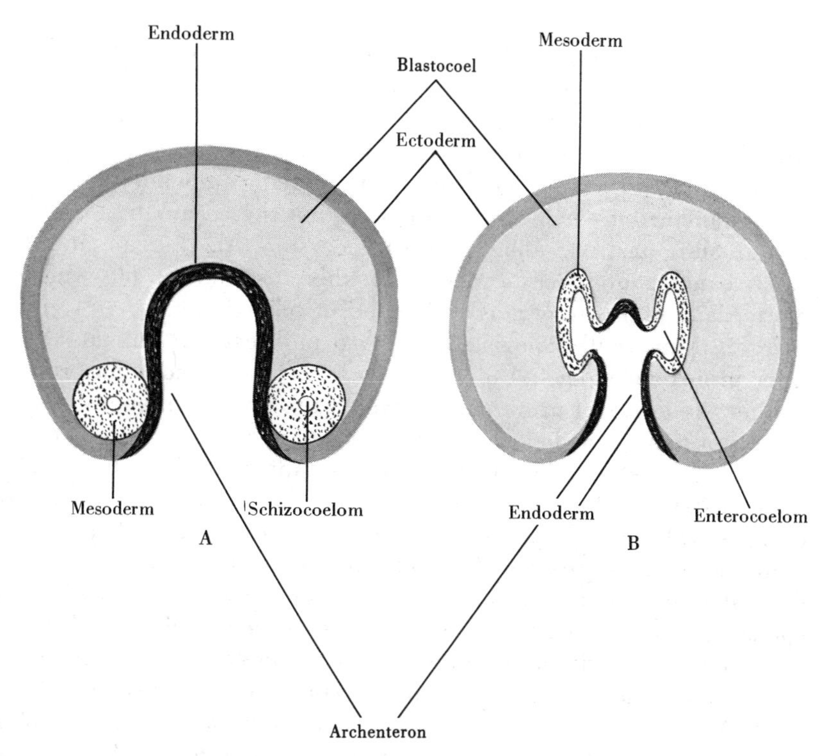

Fig. 1-18 *The (A) schizocoelous and (B) enterocoelous methods of coelom formation. A schizocoel arises from splits in originally solid clusters of mesoderm cells, whereas an enterocoel forms from cavities in mesodermal outpocketings from the endodermal wall of the archenteron.*

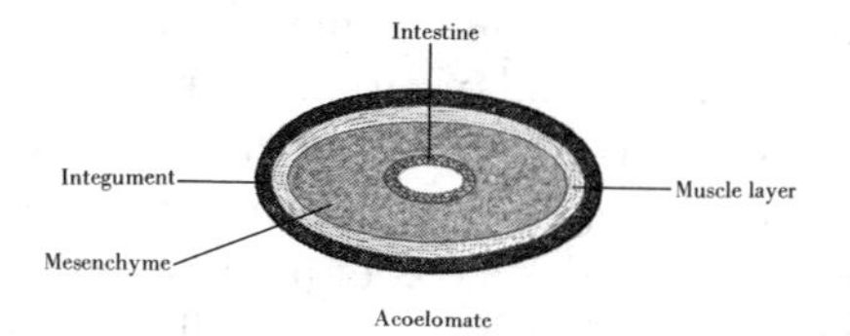

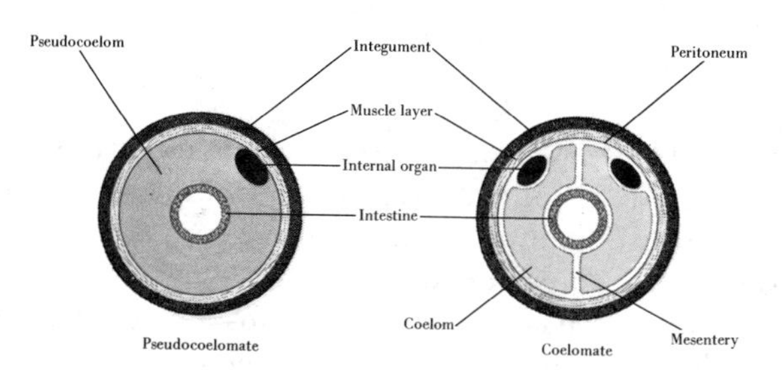

Fig. 1-19 *The acoelomate, pseudocoelomate, and coelomate body plans as seen in flatworms, nematodes, and annelids, respectively. (Modified from Wilson, Eisner, Briggs, Dickerson, Metzenberg, O'Brien, Susman, and Boggs,* Life on Earth, *Sinauer Associates, Inc. 1973.)*

The evolution of a cavity between the body wall and the digestive tract was an important advance over the solid type of body construction found in the flatworms. When the digestive tract was freed from encirclement by mesenchyme, the way was paved for the evolution of a more efficient digestive system. Organs embedded in solid tissue are distorted when the body moves, but the organs in a body cavity can slide past one another when necessary. Also, organs in a body cavity can expand when necessary, as when the stomach fills with food, more easily than can organs embedded in solid tissue.

Almost all biologists agree that multicellular animals evolved from protozoan stocks. Unfortunately, the fossil record has revealed neither the group of protozoans to which the protozoan ancestor or ancestors of metazoans belonged nor the mode of origin of metazoans from the protozoan stock. Theoretically, metazoans could have had a monophyletic origin, evolving from a single, common protozoan stock. However, as will be explained in Chapter 3, it seems more likely that the metazoans had at least a diphyletic origin, at least two lines having evolved independently from protozoans, one line developing into the sponges, which appear to have been an evolutionary dead end, with at least one other giving rise to the rest of the metazoans. There are two theories that can explain how a protozoan could have become a metazoan. According to the *syncytial theory*, a *syncytium*, or multinucleate cell, developed cell membranes that separated the nuclei from one another, thus producing a multicellular organism. The second theory, the *colonial theory*, states that the individuals in a *colony* (group of organisms of the same species living together) of protozoans gradually became interdependent as a result of cellular specialization, thereby producing a multicellular organism. The colonial theory has received more general acceptance because cellularization of a syncytium does not occur during the development of any of the lower metazoans, whereas some colonies of protozoans are hollow spheres which are suggestive of a hollow blastula. However, on the basis of our present knowledge, neither theory can be ruled out. These theories will be referred to again later in more detailed discussions.

THE FOSSIL RECORD

The fossil record, an important aspect of animal diversity, introduces the variable of "time" into the study of the evolutionary process and the determination of phylogenetic relationships, enabling us to follow the changes that occurred throughout the past. Study of the fossil record reveals that animals have gradually changed with the passage of time, and that new groups have evolved while others have become extinct. In some phyla, the echinoderms for example, the fossil record has proved very useful by revealing how the body form of a living species evolved from the ancestral type. However, the fossil record has been of little value in determining the origins of the various phyla.

Geologists have found particular fossils associated with specific geologic strata, and, using them, have constructed a geologic time table (Table 1-4). The rock record, with its included fossils, is well enough known that good estimates can be made as to when particular animal groups made their first appearance. The Archeozoic Era began with the formation of the earth's crust, but life did not begin until the primordial seas were formed, and spontaneously synthesized organic

Table 1–4 The Geologic Time Table

Era	*Period*	*Time from Beginning of Period to Present (Millions of Years)*	*Important Events*
Cenozoic	Quarternary	2	Appearance of *Homo sapiens*
	Tertiary	63	Dominance of land by mammals
Mesozoic	Cretaceous	135	Last of the dinosaurs
	Jurassic	181	First birds
	Triassic	230	First dinosaurs and first mammals
Paleozoic	Permian	280	Great expansion of reptiles, decline of amphibians
	Pennsylvanian	320	First reptiles
	Mississippian	345	Spread of amphibians
	Devonian	405	First insects and first amphibians
	Silurian	425	Invasion of land by plants and arachnids
	Ordovician	500	Invertebrates become more diversified, coral reefs begin to form
	Cambrian	570	All major phyla represented in the fossil record
Proterozoic	Not divided into periods	2,500	Earliest known fossils of animals
Archeozoic	Not divided into periods	4,500	Origin of life

molecules combined in these seas to form self-replicating entities. Fossil bacteria have been found in rocks from the Archeozoic Era, along with spheroids and filaments, which resemble modern blue-green algae.

The earliest fossils revealing the existence of metazoan animal life have been found in rocks from the late Proterozoic Era. They represent the probable remains of jellyfishes and annelids. In addition, there are questionable reports of mollusks and arthropods. However, all of the extant major phyla are represented in the fossil record of the first portion of the next era—the Cambrian Period of the Paleozoic Era. All of the Precambrian and Cambrian animals were marine organisms. Animals may have migrated from the seas and taken up a freshwater existence as early as the Ordovician Period (freshwater animals did indeed exist during the next period, the Silurian Period), but not until the Devonian Period did colonization of the freshwater environment gather much momentum. The Cambrian Period is often referred to as the "Age of Invertebrates" because invertebrates were the dominant animals at that time. The earliest vertebrate fossils discovered are those of the now extinct ostracoderms, fishes that lacked jaws. These ancient jawless fishes probably fed by filtering food particles from a water current that entered the mouth and exited after passing through the gill system.

During the Ordovician Period of the Paleozoic Era, the invertebrates exhibited much diversification, many new invertebrate species having evolved. Corals, for example, first appeared during this period, and the first coral reefs were formed.

The Silurian Period is historically important, because during this period, the first colonization of land by plants and animals occurred. The first terrestrial animals were arachnids that resembled present-day scorpions; scorpions are an order of arachnids, and arachnids are a class of arthropods. Obviously the transition from an aquatic to terrestrial existence was an important landmark in the history of life on earth, but it was not accomplished easily. A terrestrial existence presented new hazards. Because these colonizers of land were no longer immersed in water, they required, for example, mechanisms to conserve water and prevent desiccation. In the subsequent chapters that deal with terrestrial organisms we shall see how, through the process of natural selection, the problems of terrestrial existence were overcome.

The Devonian Period is often referred to as the "Age of Fishes" because fishes then were so abundant and varied. This period witnessed the extinction of the ostracoderms and the evolution from them of a number of different kinds of fishes that were the ancestors of present-day fishes. The first amphibians also evolved during the Devonian Period. Amphibians never became completely terrestrial. Most amphibians return to an aquatic habitat to breed, although some utilize very moist places on land; otherwise, their eggs would become desiccated. Reptiles did not make their appearance until the Pennsylvanian Period, but are an important group if only because they were the first vertebrates to become truly terrestrial. They developed a dry, scaly skin, which minimized water loss by evaporation, and a shelled egg, which completely freed all of them from returning to an aquatic environment to breed. The Mississippian, Pennsylvanian, and Permian Periods (together forming the late Paleozoic Era) are often collectively called the "Age of Amphibians" because amphibians were the dominant group then. When amphibians first evolved, they faced no competition on land from any other vertebrates. However, during the Permian Period the amphibians went into decline. They were probably unable to compete successfully with the more highly evolved members of the new class, the reptiles.

The Mesozoic Era is often referred to as the "Age of Reptiles" because reptiles dominated the scene then. Late in the Triassic Period of the Mesozoic Era, mammals evolved, and birds during its Jurassic Period. The most prominent reptilian species during the Mesozoic Era were the dinosaurs, which first appeared during the Triassic Period of this era. *Tyrannosaurus rex* lived during the Cretaceous Period and was undoubtedly the most spectacular predator ever to walk the earth. However, the dinosaurs and many other reptiles became extinct by the end of the Cretaceous Period. Several hypotheses have been proposed to explain the extinction of so many reptiles, especially the dinosaurs, at that time. These include deterioration of the climate (becoming too hot or cold), drainage of swamp and lake environments, infestation by parasites, and anatomical and metabolic disorders. Competition with the mammals is not an adequate explanation for their disappearance, because at that time the mammals were small and few in number.

The evolution of birds from reptiles was attested to by the discovery of a fossil bird, *Archaeopteryx*, which, as do reptiles, had teeth and a long, jointed tail (present-day birds have no teeth and only a much shortened tail). *Archaeopteryx* had feathered wings; since feathers are found only on birds, it is clearly a link between modern birds and their reptilian ancestors. The number of such clearly intermediate or transitional species in the fossil record is unfortunately small.

The Cenozoic Era is usually referred to as the "Age of Mammals" because mammals have become the dominant animals on earth. The relatively rapid increase in the number of mammalian species in the Cenozoic Era was probably, to a large extent, the result of the earlier decline of the reptiles. The extinction of many species of reptiles late in the Mesozoic Era left the evolving mammals without serious competitors for the available food and space on land. Consequently, the mammals were able to spread and take advantage of this void early in the Tertiary Period. In any case, mammals, with their superior intelligence, would probably have ultimately prevailed over the reptiles, even without the disappearance of the dinosaurs.

The Cenozoic Era has also been the time marked by the evolution of a wide variety of flowering plants, insects, and birds. In fact, one could argue with justification that this era should be called the Age of Flowering Plants, Insects, Birds, and Mammals, not merely the Age of Mammals.

FURTHER READING

Ayala, F. J., ed., *Molecular Evolution.* Sunderland, Mass.: Sinauer Associates, 1976.

Barnes, R. D., *Invertebrate Zoology,* 4th ed. Philadelphia, Saunders College/HRW, 1980.

Carter, G. S., "Phylogenetic Relations of the Major Groups of Animals," in *Ideas in Modern Biology, Vol. 6, Proceedings, XVI, International Congress of Zoology,* J. A. Moore, ed. Garden City, N.Y.: The Natural History Press, 1965, p. 427.

Clarke, B., "The Causes of Biological Diversity," *Scientific American,* vol. 233:2 (1975), p. 50.

Cloud, P. E., Jr., "Pre-Metazoan Evolution and the Origins of the Metazoa," in *Evolution and Environment,* E. T. Drake, ed. New Haven: Yale University Press, 1968, p. 1.

Dobson, E. O., "The Kingdoms of Organisms," *Systematic Zoology,* vol. 20 (1971), p. 265.

Dobzhansky, T., F. J. Ayala, G. L. Stebbins, and J. W. Valentine, *Evolution.* San Francisco: W. H. Freeman and Co., 1977.

Editors of Time-Life Books, *Life Before Man.* New York: Time-Life Books, 1972.

Fretter, V., and A. Graham, *A Functional Anatomy of Invertebrates.* New York: Academic Press, 1976.

Grant, V., *Organismic Evolution.* San Francisco: W. H. Freeman and Co., 1977.

Grassé, P.-P., *Evolution of Living Organisms.* New York: Academic Press, 1977.

Greenberg, M. J., "Ancestors, Embryos, and Symmetry," *Systematic Zoology,* Vol. 8 (1950), p. 212.

Hanson, E. D., *The Origin and Early Evolution of Animals.* Middletown, Conn.: Wesleyan University Press, 1977.

Hyman, L. H., *The Invertebrates: Protozoa through Ctenophora,* Vol. I. New York: McGraw-Hill, 1940.

Knoll, A. H., and E. S. Barghoorn, "Archean Microfossils Showing Cell Division from the Swaziland System of South Africa," *Science,* vol. 198 (1977), p. 396.

Lewontin, R. C., "Adaptation," *Scientific American,* vol. 239:3 (1978) p. 212.

Marcus, E., "On the Evolution of the Animal Phyla," *Quarterly Review of Biology,* vol. 33 (1958), p. 24.

Mayr, E., *Evolution and the Diversity of Life.* Cambridge, Mass.: Belknap Press of Harvard University Press, 1976.

Mavr. E.. "Evolution," *Scientific American,* vol. 239:3 (1978), p. 46.

Meglitsch, P. A., *Invertebrate Zoology*, 2nd ed. New York: Oxford University Press, 1972.

Prosser, C. L., ed., *Comparative Animal Physiology*, 3rd ed. Philadelphia: W. B. Saunders, 1973.

Repetski, J. E., "A Fish from the Upper Cambrian of North America," *Science*, vol. 200 (1978), p. 529.

Valentine, J. W., "The Evolution of Multicellular Plants and Animals," *Scientific American*, vol. 239:3 (1978), p. 140.

Whittaker, R. H., "New Concepts of Kingdoms of Organisms," *Science*, vol. 163 (1969), p. 150.

chapter 2

Protozoa

The phylum Protozoa [Gr. *protos*, first + *zoon* animal] consists of the unicellular animals (see Fig. 1-4). Although every animal has a stage in its life cycle when it is unicellular—the fertilized egg—only protozoans remain unicellular throughout life. Some species of protozoans form colonies, but the members of the colony retain sufficient independence so that considering such a colony a multicellular organism is not justifiable. It is erroneous, however, to assume that protozoans are simple animals merely because they are single-celled, microscopic organisms. T. M. Sonneborn has stated that a paramecium is "far more complex in morphology and physiology than any cell of the body of man."[1] Aside from their unicellularity, there is much diversity among the protozoans. Whereas in multicellular animals, the complexity of their organs reveals their evolutionary level of attainment, among

[1]T. M. Sonneborn, "*Paramecium* in Modern Biology," *Bios*, vol. 21 (1950), p. 31.

protozoans this is reflected in the number and kinds of organelles that exist in or on the single cell that constitutes the entire organism.

Protozoans are the subject of active investigation in many laboratories. The studies include such topics as nutritional requirements, mechanisms of locomotion, life cycles, genetics, and the treatment of infections caused by pathogenic protozoans. An understanding of protozoans is of prime importance to humans because of the devastating parasitic species that infect us and our domestic animals. Amoebic dysentery, sleeping sickness, and malaria affect many millions of humans. Malaria is probably responsible for the deaths of more people than any other illness caused by a pathogen. In tropical Africa alone, more than a million children under age 14 die each year as a consequence of malaria.

CLASSIFICATION The phylum Protozoa has been divided into four classes. The divisions are based primarily on the presence and type or absence of structures that can provide for locomotion.

The class Mastigophora consists of those protozoans which possess, as locomotor organelles, one or more "whiplike" structures called *flagella* (Fig. 2-1). Consequently, mastigophorans are commonly known as *flagellates*. *Trypanosoma gambiense*, the causative agent of African sleeping sickness, is a flagellate.

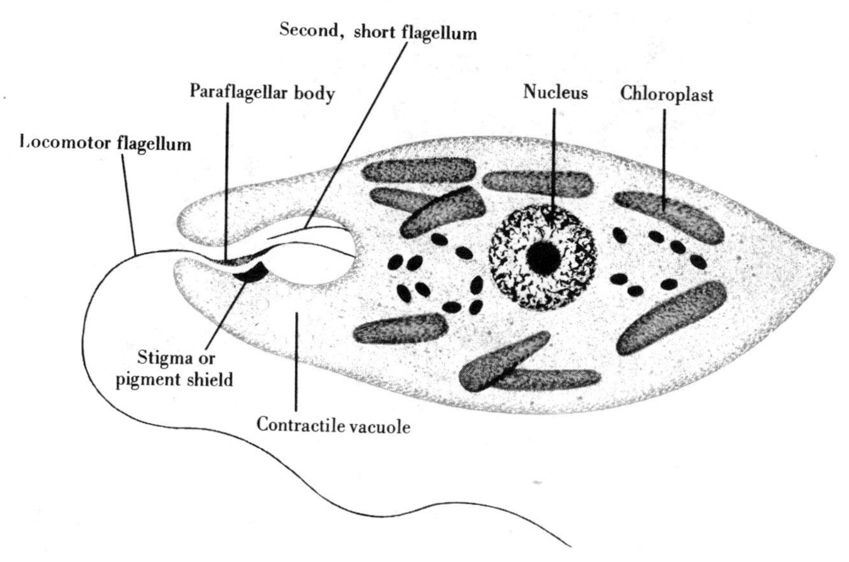

Fig. 2-1 Euglena *sp.*

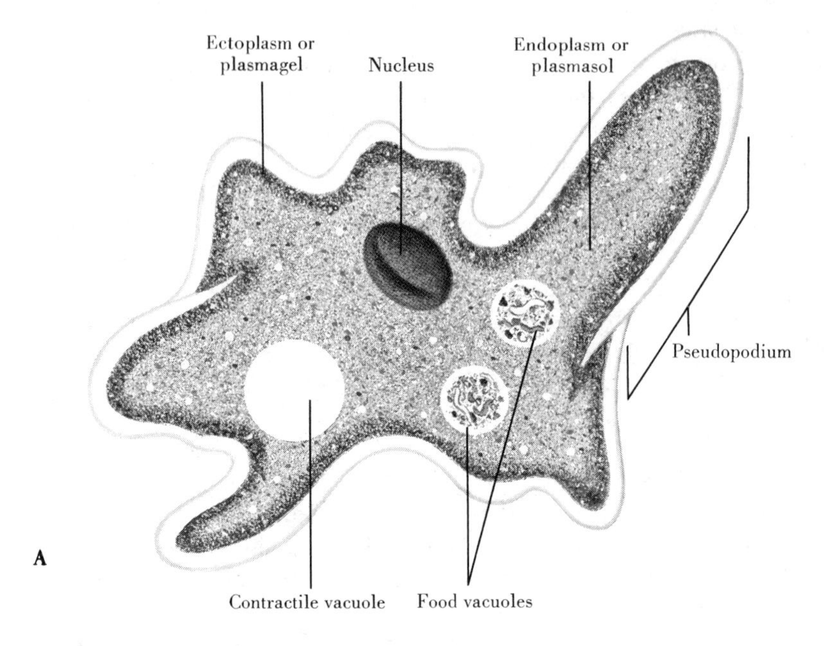

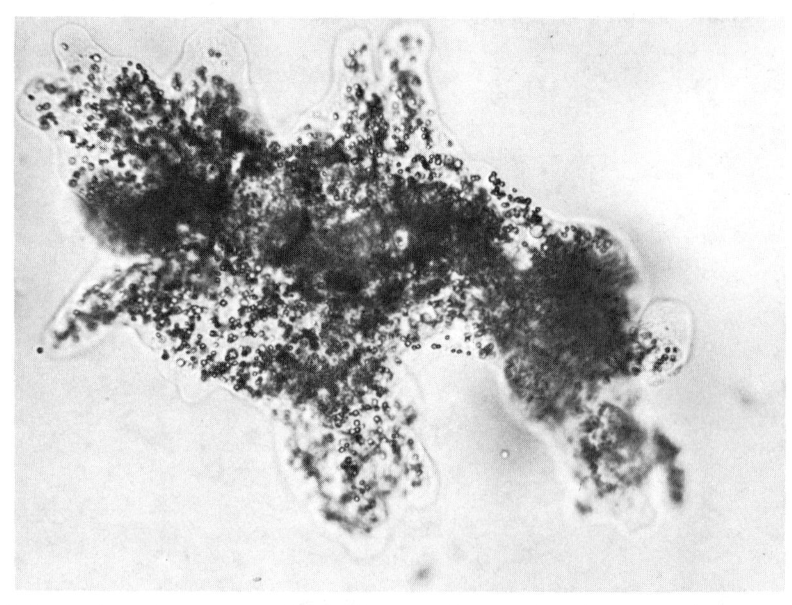

Fig. 2-2 Amoeba proteus. *(A) A diagram illustrating the parts of the organism. (B) Photograph of a living specimen. [(B) Courtesy of Ward's Natural Science Establishment, Inc.]*

The class Sarcodina includes those protozoans which move by means of *pseudopodia,* or flowing protoplasmic extensions. In addition to free-living species (Fig. 2-2) this class also contains *Entamoeba histolytica,* the causative agent of amoebic dysentery in humans.

The class Sporozoa consists of the protozoans which lack locomotor structures at maturity. All members of this class are parasitic. Malaria in humans is caused by four species of the genus *Plasmodium, Plasmodium vivax,* for example.

The class Ciliata is characterized by the possession of *cilia,* or short, hairlike organelles. Cilia occur in large numbers on the surfaces of the members of this class at some stage during their life cycles and serve as their locomotor organelles. *Paramecium caudatum* (Fig. 2-3) is a ciliate commonly found in fresh water.

MORPHOLOGY AND PHYSIOLOGY

LOCOMOTION

The question of how cells move from place to place has concerned biologists for many years, and the answers are still incomplete. Many cells move along by virtue of changes in cell shape that are associated with flexible cell membranes and the extension of

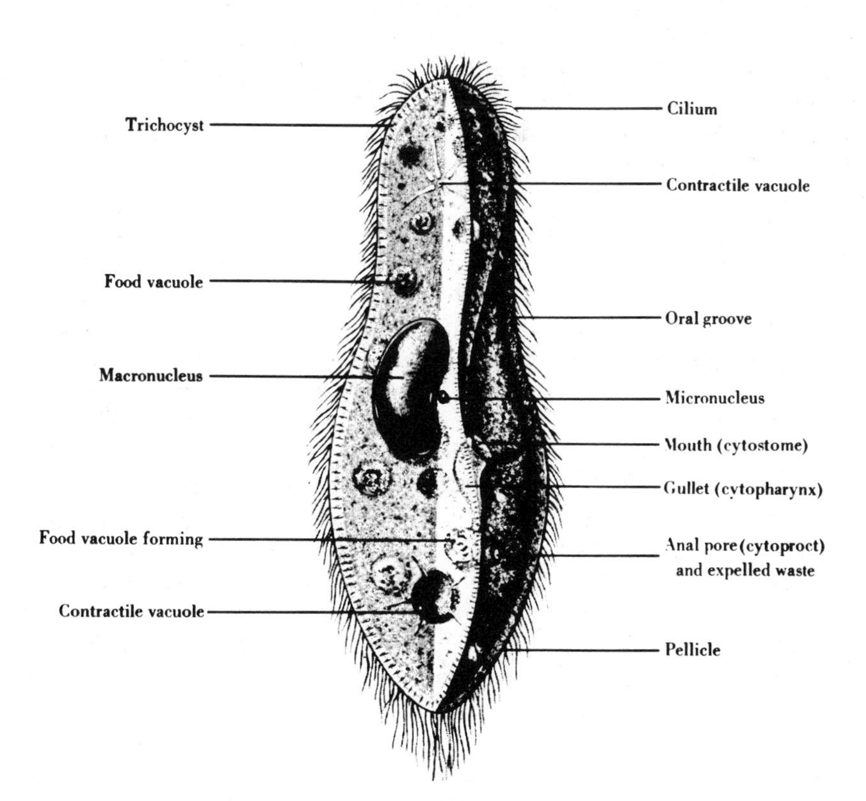

Fig. 2-3 Paramecium caudatum. *(After Johnson, Laubengayer, DeLanney, and Cole,* Biology, *3rd ed., Holt, Rinehart and Winston, 1966.)*

pseudopodia. Pseudopodia vary in shape; generally, they are either broad with rounded tips or threadlike with pointed tips. Locomotion across a substrate by the extension of pseudopodia is called *amoeboid movement*. It is used by many cells in metazoans (such as the white blood cells of vertebrates) as well as by sarcodines. The core of an amoeboid cell consists of a low-viscosity *endoplasm*, or *plasmasol*, in which the cytoplasmic granules flow freely. Surrounding this core is the relatively high viscosity *ectoplasm*, or *plasmagel*. The viscosity of a substance is a measure of the degree to which its molecules associate with one another. Amoeboid cells differ markedly in cell form. Some move by use of only a single large pseuopodium, whereas others use several small ones. The initial event in amoeboid movement is that the cell attaches itself to the substrate, the strongest attachment being at the tips of pseudopodia as they are extended. Then, as an amoeboid cell begins to move forward, the ectoplasm in its posterior region is converted into the more fluid endoplasm and flows forward into any forming pseudopodium, where it is again converted into the more viscous ectoplasm. Several theories have been proposed to account for the forward flow of the endoplasm. The following three theories are now the most popular ones. The endoplasm may be pulled forward by contraction of ectoplasm at the leading end; it may be squeezed forward by contraction of ectoplasm at the trailing end; or chemical "ratchets" on the inner edge of the ectoplasm may push forward individual molecules of endoplasm, which in turn drag along the endoplasm molecules near the center. It is unlikely that only one of these theories is applicable to all amoeboid cells; more likely, amoeboid cells developed more than one mechanochemical mechanism of locomotion.

Cell locomotion can also be accomplished as a result of propulsion by cilia or flagella, both being hairlike motor organelles at the cell surface. Characteristically, ciliated cells have many more cilia than flagellated cells have flagella. Cilia have a to-and-fro bending movement, whereas flagella usually have an undulatory motion. When viewed with an electron microscope, the only structural difference that is seen between flagella and cilia is that flagella are much longer than cilia. The structure of both organelles is, with very few exceptions, the same in all eukaryotic cells (Fig. 2-4). A typical cilium or flagellum is bounded by a membrane, within which is a circular array of nine double microtubules and, at the center, two single *microtubules*. Microtubules are minute, elongated, hollow cylinders (about 25 nm in diameter) which are found only in eukaryotic cells. Located at the base of each cilium and flagellum is a *basal body*. In the outer wall of the basal body are embedded nine groups of three microtubules.

This nine triplet pattern is also seen in *centrioles*. Centrioles and basal bodies seem to be the same structure put to different uses. In fact, in some cells that have a single flagellum, a centriole actually migrates toward the cell surface and becomes the basal body of the flagellum. The function of the basal bodies is to regulate synthesis and

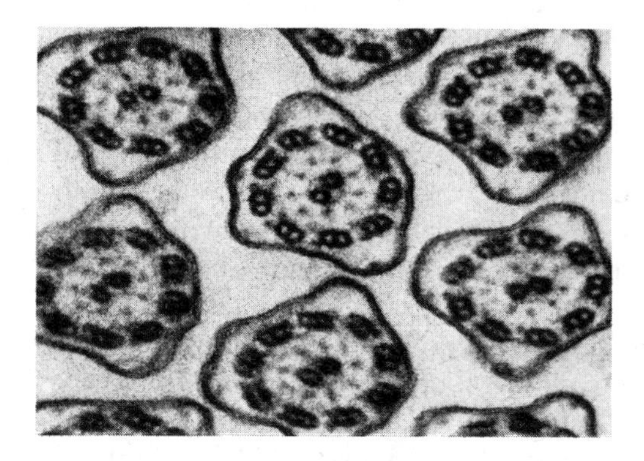

Fig. 2-4 *The structure of flagella from* Pseudotrichonympha *sp. as seen in cross section by means of the electron microscope. (Courtesy of I. R. Gibbons, Harvard University.)*

aggregation of the protein molecules required for the formation of the microtubules in cilia and flagella. The mechanism providing for movement of cilia and flagella is not fully understood, but modern theories favor the view that the moving force is in the organelles themselves, the double microtubules having the ability to slide past each other and thereby producing the movement. In contrast, the older, less likely view is that the moving force is in the cell body proper and these organelles act only passively. One microtubule of each doublet around the periphery of the organelle has two rows of arms consisting of a protein *(dynein)* that has enzymatic activity, splitting the energy-yielding molecule adenosine triphosphate. The enzyme is there presumably to release energy to be used in the movement of these organelles.

NUTRITION AND FEEDING

An active and important area of research with protozoans is the study of their nutritional requirements. Efforts have been made to culture many of them in a completely defined medium, one consisting of substances of known chemical composition. Probably the best known example is the study performed with the ciliate *Tetrahymena pyriformis*. G. W. Kidder and his associates obtained an excellent rate of growth of the population in their culture of this ciliate with a medium consisting of 16 amino acids, 10 vitamins, glucose, and other compounds, both organic and inorganic. Omission of any one of these substances resulted in a decline of the growth rate. Once a defined medium is established, it is possible to study the biochemistry of the organism in a depth greater than would be possible otherwise.

Protozoans have three types of nutrition. The chlorophyll-bearing flagellates are capable of photosynthesis, using energy from the sun, and manufacture their own organic molecules and are therefore called *autotrophs;* but all other protozoans derive their energy from organic compounds in their diet. Although these autotrophic protozoans have a plantlike nutrition, they have several counterbalancing animal-like characteristics such as locomotion by means of flagella and the presence of a contractile vacuole (an organelle for salt and water balance that will be discussed below); in most such species there is no rigid cell wall, and many of them have a light-sensitive structure that functions as an eye. The presence of such animal-like characteristics among these autotrophic unicellular organisms seems to be adequate justification to treat them as members of the phylum Protozoa. Some colorless (that is, lacking chlorophyll) flagellates and all parasitic protozoans are *saprozoic*, that is, they do not ingest solid food but subsist entirely on dissolved substances actively absorbed from the medium. In the case of an internal parasite, the medium is, of course, the host. But some flagellates can change their mode of nutrition, being autotrophs or saprozoites depending upon environmental conditions such as available light. The rest of the protozoans are *holozoic:* they subsist on other living organisms such as bacteria, small algae, other protozoans, and even small metazoans. Species that are capable of amoeboid movement ingest the organisms on which they feed by encircling them with pseudopodia, a process called *phagocytosis*. Once the food is surrounded, the tips of the pseudopodia fuse and the cell membrane forms a food vacuole, which is moved into the endoplasm. Digestive enzymes are then released into the vacuole. After digestion has been completed, the nutrients are absorbed from the vacuole. With its residual undigestible matter the vacuole is then moved to any part of the cell surface, and the contents are emptied to the outside. Holozoic flagellates commonly engulf food in the depression where the flagella attach to the body. Dissolved substances, which are too large to pass through the cell membrane into the cell by means of the ordinary passive and active transport mechanisms, can nevertheless be ingested by saprozoic protozoans by use of a process called *pinocytosis* (cell drinking), which does not involve the formation of pseudopodia. Instead, the material becomes absorbed onto the outer surface of the cell membrane. Then the cell membrane folds inward, and a small vesicle containing the material is subsequently pinched off into the cytoplasm.

Free-living ciliates have the most advanced feeding apparatus of all protozoans. A mouth *(cytostome)* is ordinarily present, and this leads into a gullet *(cytopharynx)* at the end of which food vacuoles form. Quite commonly there is also an oral groove, as in *Paramecium*, that leads from the body surface to the cytostome. Certain of the cilia beat in such a manner as to direct food toward the cytostome. In both ciliates and flagellates, undigested food is egested at a specific point on the body, the *cytoproct*.

All ciliates, a few sarcodines, and many flagellates have a *pellicle*. A pellicle is a living, membranous structure that is a modification of the usual three-ply cell membrane, in that it is thicker and consequently is able to provide more protection to the cell than can a simple cell membrane alone. Associated with the pellicle of some ciliates are oval or rodlike organelles, called *trichocysts*, that are used for capturing food, for anchoring, and for defense. There are three types of trichocysts. The type found in *Paramecium* and some other ciliates consists in the undischarged state of a capsule with a dense, thornlike tip. Even the electron microscope does not reveal the presence of a tube in the capsule; but when this type of trichocyst discharges, the capsule's contents are rapidly extruded, forming in the process an elongated tube at the free end of which is the thornlike tip. Still other ciliates, including *Dileptus anser*, possess a second type of trichocyst. It contains, even in the undischarged state, a tube that has an open tip. During the discharge of this second type the tube turns itself inside out, like pushing out the finger of a glove. From the open tip of the elongated tube is then released a fluid that has a paralyzing effect on other protozoans. The third type of trichocyst, found in *Dileptus gigas* and still other ciliates, consists of a fluid-filled capsule the contents of which are released upon stimulation. No tube is present nor is one formed. As in the second type, the contents of the third type are paralytic. Little or nothing is known of the mechanism of trichocyst discharge. Discharge probably involves either (a) the hydration and consequent swelling of a protein within the trichocyst, perhaps in response to a pH change in the trichocyst or (b) the simple osmotic flow of water into the trichocyst. In either case, swelling would be induced within the trichocyst, and the resultant increase in pressure within the trichocyst evokes extrusion or release of the contents. It appears that the type found in *Paramecium* functions as an organelle of attachment for temporary anchorage and for food capture, whereas the second and third types seem to provide for protection of the bearer.

SHELLS AND SKELETONS

Some of the sarcodines have either a protective shell or an internal skeleton. The animal forms a shell either by secreting one around itself or by cementing together particles such as sand grains from the environment. There is at least one opening in the shell through which pseudopodia can protrude. The shells of the group of sarcodines called foraminiferans are almost always secreted and most formaniferan shells are composed of calcium carbonate. The type of shell formed by cementing foreign particles together is formed by some other sarcodines such as *Difflugia*. Its shell is egg-shaped, formed of sand grains, and has an opening at one end through which pseudopodia can project. Skeletons are found in still other sarcodines, for example radiolarians, whose skeletons are usually

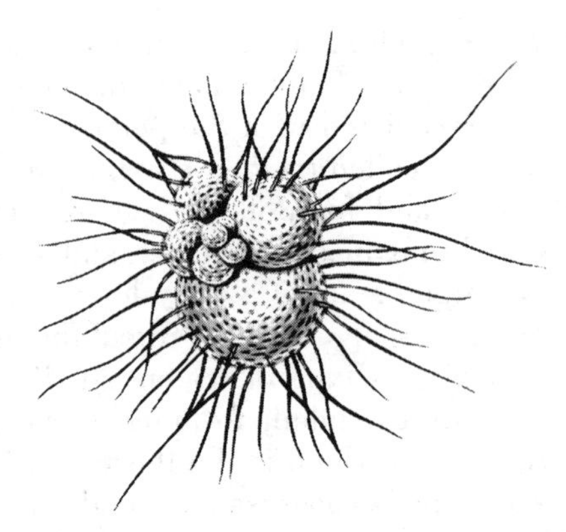

Fig. 2-5 Globigerina *sp. Note the threadlike pseudopodia and the different-sized chambers of the shell.*

composed of silicon dioxide. Large deposits of foraminiferan shells have accumulated through the ages, resulting in limestone beds, such as the White Cliffs of Dover. *Globigerina* (Fig. 2-5) is an example of a foraminiferan. *Globigerina* begins life in a single chamber. As it grows, it overflows through an opening in the wall of the chamber and then secretes another larger compartment. This process continues throughout the life of the individual. Ultimately, the shell consists of a series of interconnected chambers with a spiral-like arrangement, and it has many fine pores through which numerous threadlike pseudopodia project. At ocean depths of 2,500–4,500 meters, empty foraminiferan shells occur in great quantities in the mud, which is called *Globigerina ooze* because of the prevalance of this genus. At greater depths, the calcareous shells of foraminiferans dissolve because the increased capacity of the sea water to hold carbon dioxide at the prevailing pressure results in the deeper waters becoming less alkaline than the surface waters. There the bottom mud consists largely of siliceous skeletons and is consequently called *radiolarian ooze*.

Geologists have great interest in the fossil remains of foraminiferans and radiolarians for several reasons. First, both types of fossils serve as a means of determining the age of deposits. Second, geologists can make use of the fact that some species of radiolarians are characteristic of warm waters and others of cold waters, in determining the ancient temperatures of bodies of water. Third, of interest to oil geologists in particular is the fact that certain species of fossil foraminiferans are associated with oil-containing strata.

EXCRETION AND RESPIRATION Excretion, in addition to the elimination of metabolic waste products, particularly nitrogenous ones, includes the regulation of the salt and water concentrations in an animal. Freshwater protozoans are *hyperosmotic* to their environment, that is, the salt concentration within the cell is greater than that of the environment. Consequently, water constantly enters through the cell surface by osmosis and threatens to cause swelling and eventual rupture of the cell unless the excess is excreted. Animals that have the capacity to maintain a relative constancy of their body fluids when under environmental osmotic stress are known as *osmoregulators*, and the regulatory process itself that is used to accomplish this is termed *osmoregulation*. All freshwater metazoans are likewise hyperosmotic to their environment.

Because the bloods and tissue fluids of metazoans are similar to seawater in their relative ionic compositions, it is generally accepted that life evolved in the sea. In order for animals to have colonized the freshwater environment successfully, they had to have a means of eliminating the excess water load. It is the function of the protozoan *contractile vacuole* to accomplish this task. The contractile vacuole becomes filled with the excess water and then expels its contents into the environment. Contractile vacuoles are found in all freshwater protozoans and, in addition, in many marine and a few parasitic protozoans. In marine and parasitic protozoans, the solute concentration in the cytoplasm is ordinarily the same as that of the surrounding medium; that is, these organisms are ordinarily *isosmotic* with their environment and do not have the problem of osmotic swelling caused by water entering through the body surface. The marine and parasitic species that have contractile vacuoles use them to eliminate water derived from nonosmotic sources (food and metabolism). The electron microscope has revealed that contractile vacuoles are usually formed initially by the fusion of numerous small vesicles. Then the contractile vacuole grows as additional vesicles fuse with it. In ciliates, the contractile vacuole forms in association with a permanent, surrounding system of canals.

Several theories have been proposed for the mechanism of filling the contractile vacuole. According to an older theory, water is actively secreted directly into the contractile vacuole. But a variety of experiments have led many physiologists to doubt seriously that active transport of water ever occurs in any animal. Consequently, mechanisms based on solute transport, such as that proposed by D. H. Riddick on the basis of experiments he performed with the freshwater sarcodine, *Pelomyxa carolinensis*, are now favored. He found that the fluid in the contractile vacuole not only has a constant concentration, but also is always *hyposmotic* to (less concentrated than) the surrounding cytoplasm, no matter how small the vacuole, and hypothesized that the vesicles that will fuse to form a new contractile vacuole, or fuse with it to enlarge it, are initially not only fluid-filled but also isosmotic with the cytoplasm. But prior to any fusion, the vesicles will lose solute, becoming hyposmotic, and then they will fuse to form a new contractile vacuole or fuse with the contractile vacuole to enlarge it, discharging their contents into the vacuole, but not until they have become hyposmotic to the surrounding cytoplasm. If solute were being pumped from the contractile vacuole itself after fusion of the vesicles, we would expect the vacuole to be more concentrated when it is first formed than later, but at least in *Pelomyxa carolinensis* this is not the case. Different species of protozoans may, of course, ultimately be shown to have different mechanisms for filling their contractile vacuoles. In *Paramecium*, for example, the canals surrounding the contractile vacuoles appear to empty their contents into the vacuoles. How these canals fill is not known.

Because of their small size, protozoans are able to carry on respiration and eliminate their nitrogenous wastes by diffusion alone; that is, the force moving the molecules across the cell membrane is simply their random thermal movement. Calculations based on the rate of diffusion of oxygen through tissues have revealed that even in an organism with a fairly high metabolic rate an adequate supply of oxygen can be delivered to the cells by diffusion alone, provided the body shape is such that every part is no more than 0.5 mm from an oxygen supply. Only when an organism attains a larger size must special mechanisms be developed to transport oxygen, unless the metabolic rate of the organism is very low. Larger protozoans tend to be elongated rather than spherical, thereby assuring an efficient exchange of materials between the environment and all parts of the cell. This is true not only because the central point within the elongated cell is closer to the surface than it would be in a spherical organism of the same volume, but also because the surface-to-volume ratio is larger. The main nitrogenous waste product of protozoans appears to be ammonia. In fact, ammonia is the main nitrogenous waste product of all marine and freshwater invertebrates.

REPRODUCTION Although some flagellates and some sarcodines can only reproduce asexually, the rest of the protozoans can reproduce both asexually and sexually. The usual method of asexual reproduction is *binary fission,* whereby the organism divides into two approximately equal parts through an ordinary cell division involving mitosis. Where there is a definite body axis, as in flagellates and ciliates, fission occurs in a predictable plane. In flagellates the plane of splitting runs from the anterior end to the posterior, while in most ciliates the split is transverse.

What is the biological significance of sexual reproduction? Briefly, it provides not only for an increase in the number of individuals but also for a change in their genetic makeup. Through a sexual process genetic variation can express itself, whereas in asexual reproduction the daughter cells have the same genetic constitution as did the parent. In many flagellates, some sarcodines, and all sporozoans, sexual reproduction involves fusion of two individuals, followed by meiosis to produce offspring with new genetic constitutions. Characteristic of ciliates is a specialized sexual process, *conjugation*, in which two organisms form a lateral attachment, exchange nuclear material, separate, and subsequently undergo a series of divisions involving at times only nuclei and at other times the entire animal (Fig. 2–6).

Many protozoans, especially freshwater and parasitic species, can encyst during unfavorable conditions such as drought and lack of food. This is called "protective encystment" and involves formation of a gelatinous, protective covering, loss of cilia and flagella, if either had been present, and decrease in size due to expulsion of water. However, encystment can also be part of the normal reproductive cycle, and the cysts formed then are called reproduction cysts. An interesting example of reproductive encystment is that of the flagellate genus *Trichonympha*, seven species of which live in the digestive tract of the cockroach, *Cryptocercus punctulatus*, which in nature feeds only on wood. The cockroach is incapable of digesting the wood and needs the flagellate to provide cellulase, a wood-digesting enzyme. After the wood has been digested, the cockroach has available to it glucose, a carbohydrate that it is capable of metabolizing. The protozoans in turn contain intracellular bacteria, and it may be the bacteria, rather than the protozoans themselves, that produce the cellulase. When the immature cockroach undergoes the molting process (which includes not only the formation of a new external skeleton but also *ecdysis*, or the shedding of the old one) in order for it to grow, it not only sheds the old skeleton but also loses part of the lining of its digestive tract, both anteriorly and posteriorly. At the posterior end, the lining of the posterior portion of the intestine (rectum) is lost. About four days prior to ecdysis the flagellates encyst. When, at ecdysis, the lining and contents of the rectum are eliminated, many, but not all, of the encysted flagellates pass out into the environment, where, if they had not encysted, they would not survive for long because of desiccation. Those flagellates that remain in the cockroach soon emerge from their cysts. Newly hatched cockroaches then ingest cysts that were expelled by older roaches to obtain the flagellate without which they could not survive. Encystment provides resistant organisms in the environment to infect newly hatched cockroaches which at hatching are apparently always uninfected; the purpose is not to maintain the infection in the cockroach from which they came. Molting of the cockroach is induced by a hormone, ecdysone, and according to L. R. Cleveland and his associates, this same hormone causes encystment of *Trichonympha*. These findings not only illustrate a flagellate's sensitivity to an insect hormone, but also show the

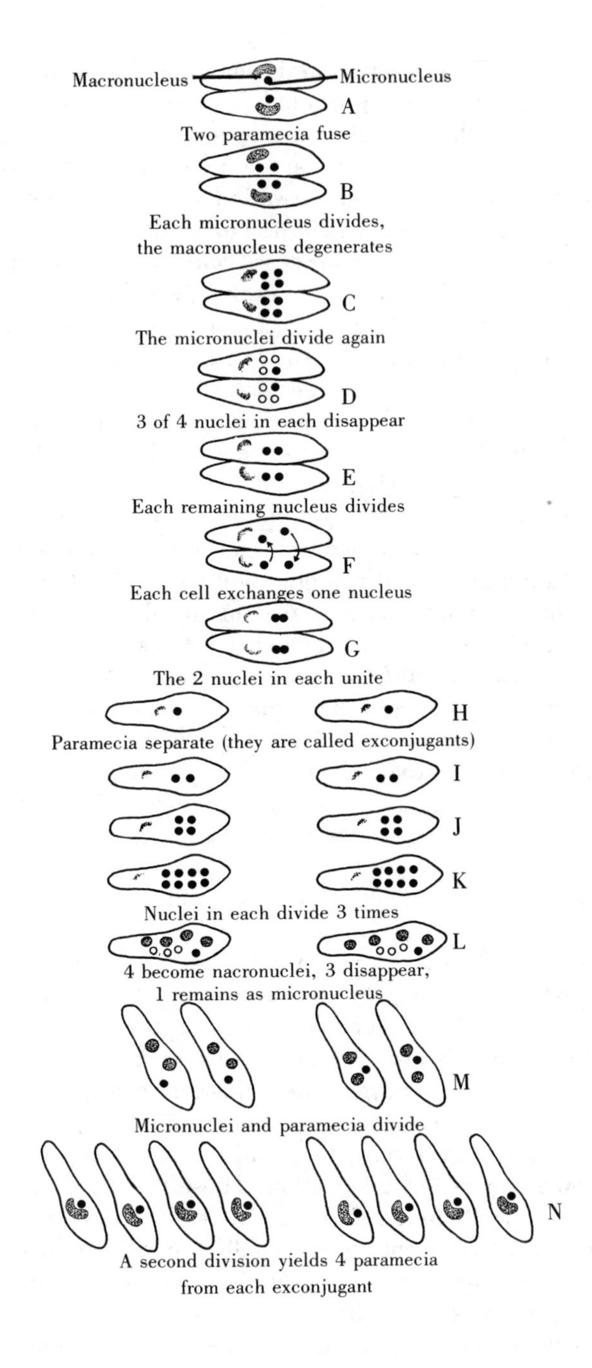

Fig. 2-6 *Conjugation in* Paramecium caudatum. *(Modified from Storer, Usinger, Nybakken, and Stebbins,* Elements of Zoology, *4th ed. Copyright © 1977, McGraw-Hill Book Company. Used by permission.)*

evolution of a remarkable relationship between two organisms. See Chapter 9 for a further discussion of insect hormones, including ecdysone.

Another example of encystment can be seen in the life cycle of the sporozoan *Plasmodium*, which causes malaria. Malaria is transmitted by female mosquitoes of the genus *Anopheles*. When an infected mosquito bites a human and sucks the blood, the infective stage, a *sporozoite*, of *Plasmodium* enters the human's bloodstream. Only female mosquitoes are blood-sucking. Male mosquitoes cannot suck blood because they lack the appropriate mouth parts needed to pierce a human's skin and, therefore, do not transmit malaria. The sporozoites enter the wound with saliva of the mosquito that is released when the mosquito starts to suck the blood. The mosquito's saliva contains an anticoagulant that prevents clotting of the blood. The sporozoites then enter red blood cells. There they reproduce asexually, each sporozoite forming from 6 to 36 (the number depending upon the species of *Plasmodium*) individuals of another stage in the life cycle, *merozoites*. The infected red blood cells burst at regular intervals (48 or 72 hours depending upon the species of *Plasmodium*), releasing the merozoites which then enter other red blood cells and reproduce asexually again, each merozoite producing 6 to 36 more merozoites. Some of the merozoites, while inside the blood cells, eventually develop into *gametocytes*, the next stage in the life cycle. There are two types of gametocytes, small ones called "male gametocytes" and large ones called "female gametocytes." Gametocytes become liberated by destruction of the red blood cells while still in the blood of the human, but gametocytes can undergo no further development in a human. However, when a mosquito bites an infected human, the mosquito will suck up some blood containing these gametocytes. In the stomach of the mosquito the male and female gametocytes then undergo a maturation process forming male and female *gametes*, mature reproductive cells. A male gamete then fuses with a female gamete. As with the gametocytes, these male gametes are smaller than the female gametes. The cell formed by the union of the two gametes then invades a cell in the wall of the mosquito's digestive tract, where it encysts and then undergoes a series of divisions, producing many new sporozoites. When the cyst ruptures, the sporozoites escape and migrate to the salivary glands, where they are ready to invade another human and repeat the life cycle.

LIGHT AND CHEMICAL SENSITIVITY

Protozoans can respond to light. Most simply exhibit a general light sensitivity of the body surface, but in some flagellates there is a localized light sensitive structure. In *Euglena*, the primary structure for light reception appears to be the paraflagellar body which is contained within the long flagellum (Fig. 2-1). The adjacent stigma (or pigment shield), which is composed of orange-red granules, is thought to function as an inert shading device only, thereby enabling the photoreceptor to function as a directional detector. The paraflagellar body and stigma of *Euglena* form a functional unit, a localized photoreceptor that is entitled to be called an "eye." In general, protozoans avoid bright light and complete darkness, tending to select a moderate light intensity. However, as will be seen below, the response to light can vary depending upon the time of day. Protozoans are sensitive also to chemicals in the environment, being attracted to some and avoiding others.

THE BIOLOGICAL CLOCK

The ability of plants and animals to "mark off" periods of time repeatedly with remarkable precision, independent of environmental variables such as light and temperature, is now well documented. This time-keeping ability is due, at least in part, to an internal timer, called the *biological clock,* whose exact nature is not yet known. At present there are two theories concerning the clock mechanism. First, the endogenous theory states that it is a self-sustaining oscillator: that is, the organism is able to mark off periods of time independently of any environmental factors. Second, the exogenous theory holds that the biological clock within the organism is dependent for its time-keeping ability upon subtle, external rhythmical geophysical forces. Because protozoans offer an investigator the opportunity to study the clock at the cellular level, they are being used for just this purpose in several laboratories. *Paramecium bursaria* shows an interesting behavior pattern, in which the population typically mates during the daytime, and rarely or never at night. *Euglena gracilis* is attracted to a particular intensity of illumination by day but evidences little or no attraction to light of the same intensity at night. These regularly recurring changes are known as *biological rhythms.* They are the overt expression of the biological clock upon which they depend. The rhythm of phototactic sensitivity exhibited by *Euglena gracilis* may have evolved as a survival mechanism that ensures that this photosynthetic organism is attracted to light during the period of time when it would normally be exposed to illumination in nature. When cultures of these organisms are maintained in the laboratory in constant darkness, both rhythms persist (for at least a few days) in close synchrony with the external environmental daytime–night illumination changes of the solar day. Solar-day rhythms such as these are also known variably as 24-hour, diurnal, and *circadian* rhythms. The latter term, circadian (meaning "about a day"), refers to the fact that when animals are kept under conditions of constant light and temperature which deprive them of variations in their two major phase-setting periodic environmental cues, the recurring cycles may become slightly longer or shorter than 24 hours. One of the basic properties of the biological clock is a virtual temperature-independence, that is, the period of the biological clock is essentially unaffected by the daily fluctuations in environmental temperature. For example, the clock in *Euglena* is virtually unaffected by changes in temperature in the range 20–33°C, a phenomenon remarkable in an organism whose other physiological processes are markedly slowed by cold and accelerated by heat. If the speed at which the clock ran were as sensitive to these temperature fluctuations as are other physiological processes, it would be unreliable at best, and the organism would probably be better off without it.

PHYLOGENY There is general agreement that the flagellates are the most primitive protozoans, approximating the ancestral protozoans more closely than any other living protozoans. Furthermore, because there is so much diversity among the protozoans, it is most likely that they had a polyphyletic origin. That is, the living protozoans probably evolved from several different ancestral forms, which evolved into different stocks of protozoan flagellates. These newly evolved flagellate stocks then gave rise to different groups among the sarcodines, sporozoans, and ciliates. There was apparently not a single common ancestor for all the protozoans that gave rise to all the flagellates and through them to the rest of the phylum as well. As already noted, some species of flagellates have the plantlike ability of being able to carry on photosynthesis through the use of chloroplasts. Such flagellates appear to represent a transition group between those unicellular organisms which are strictly plantlike and the strictly animal-like protozoans. Quite likely, such transitional flagellates gave rise to the higher plants, to strictly animal-like flagellates, and through the latter to the rest of the animal kingdom as well. To retrace one step further, the flagellates themselves, which are eukaryotes, probably arose from prokaryotes. One theory suggests that eukaryotes arose from combinations of prokaryotes, one prokaryotic cell coming to live inside a larger host prokaryotic cell. According to this theory, the organelles, such as mitochondria and chloroplasts, that are present in eukaryotic cells but not in prokaryotic cells were at one time independent organisms. Prokaryotes, in addition to lacking a membrane-bound nucleus, also have no membrane-bound organelles such as mitochondria or chloroplasts.

A close relationship between the flagellates and sarcodines is evidenced by a group of protozoans, the order Rhizomastigida (Fig. 1-14B), which are amoeboid but possess at the same time one or more flagella. This group has been arbitrarily classified with the flagellates. Pseudopodia probably evolved as a supplementary feeding-mechanism in the flagellates and eventually became a new means of locomotion as

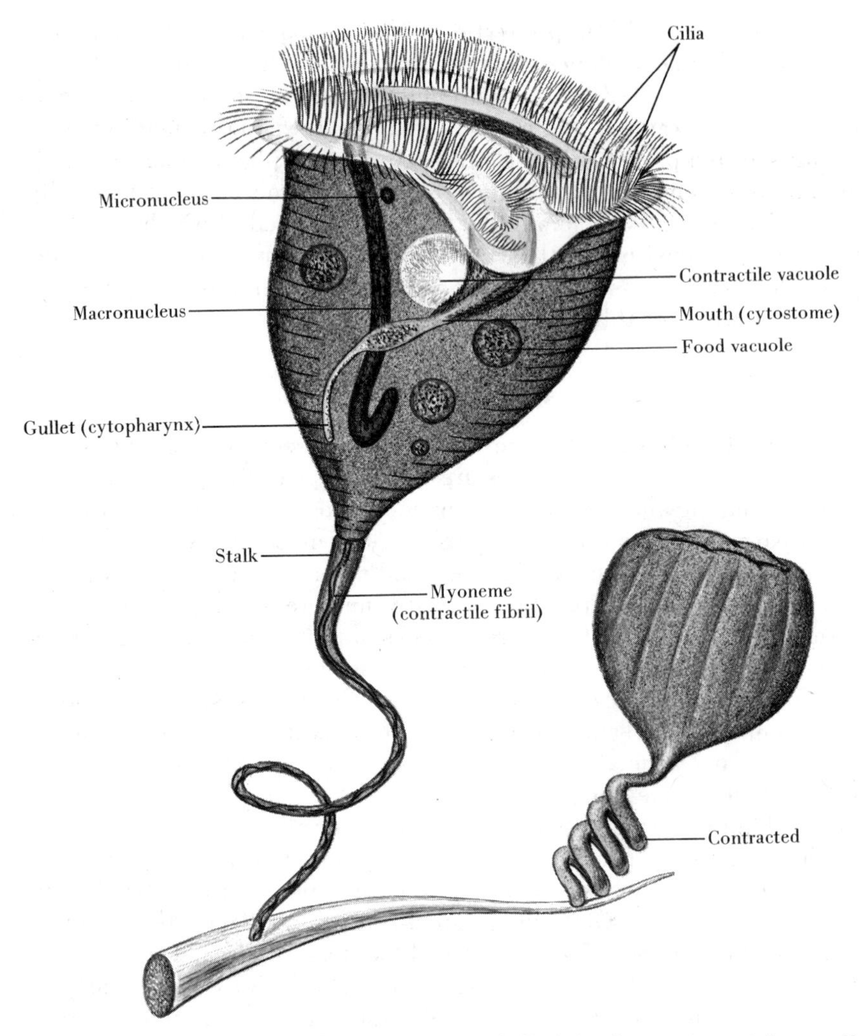

Fig. 2-7 Vorticella *sp. This ciliate is a bell-shaped organism with a stalk containing contractile fibril. The animal becomes attached to the substrate by means of the stalk. When the stalk contracts, the ciliature around the mouth region becomes folded inward. (After Mackinnon and Hawes,* An Introduction to the Study of Protozoa, *The Clarendon Press, 1961; and Sherman and Sherman,* The Invertebrates: Function and Form, *Macmillan, 1970.)*

well. This led ultimately to the loss of the flagellum and to the evolution of the sarcodines. The sporozoans are closely related to both the flagellates and the sarcodines because immature sporozoans may possess flagella or pseudopodia. The ciliates are a more distantly related group.

Ciliates are the most highly evolved protozoans. They provide an excellent illustration of how complex an organism can become and yet remain a single cell (Fig. 2-7). For example, unlike the rest of the protozoans, each ciliate characteristically has two types of nuclei, a macronucleus concerned with vegetative functions ("everyday housekeeping") and a micronucleus that is essential for sexual reproduction (Fig. 2-6). However, a unicellular organism is limited in its potentialities. It is, for example, limited in size, because a single cell cannot provide the necessary respiratory and circulatory mechanisms that would be required in larger animals. Only through multicellularity was there an opportunity for a great step forward. Specialization of different cell groups for specific tasks, each group contributing to the survival of the whole, resulted in an animal that had distinct advantages over protozoans. No longer would a cell have to be a "jack-of-all-trades." A cell could become specialized for doing a single job well, leaving other specialized functions to different cells in the animal. Multicellular animals are treated in the following chapters.

FURTHER READING

Anderson, J. D., Amoeboid Movement, in *Comparative Animal Physiology*, 3rd ed., C. L. Prosser, ed. Philadelphia: W. B. Saunders, 1973, p. 799.

Brown, F. A., Jr., J. W. Hastings, and J. D. Palmer. *The Biological Clock: Two Views*. New York: Academic Press, 1970.

Cleveland, L. R. "Correlation between the Molting Period of *Cryptocercus* and Sexuality in Its Protozoa," *Journal of Protozoology*, vol. 4 (1957), p. 168.

———, A. W. Burke, Jr., and P. Karlson, "Ecdysone Induced Modifications in the Sexual Cycles of the Protozoa of *Cryptocercus*," *Journal of Protozoology*, vol. 7 (1960), p. 229.

Elliott, A. M., *Biology of Tetrahymena*. Stroudsburg, Pennsylvania: Dowden, Hutchinson, and Ross, 1973.

Grell, K. G., *Protozoology*. New York: Springer-Verlag, 1973.

Kidder, G. W., and V. C. Dewey, "The Biochemistry of Ciliates in Pure Culture," in *Biochemistry and Physiology of Protozoa*, vol. 1, A. Lwoff, ed. New York: Academic Press, 1951, p. 323.

Pappas, G. D., and P. W. Brandt, "The Fine Structure of the Contractile Vacuole in Ameba," *Journal of Biophysical and Biochemical Cytology*, vol. 4 (1958), p. 485.

Ragan, M. A., and D. J. Chapman, *A Biochemical Phylogeny of the Protists*. New York: Academic Press, 1978.

Riddick, D. H., "Contractile Vacuole in the Amoeba, *Pelomyxa carolinensis*," *American Journal of Physiology*, vol. 215 (1968), p. 736.

Satir, P., "How Cilia Move," *Scientific American*, vol. 231: 4 (1974), p. 44.

Sonneborn, T. M., "*Paramecium* in Modern Biology," *Bios*, vol. 21 (1950), p. 31.

chapter 3

Porifera

From the previous chapter it should be evident that although unicellular organisms are capable of performing those functions required to survive and perpetuate their species, unicellularity limits size. Consequently, large animals must be multicellular. The phylum Porifera [L. *porus*, pore + *ferre*, to bear], the sponges, represents one way this limitation was overcome. The phylum is so named because the bodies of these animals are perforated with many small pores (See Fig. 1-5), through which passes a constant current of water. In spite of the fact that sponges are metazoans, they are still at the cellular grade of construction, that is, their cells have maintained a considerable degree of independence, instead of being organized into tissues and organs. Nevertheless, there is specialization among the cells for different functions and some cooperation among the cells to produce a functional whole; thus, a sponge is more advanced than a protozoan colony. Some colonial protozoans do, however, approach the multicellular condition because of the degree of cellular interdependence they have attained. In a colony of the green flagellate *Volvox*, for example, the function of reproduction is carried out only by a few of the cells. Adult sponges are *sessile*. A sessile animal is either fixed in place or sedentary. Adult sponges have no means of locomotion. They are almost always attached to rocks, shells, or other objects. Although some sponges are radially symmetrical, the majority of sponges are asymmetrical. Sponges vary in size from a few millimeters to about 2 meters in diameter.

CLASSIFICATION

Sponges have been separated into four classes on the basis of their skeletal structures.

The class Calcarea consists of those sponges with a skeleton of separate crystal-like elements, called *spicules*, of calcium carbonate. Consequently, members of this class are known as calcareous sponges. Syconoid, leuconoid, and all the asconoid sponges are members of this class. The terms "syconoid," "leuconoid," and "asconoid" refer to the three structural types of sponges. These types will be described in the next section. *Leucosolenia botryoides* is a species of calcareous sponge that is commonly used in biology courses as a representative of this class. This class is exclusively marine.

The class Hexactinellida consists of sponges with a skeleton of siliceous, six-pointed spicules. Hexactinellids, commonly known as glass sponges, have syconoidlike and simple leuconoid body forms. They are all marine and mainly deep-water forms. The genus *Euplectella* (Venus's-flower-basket) is a member of this class; *Euplectella suberea*, for example, is found in the West Indies.

The class Desmospongiae consists of leuconoid sponges having a skeleton of siliceous spicules other than the six-pointed variety, or spongin fibers, or both. This class contains most of the species of sponges as well as the largest sponges. One family of this class consists of fresh-water sponges; the rest are marine. The spongin skeletons of *Hippospongia gossypina*, which are found near Florida and the Bahamas, have commercial importance as a bath sponge.

The class Sclerospongiae consists of a small number of leuconoid sponges, the skeletons of which consist of spongin fibers, siliceous spicules, and a basal mass of calcium carbonate. The living cells form a thin veneer over the calcareous mass. Spongin fibers and siliceous spicules lie among the living cells. As the sponge grows upward and outward, and the calcareous mass increasingly thickens, some fibers and

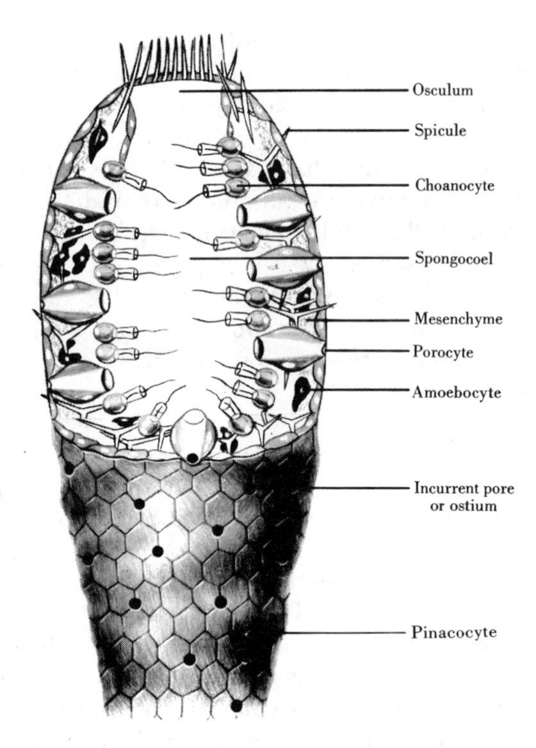

Fig. 3-1 *Diagram of an asconoid type of sponge. (Modified from Buchsbaum,* Animals without Backbones, *2nd ed., University of Chicago Press. Copyright © 1948. Used by permission.)*

spicules become trapped in the calcareous mass. *Ceratoporella nicholsoni* from the Jamaican coral reefs is an example. It can be up to 1 meter wide.

MORPHOLOGY AND PHYSIOLOGY

SPONGE STRUCTURE

Sponges can be readily divided into three structural types. The simplest and most primitive (Fig. 3-1, Fig. 3-2A) is the *asconoid* type. Asconoid sponges are all small, with radially symmetrical, vaselike bodies. A central cavity *(spongocoel)* opens to the exterior through a large opening at the top, the *osculum*. Sponges have neither a mouth nor digestive cavity. Structures peculiar to asconoid sponges alone are *porocytes* (tubular cells), which extend from the outer to the inner surface and are scattered throughout the body wall. The outer surface, except where the porocytes are located, is covered by a layer of flattened cells called *pinacocytes*. The entire layer is called the *pinacoderm*, a name given specifically to the epidermis of sponges. The

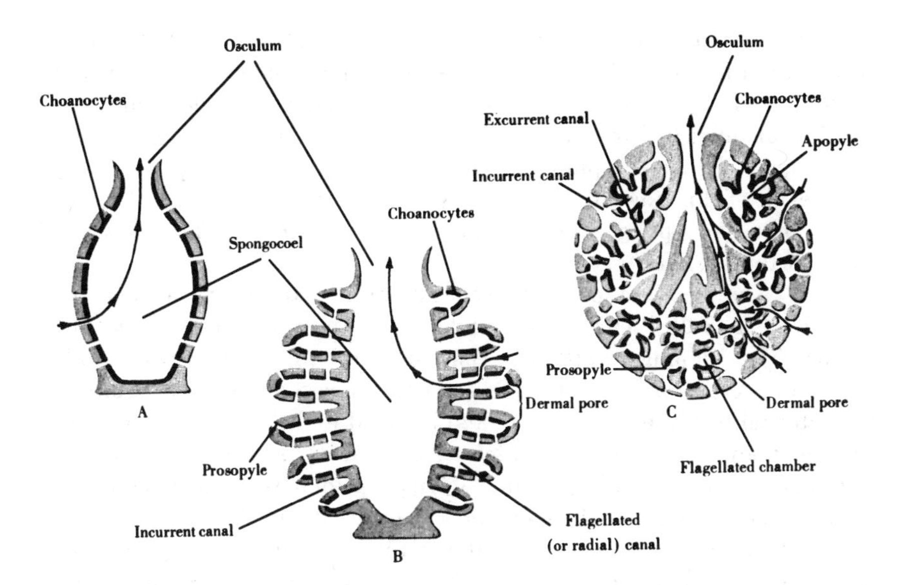

Fig. 3-2 *The three structural types found among sponges: (A) asconoid; (B) synconoid; (C) leuconoid. Arrows indicate the direction of the water current. (After Hyman,* The Invertebrates, *Vol. 1. Copyright © 1940, McGraw-Hill Book Company. Used by permission.)*

epidermis is the outermost cellular layer of the body wall of a metazoan. The spongocoel of asconoid sponges is lined with a single layer of *choanocytes* (flagellated collar cells). The lining of choanocytes is interrupted only by the inner ends of the porocytes. The electron microscope has revealed that the collar is not a homogeneous cylinder but actually consists of about 30 to 40 circularly arranged filaments projecting from the cell body. A layer of mucus helps to bind each filament to its neighbors, and, as will be described later, also assists in food capture. The beating of the flagella creates a current of water that enters the sponge through the porocytes and leaves *via* the osculum. The lumen of a porocyte is called an *incurrent pore,* or *ostium.* The middle layer of the body wall of a sponge is composed of mesenchyme.

In the asconoid type, the spongocoel contains such a large volume of water relative to the total volume of the sponge that the flagella are only able to force the water out through the osculum slowly. This situation has limited asconoid sponges to a small size. A simple increase in size without any compensating changes in the body plan would only have magnified the problem; as is clear from the surface-to-volume relationship, the volume of the spongocoel would increase at a faster rate than would the internal surface area that could accommodate additional choanocytes. The problem was surmounted when, through natural selection, the body wall became folded, resulting in a reduction of the volume of the spongocoel (decreasing the volume of water that must be circulated) and an increase of the internal surface area used to house the pumping cells (the choanocytes) (Fig. 3-2B, C). The result of this new organization, which presumably allowed the evolution of larger sponges, was a more efficient, greater flow of water through the sponge.

The first stages of body wall folding are seen in *syconoid sponges* (Fig. 3-2B). The external folds form the walls of the *incurrent canals.* The openings on the external surface into the incurrent canals are called *dermal pores.* Choanocytes became restricted to the walls of the internal folds. The space in each internal fold that is encompassed by choanocytes is called a *flagellated,* or *radial, canal.* The thick, horizontally folded walls contain alternating incurrent and flagellated canals. The incurrent and flagellated canals are connected by very small intercellular canals *(prosopyles);* the porocytes were lost. Water flows from the flagellated canals into the reduced spongocoel. Syconoid sponges, while retaining the radial vaselike shape of the asconoids, differ from them in two important aspects; first, in the thick folded walls containing the alternating incurrent and flagellated canals, and, second, in the breaking up of the choanocyte layer, which no longer covers the entire interior but is limited to the flagellated canals.

The most complex structural type is the *leuconoid* plan (Fig. 3-2C), where there is even further folding and thickening of the body wall than in the syconoid type. Almost all leuconoid sponges lost the radial symmetry of the asconoid and syconoid sponges, becoming asymmetrical. In the leuconoid type, the choanocytes are even further restricted, being found only in the walls of chambers *(flagellated chambers)* that were formed by evagination of the walls of the elongated flagellated canals. Water enters the sponge through the intercellular dermal pores on the external surface, passes through incurrent canals, and then into the flagellated chambers. The very small canals leading from the incurrent canals into the flagellated chambers are here also termed *prosopyles;* the apertures from these chambers are called *apopyles.* The water passes through the apopyles into small *excurrent canals,* which unite to form larger and larger excurrent canals, the largest leading to the osculum. Unlike asconoid and syconoid sponges, leuconoid sponges do not have a single large cavity that can be identified as the spongocoel. Among leuconoid sponges the spongocoel has become so reduced that all that remains of it are the excurrent canals leading from the flagellated chambers to the osculum. Because most species of sponges, as well as the largest sponges, are leuconoid, it is logical to conclude that this body plan is the most efficient of the three for meeting the needs of sponges.

A skeleton, secreted by amoebocytes, that strengthens and helps support the body is located primarily in the mesenchyme (although the pinacoderm is often pierced by spicules). It consists of (a) *calcareous spicules*, or (b) *siliceous spicules*, or (c) *spongin fibers*, or (d) a combination of (b) and (c) or (a), (b), and (c) with the spongin binding together siliceous spicules. Spongin is a fibrous protein related to collagen. Spicules of sponges are microscopic crystal-like structures that come in a great variety of shapes (Fig. 3-3). In some sponges (Venus's-flower-basket, for example), the individual spicules fuse into a continuous network that conforms to the general body shape. Bath sponges, family Spongiidae, which have a skeleton consisting of only spongin fibers, are collected for their skeletons. These sponges are most abundant in the Mediterranean Sea, the Gulf of Mexico, and the Caribbean Sea. After a bath sponge has been brought up from the bottom, the cellular material is allowed to decompose, and the skeleton then is cleaned and freed from the once living matter. The water-holding capacity of the skeleton is due to the capillary forces of the network of spongin fibers.

FEEDING AND DIGESTION

Sponges are *filter feeders;* that is, they produce a water current that carries to them suspended food particles, which they are able to trap and ingest. In sponges specifically, the water current is created by the flagella of the choanocytes, which also ingest most of the food. The effective stroke of the flagella is from the base to the tip, causing water to be drawn in through the gaps between the filaments that form the collar of the cell. Then the filtered water is caused to flow toward the top of the flagellum because of the direction of the flagellar effective stroke. The food particles (detritus, which are particles of nonliving organic material, and microscopic organisms) are trapped on the sticky mucus-coated outer surface of the collar as the water flows through the gaps in the collar and then passed downward to the cell body, where they are engulfed. However, wandering amoebocytes that come in contact with the water current can also ingest food. Furthermore, if the choanocytes are small, they may not themselves digest the food they have ingested but pass it to an amoebocyte for digestion. Digestion in sponges is protozoan-like, being completely intracellular in food vacuoles. The end-products of digestion then diffuse throughout the body, providing nutrients to the cells that do not ingest food. The sponge body-plan does not include a mouth or digestive cavity.

EXCRETION AND RESPIRATION

Sponges seem to have retained the mechanisms found in protozoans for excretion and respiration. Contractile vacuoles have been

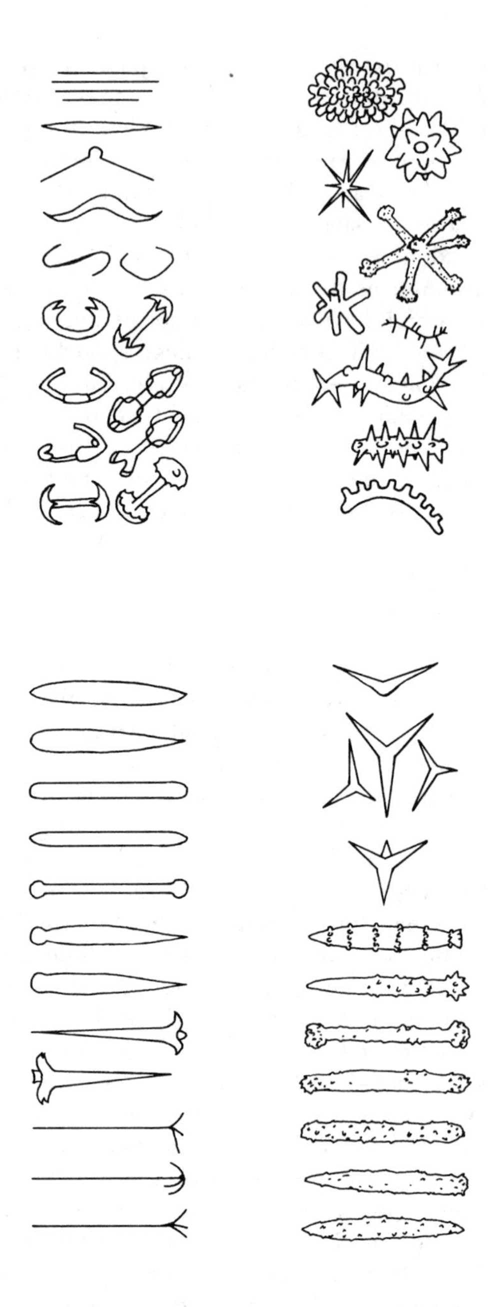

Fig. 3-3 *Common types of sponge spicules. They occur in a wide variety of shapes, and are often used as a basis for identifying individual species of sponges. (From de Laubenfels,* A Guide to the Sponges of Eastern North America. *Copyright © Rosenstiel School of Marine and Atmospheric Science, University of Miami. Used by permission.)*

seen in choanocytes, amoebocytes, and pinacocytes of freshwater sponges, and are used to eliminate the excess water load, as does the contractile vacuole of protozoans. As with protozoans, sponge contractile vacuoles appear to form and enlarge by the fusion of vesicles. Marine sponges do not have contractile vacuoles. Excretion of nitrogenous wastes, primarily ammonia, and respiratory exchange are performed by individual cells. The water current provides a constant supply of oxygen, as well as a vehicle for waste removal. The fact that the choanocytes alone produce the water current that brings in this fresh supply of oxygen for all the cells provides an excellent example of the intercellular cooperation that distinguishes a multicellular organism from a typical colony of unicellular organisms.

When toxic substances are present in the environment, the volume of water being pumped decreases. This is due to constriction of the osculum, which is surrounded by spindle-shaped contractile cells, called *myocytes*, that appear to contract in direct response to such environmental stimuli. In metazoans other than sponges, muscle cells are controlled by the nervous system. Sponges, however, do not seem to have nervous control of their muscle cells, which is not overly surprising in view of the degree of independent activity that sponge cells exhibit. Cells with long processes and staining properties similar to those of neurons (nerve cells) have been reported, but there is no physiological evidence that they are neurons. These cells may simply be amoebocytes that, by coincidence, stain in a fashion similar to nerve cells.

REPRODUCTION Sponges reproduce sexually and asexually. Gametes (both eggs and sperm) develop from amoebocytes, but choanocytes have also been reported to transform into sperm. Most species are *hermaphroditic*, or *monoecious*, eggs and sperm being formed by the same individual. A monoecious sponge does not, however, produce eggs and sperm at the same time, thus ensuring that crossbreeding will take place. Sperm leave one sponge and enter another in the water current. A newly arrived sperm enters either a choanocyte or an amoebocyte. These cells act as carriers, transporting sperm to the egg. The carrier cell then fuses with the egg and transfers the sperm to it.

The early blastulas of practically all calcareous sponges and a few in the class Desmospongiae undergo a process of inversion that does not occur in any other metazoans. This inversion process, whereby the embryo turns itself inside out, results in the formation of a late blastula stage called an *amphiblastula* (Fig. 3-4A). Within the mesenchyme of the parent, the fertilized egg first develops into the early blastula stage, the cells at one end having flagella that project into the blastocoel with an opening at the opposite end. The embryo then turns itself

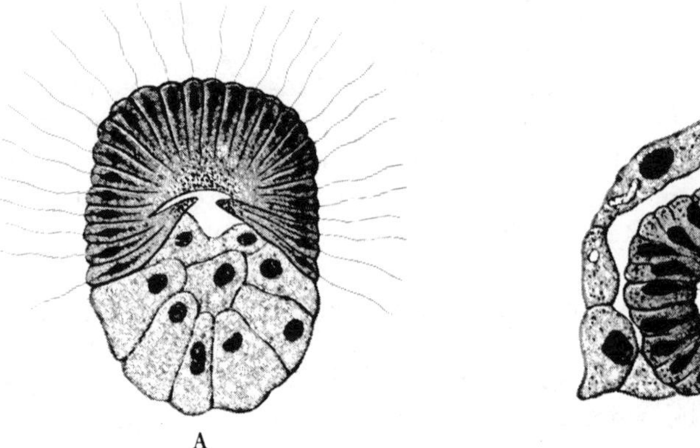

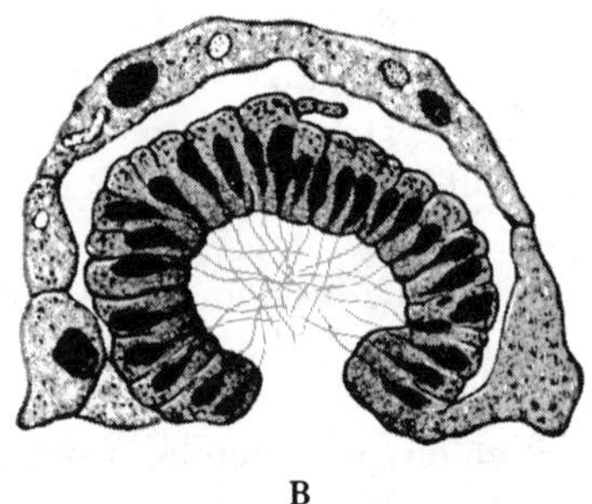

Fig. 3-4 *The (A) amphiblastula and (B) gastrula of a calcareous sponge,* Sycon *sp. (After Hyman,* The Invertebrates, *Vol. I. Copyright © 1940, McGraw-Hill Book Company. Used by permission.)*

inside out through the opening: the flagellated cells are now on the external surface, and the embryo has become an amphiblastula, which then breaks out of the mesenchyme and escapes through the osculum of the parent. After escaping, the amphiblastula has a brief free-swimming existence, settles to the bottom, becomes attached there at its blastopore end, and undergoes gastrulation. Gastrulation occurs by invagination of the flagellated half or overgrowth of the flagellated cells by the nonflagellated cells, resulting in a gastrula with a blastopore, a nonflagellated outer layer, and a flagellated inner layer (Fig. 3-3B). The central flagellated cavity is the beginning of the spongocoel. The gastrula then develops into a young sponge, an osculum breaking through at the free end.

The majority of sponges, however, do not form an amphiblastula and a hollow gastrula, but instead produce a solid gastrula directly from a solid blastula, neither the blastula nor gastrula having an internal cavity (the gastrula also has no blastopore). This solid gastrula of sponges is called a *parenchymula.* The surface of a parenchymula is almost completely covered with flagellated cells. Like the amphiblastula, the parenchymula develops in the mesenchyme of its parent. After leaving its parent, the parenchymula swims about freely for a short while and then settles, attaches, and becomes a young sponge, its interior hollowing out to form the spongocoel. Internal cells move to the outer surface to form pinacocytes; flagellated cells move to the inside to form choanocytes, and an osculum forms.

Asexual reproduction occurs by budding and by a variety of other processes that involve the production and liberation of cell aggregates capable of developing into a new sponge. The buds can either separate from the parent or remain attached. *Gemmules,* asexual repro-

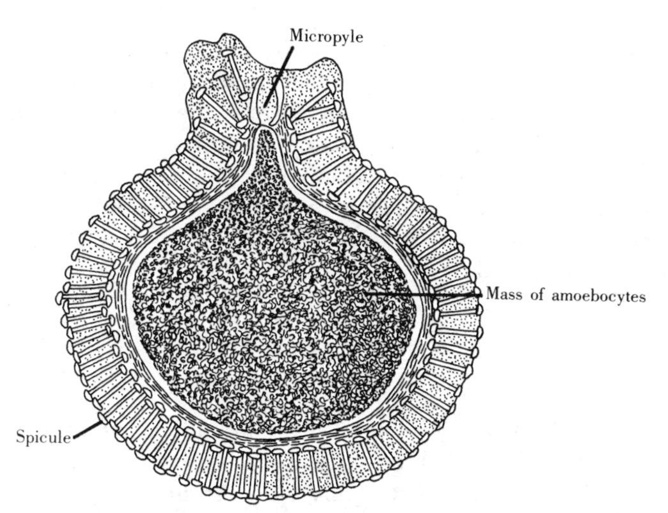

Fig. 3-5 *A gemmule of the freshwater sponge,* Ephydatia blembingia. *(After Evans, Quarterly Journal of Microscopical Science, vol. 44, 1901. Used by permission.)*

ductive bodies, are regularly formed as part of the normal life cycle of all freshwater sponges and some marine species (Fig. 3-5). Gemmules consist of an aggregation of amoebocytes enriched with food material which is covered by other amoebocytes. In freshwater sponges, these covering amoebocytes deposit a thick, hard, spicule-containing coat on the gemmule. Gemmules are formed primarily in the autumn, and the parent sponge subsequently disintegrates when winter arrives, releasing them. The gemmules are able to survive freezing during the winter and hatch the following spring, the amoebocytes emerging through an opening, the *micropyle,* and then develop into adult sponges. Another asexual reproductive cell aggregate is the *reduction body.* Unlike gemmules, which are formed regularly as part of the normal life cycle, reduction bodies are formed by many sponges, freshwater and marine, under various adverse conditions such as drought or low temperatures. The sponge disintegrates, leaving the reduction bodies, which consist of an internal mass of amoebocytes covered by pinacoderm. Upon the return of favorable conditions, the reduction body can develop into a complete sponge.

REGENERATION Sponges have a remarkable ability to regenerate. Any piece containing amoebocytes and choanocytes is capable of growing into a complete sponge. In a now-classic experiment, the regenerative capability of sponges was investigated by H. V. Wilson in 1907. He observed that when he dissociated a sponge into individual cells and cell clusters by squeezing it through a fine silk cloth, the cells would begin to migrate by amoeboid movement and reaggregate, reconstructing a sponge. For the reaggregating cells to adhere to each other, a macromolecule, secreted by the sponge, that holds the cells together is required along with calcium and magnesium ions. Biochemical tests suggest that this macromolecule consists predominantly of protein and polysaccharide. In the absence of calcium and magnesium, this aggregation factor, as it has been called, dissolves from the cell surface and leaves the cells nonadhesive.

PHYLOGENY Sponges give no indication of being anything other than an evolutionary dead end; that is, there is no evidence that any other group evolved from them. Their unique system of water canals proved satisfactory for these sessile organisms, but is not capable of providing the greater amounts of nutrient that active animals of the same size would require. The lack of a mouth and a digestive cavity restricts sponges to a protozoan type of nutrition whereby only particulate food can be consumed. Once the basic asconoid body plan had been established, the only result was the evolution of new sponges, which were more efficient in creating a water current.

As stated earlier, almost all biologists agree that metazoans evolved from protozoans, but disagreement arises in the choice of the particular ones. It seems more likely that sponges, and the rest of the metazoans as well, evolved by colonial integration of flagellated protozoans (recall the colonial theory of Chapter 1) and not from any other class of protozoans. Evidence in favor of this statement consists in part of the following facts: (1) Flagellated sperm cells are almost universal among metazoans. Such a spermatozoon may be regarded as a modified flagellate. (2) Flagellated body cells occur in sponges and cnidarians. (3) Some colonial flagellates, such as *Volvox*, have an organization approaching that of a multicellular animal. (4) The spherical, hollow colonies of protozoans such as *Volvox* are suggestive of a hollow blastula.

There seems to be no dissent over the theory that sponges originated from a colony of flagellates, but with respect to the origin of the rest of the metazoans opinions are more divided. The current proponents of the syncytial theory (Chapter 1) not only do not accept the colonial theory but also contend that a multinucleate, ciliated protozoan was the ancestor of all metazoans other than sponges. However, with a ciliate as an ancestral form, it is difficult to explain away the facts in favor of flagellates as ancestors for all multicellular animals. Furthermore, as mentioned in the brief discussion of these theories in Chapter 1, probably the major objection to the syncytial theory is the fact that none of the lower metazoans develops by way of a syncytium;

all undergo cleavage. The major criticism leveled against the colonial theory as applied to metazoans other than sponges appears to be that the flagellates living today that would appear to be most closely related to the ancestral colonial flagellate of this theory are organisms that have plantlike characteristics (such as chlorophyll and photosynthetic ability) not found in living metazoans. However, those flagellates that would have given rise to the sponges and the rest of the metazoans surely would have lost their plantlike characteristics, becoming strictly animal-like, before they embarked on the road to multicellularity.

Two types of flagellates have been suggested as the possible ancestor of sponges. Some investigators choose a choanoflagellate, a protozoan very similar in appearance to a choanocyte. Some choanoflagellates form colonies that could have evolved into a sponge. The chief criticism of this theory is that larval flagellated cells in sponges are without collars; they are not choanocytes. The second type of flagellated protozoan that has been suggested would have been an animal-like flagellate (but not a choanoflagellate) that formed hollow, spherical colonies similar to those of the green flagellate, *Volvox*. Support for the second type of flagellate as the ancestor of the sponges seems to be greater then for a choanoflagellate for the following reasons. The criticism of the choanoflagellate theory does not apply to the second theory; the amphiblastula bears a resemblance to a colony of *Volvox;* and the inversion (turning inside out) that leads to the formation of the amphiblastula is similar to the inversion process that normally occurs in the development of a colony of *Volvox*, the young colony turning itself inside out through an opening in the sphere so that the flagella which are initially pointing into the hollow center of the sphere assume their normal position, pointing outward.

If flagellates did give rise to all of the metazoans, (a) did one ancestral flagellate stock give rise to all the metazoans, with the sponges diverging early from the main line of metazoan evolution, or (b) did sponges evolve from a flagellate stock different from that which gave rise to the rest of the metazoans? Because sponges are so different from all other metazoans, alternative (b) seems more likely.

FURTHER READING

Bergquist, P. R., *Sponges*. Berkeley: University of California Press, 1978.

Brauer, E. B., "Osmoregulation in the Fresh Water Sponge, *Spongilla lacustris*," *Journal of Experimental Zoology*, vol. 192 (1975), p. 181.

Brauer, E. B., and J. A. McKanna, "Contractile Vacuoles in Cells of a Fresh Water Sponge, *Spongilla lacustria*," *Cell and Tissue Research*, vol. 192 (1978), p. 309.

Bidder, G. P., "The Relation of the Form of a Sponge to its Currents," *Quarterly Journal of Microscopical Science*, vol. 67 (1923), p. 293.

Burbanck, W. D., "Porifera," in *Selected Invertebrate Types*, F. A. Brown, Jr., ed. New York: Wiley, 1950, p. 72.

Dougherty, E. C., ed., *The Lower Metazoa: Comparative Biology and Phylogeny*. Berkeley: University of California Press, 1963.

Ebert, J. D., and I. M. Sussex, *Interacting Systems in Development*, 2nd ed. New York: Holt, Rinehart and Winston, 1970.

Fjerdingstad, E. J., "The Ultrastructure of Choanocyte Collars in *Spongilla lacustris*," *Zeitschrift für Zellforschung*, vol. 53 (1961), p. 645.

Fry, W. G., *The Biology of the Porifera*. New York: Academic Press, 1970.

Giese, A. C., and J. S. Pearse, eds., *Reproduction of Marine Invertebrates, Vol. 1, Acoelomate and Pseudocoelomate Metazoans*. New York: Academic Press, 1974.

Harrison, F. W., and R. R. Cowden, eds., *Aspects of Sponge Biology*. New York: Academic Press, 1976.

Humphreys, T., "Chemical Dissolution and *in vitro* Reconstruction of Sponge Cell Adhesions. I. Isolation and Functional Demonstration of the Components Involved," *Developmental Biology*, vol. 8 (1963), p. 27.

Moscona, A. A., "Aggregation of Sponge Cells: Cell-Linking Macromolecules and Their Role in the Formation of Multicellular Systems," *In Vitro*, vol. 3 (1968), p. 13.

Wilson, H. V., "On Some Phenomena of Coalescence and Regeneration in Sponges," *Journal of Experimental Zoology*, vol. 5 (1907), p. 245.

chapter **4**

Cnidaria and Ctenophora

Sponges, as explained in the previous chapter, represent in all likelihood an evolutionary blind alley. We shall now examine another of the lower phyla, the phylum Cnidaria [Gr. *knide*, nettle], and consider whether it offers a clue to the route that might have been taken by the ancestral protozoan that gave rise to the metazoans (other than sponges). Cnidarians are often still called by the common name coelenterates after their older phylum name Coelenterata [Gr. *koilos*, hollow + *enteron*, intestine]. However, because ctenophores (now in a separate phylum) were at one time included in the phylum Coelenterata, most biologists now prefer the newer name, Cnidaria, for the phylum.

The cnidarians include the hydras, jellyfishes, sea anemones, and corals. Some species, such as the hydras, are solitary; others, such as the Portuguese man-of-war, are colonial. Cnidarians exhibit considerable diversity in size, ranging from animals such as the hydras, whose body is about 1 millimeter wide, up to giant jellyfishes that are about 2 meters in diameter.

Cnidarians have advanced beyond the sponges to the tissue level of organization, for cnidarians show some division of labor among somewhat specialized groups of cells. But most cnidarian cells still lack the structural and functional specializations found in cells of higher animals. Because of this deficiency, organ development has been seriously hampered among cnidarians. The primary symmetry of the members of this phylum of invertebrates is considered by all biologists to be radial, but adult sea anemones and corals have a secondarily acquired biradial or bilateral symmetry. That is, biradial or bilateral symmetry has become superimposed upon the primary radial symmetry of these animals. Unlike the situation in sponges, cnidarians have an internal cavity, called the *gastrovascular cavity,* in which digestion of food occurs. The mouth, which is the aperture into this cavity, can open widely, and, in conjunction with the gastrovascular cavity, enables cnidarians to ingest food of a larger size than is possible for protozoans and sponges. Tentacles, which are extensions of the body wall, surround the mouth and assist in the seizure and ingestion of food. In addition, all cnidarians have cells called *cnidocytes,* or *nematocytes.* Within each of these is a *nematocyst,* an organelle containing a tube that can be discharged for use in obtaining food, for defense, or for attachment. The phylum name, Cnidaria, was coined in reference to these cnidocytes.

CLASSIFICATION Cnidarians are divided into three classes: Hydrozoa, Scyphozoa, and Anthozoa.

The class Hydrozoa includes the hydras (Fig. 1-6A), which are solitary polyps (a polyp being the cylindrical or treelike structural type of cnidarians, Fig. 4-1A); such colonial cnidarians as the Portuguese man-of-war *(Physalia pelagica)* and *Obelia commissuralis;* and the solitary medusa (the umbrella-shaped or bell-shaped structural type of cnidarian, Fig. 4-1B) *Craspedacusta sowerbyi* (Fig. 1-6B). Unlike the other cnidarian classes, the middle layer of the body wall in hydrozoans is devoid of cells. Furthermore, only hydrozoan medusae are *craspedote;* that is, they have a *velum,* a ring of tissue on the underside of the medusae that projects inward to form a shelf around the edge of the umbrella. This is the only class that has freshwater representatives. The life cycle may be exclusively polypoid, exclusively medusoid, or include both medusoid and polypoid stages, with the polypoid stage usually dominant.

The class Scyphozoa consists of cnidarians that have a medusa without a velum *(acraspedote).* The average size of scyphozoan

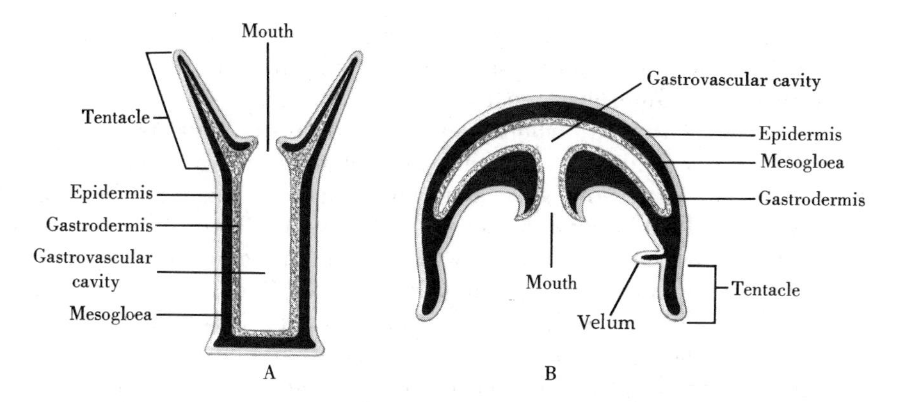

Fig. 4-1 *The two body forms of the cnidarians. (A) Polyp; (B) Medusa.*

medusae is larger than that of hydrozoan medusae. Only rarely does a scyphozoan life cycle lack a polypoid stage. But even if a polypoid stage is present, a medusa is always the dominant, conspicuous form in a scyphozoan life cycle. The middle layer of the body wall is cellular, containing scattered amoebocytes that appear to be of epidermal origin. Scyphozoans do not form colonies. *Aurelia aurita,* whose life cycle is described later in this chapter, is a member of this class.

The class Anthozoa (Fig. 1-6C) is represented by the sea anemones, which are always solitary, and the corals, which are both solitary and, more commonly, colonial. This class consists exclusively of polyps; medusae are completely absent. The middle layer of the body wall is cellular, containing amoeboid cells as the middle layer of scyphozoans. The coral, *Astrangia danae,* forms small encrusting colonies along the coast from North Carolina to Massachusetts. Anthozoans are anatomically the most complex cnidarians. These polyps are generally larger than those of hydrozoans, some sea anemones attaining 1 meter in diameter. Corals secrete a skeleton; sea anemones do not.

MORPHOLOGY AND PHYSIOLOGY

CNIDARIAN STRUCTURE

Even though all cnidarians have the same basic body plan of mouth, tentacles, and gastrovascular cavity, two different structural types have evolved in this phylum, the *polyp* and the *medusa* (Fig. 4-1). The polyp has a tubular body, with the mouth typically directed upward. Most polyps are sessile, being permanently fixed in place at the aboral end. However, although some polyps are capable of locomotion, the medusa is, in general, the active locomotory form in this phylum. Radial symmetry, as occurs in this phylum, is of particular advantage to a sessile organism because it enables the animal to meet the challenges of the environment equally well in all directions. The medusa is umbrella-shaped (or bell-shaped), with the convex side typically upward and the mouth on the concave (subumbrellar) surface.

The body wall of both polyps and medusae consists of three layers: an outer epidermis, an inner gastrodermis, and a middle layer, which can be either the gelatinous substance, mesogloea, alone or mesenchyme. In this phylum, however, the middle layer is usually referred to as the mesogloea even when it contains amoebocytes. The mesogloea is thin in polyps but thick in medusae. Because of this thick mass of mesogloea, a medusa is commonly called a jellyfish.

Colonial hydrozoans consist of more than one structural and functional type of individual; that is, they are *polymorphic*. For example, *Obelia* has two polypoid forms; the reproductive *gonangium* and the feeding *gastrozooid* (Fig. 4-2). A gastrozooid performs all functions except reproduction. By means of contraction of the body wall and flagellar activity also, the broth resulting from extracellular digestion in the gastrozooids is distributed to the gonangia through the network of living tubes *(coenosarc)* that connects all the individuals of a hydrozoan colony. These tubes are commonly covered with a nonliving *perisarc* secreted by the epidermis. Most colonial hydrozoans are anchored to the substrate by a rootlike structure. The Portuguese man-of-war is, on the other hand, a pelagic (living in the open ocean) hydrozoan colony that is

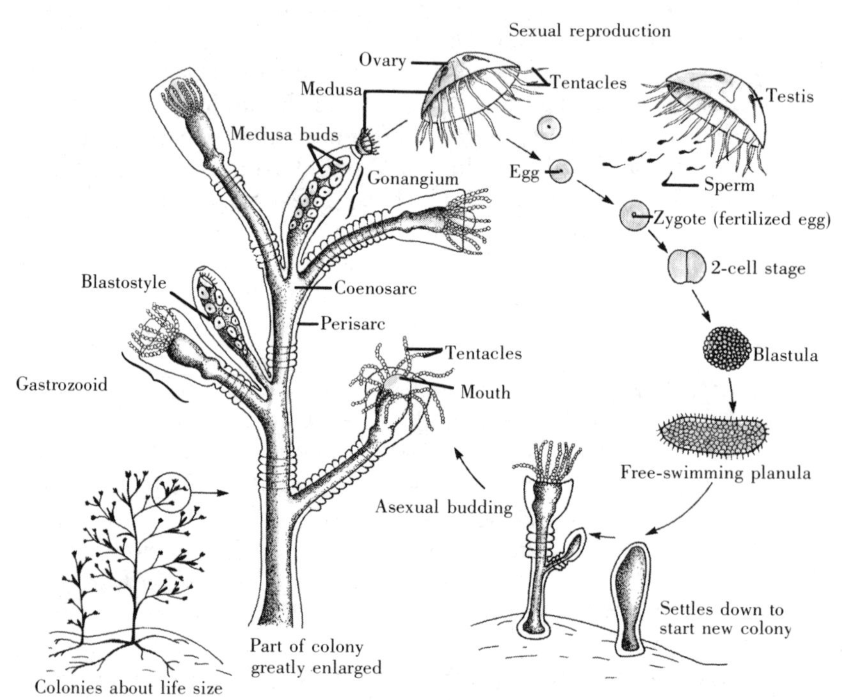

Fig. 4-2 *Life cycle of* Obelia *sp. (From Hickman, Hickman and Hickman,* Biology of Animals, *2nd ed., 1978. C. V. Mosby Company. Used by permission.)*

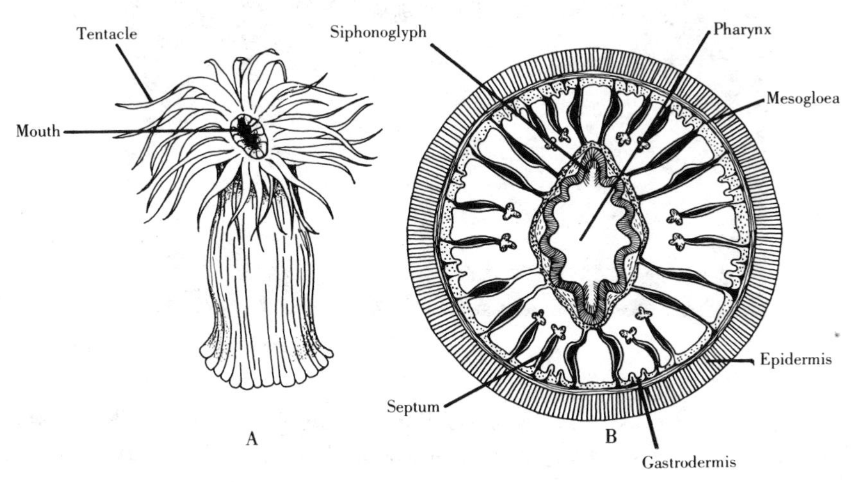

Fig. 4-3 *(A) General features of a sea anemone; (B) cross section through the pharynx of a sea anemone that has two siphonoglyphs. (After Hyman,* The Invertebrates, *Vol. I. Copyright © 1940, McGraw-Hill Book Company. Used by permission.*

carried about by currents and winds. It consists of polypoid and medusoid individuals showing extreme polymorphism, such as the polyp that becomes the air sac or float. Hydrozoan colonies form by the budding of a polyp that develops from a fertilized egg, the individuals remaining attached to each other.

Polyps of anthozoans have a tubular pharynx, which extends more than halfway from the slitlike mouth into the gastrovascular cavity (Fig. 4-3). In anthozoans other than the stony corals the slitlike mouth has a *siphonoglyph*, a ciliated groove, at one or both ends, which extends into the pharynx. Consequently, anthozoan symmetry is bilateral if one siphonoglyph is present because, as explained in Chapter 1, the organism can be cut into mirror halves in only one plane. But when there are two siphonoglyphs, or simply a slitlike mouth, as in the stony corals, the symmetry is biradial, and the organism can be cut into mirror halves in either of two planes.

Stony corals have an external skeleton, composed of calcium carbonate, which is secreted by the epidermis (Fig. 4-4). Each of their polyps is fixed in place in a depression in the skeleton. Other corals (the octocorallian corals) have an internal skeleton composed of separate or fused calcareous spicules or a horny substance. As stated in the classification section, sea anemones do not secrete a skeleton. As the result of the accumulation of layer upon layer of stony coral skeletons, a reef may be created. Although stony corals are the principal builders of a reef, a coral reef is actually an association of many different species of animals in addition to stony corals, each contributing in its own way to the structure of the reef. More than 200 species of reef-building corals

Fig. 4-4 *The skeleton of the brain coral,* Meandrina *sp. (Courtesy of Carolina Biological Supply Company.)*

alone are found in Australia's Great Barrier Reef, which is roughly 2,000 km long. Many species of invertebrates and fishes occur only in such reefs.

LOCOMOTION Cnidarians are able to move from place to place in several different ways. Those polyps, such as hydras and some of the sea anemones, that are not sessile can move about slowly by gliding on their base or aboral end. Hydras also use a method that employs looping movements. In this type of locomotion, the stalk of the body bends until the tentacles reach and attach to the substrate; the base then releases its grasp and moves to another site of attachment. Finally, the tentacles loosen their hold and the body becomes upright again. By contrast, the medusa is almost always a free-swimming form. One of the functions of the gastrovascular cavity is to act as a hydrostatic skeleton, providing an incompressible enclosed volume of fluid against which the muscle cells in the body wall and tentacles can work. A hydra can also simply float about by means of a gas bubble secreted by its base. The base detaches, secretes the bubble, and the hydra is carried to the surface. A medusa can swim by alternate contraction and relaxation of the umbrella, or it can be carried about by water currents. Medusae, however, do not direct any particular part of the body forward. Contraction of the umbrella (also called the bell) forces water out from the undersurface, moving the medusa in the opposite direction—a primitive type of jet propulsion. In contrast, the scyphozoan *Cassiopeia* has a behavior pattern that is exceptional for a medusa. It lies upside down with its convex surface attached to the bottom through a suckerlike action of a raised circular zone on the convex surface of the umbrella.

FEEDING, DIGESTION, AND NEMATOCYSTS Cnidarians are carnivorous, feeding mainly on small crustaceans (a class of arthropods), but the larger species can even capture small fishes. Food is obtained when an organism comes in contact with

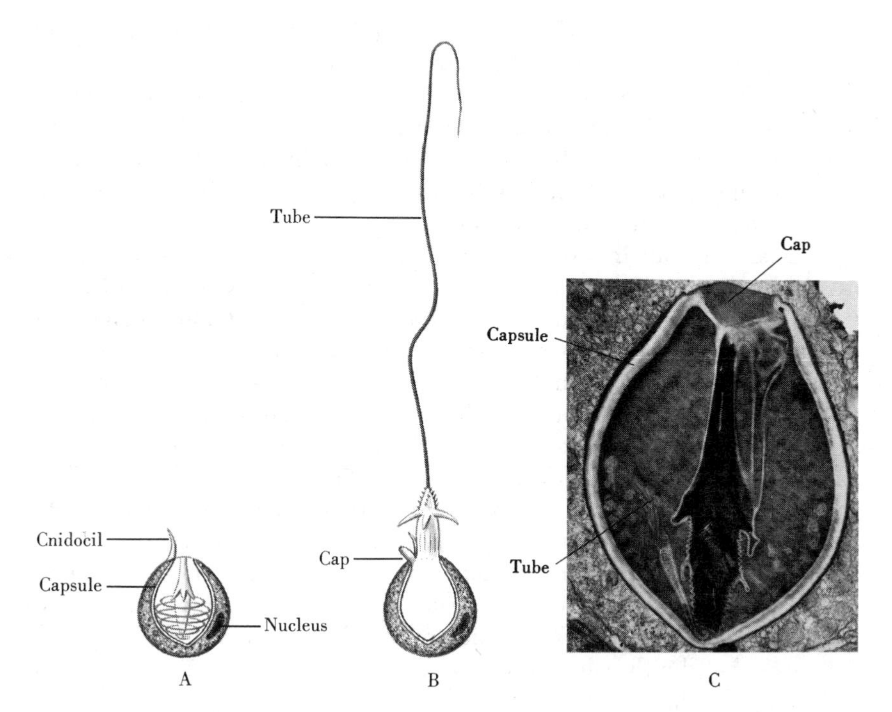

Fig. 4-5 *(A) An undischarged nematocyst inside its cnidocyte and (B) a discharged nematocyst (a penetrant); (C) Electron micrograph of an undischarged nematocyst (a penetrant) of a hydra. [(A), (B) Modified from Hyman,* The Invertebrates, *Vol. I. Copyright 1940, McGraw-Hill Book Company. Used by permission; (C) Courtesy of G. B. Chapman, from* The Biology of Hydra and of Some Other Coeloenterates, *H. M. Lenhoff and W. F. Loomis, eds. University of Miami Press, 1961.*]

the tentacles, thereby causing the discharge of the nematocysts used to capture prey. A nematocyst consists of a capsule that contains a coiled tube, a cap, and usually a cnidocil, or trigger (Fig. 4-5). The tube is released when the nematocyst discharges. The cnidocil, when present, is probably involved in the excitation process. However, this structure is not essential, as evidenced by its absence in many anthozoans. In its absence, the cnidocyte surface appears to be the receptive region for eliciting a response. In hydras and some other hydroid polyps, cnidocytes occur in clusters often referred to as "batteries."

Nematocysts fall into three categories. First, the *volvent*, which is used to capture prey, has a nonadhesive tube with a closed tip. The tube simply coils around the prey as a means of attachment. Second, the *penetrant*, also used to capture prey, has a tube, the tip of which is open; the tube penetrates the tissues of the prey and releases a toxin that paralyzes or kills the victim. Third, the *glutinant* has an open-tipped, sticky tube. This adhesive type is used to help anchor polyps capable of locomotion to the substrate and to hold prey already captured by volvents and penetrants against the tentacles, presumably aiding the tentacles in pulling the food to the mouth. Glutinants do not appear to be directly involved in prey capture.

Nematocysts are independent effectors, that is, they are not under control of the animal itself, but discharge upon receiving appropriate stimulation directly from the environment. In other words, the receptor and effector functions reside in the same cell. However, the responsiveness of nematocysts can change. They discharge more readily in starved than in well-fed individuals. Glutinants respond only to mechanical stimulation, whereas the discharge of volvents and penetrants normally requires double stimulation, chemical and mechanical; discharge of these two latter types occurs only after they have come in contact with an object that contains at least one of the lipoidal food factors known to lower the threshold of the nematocyst to mechanical stimulation. In other words, there is chemical sensitization to mechanical stimulation. In this way, the volvents and penetrants, which clearly function to capture prey, are not wasted by discharging, for example, when contact is made with a rock, nor will they discharge prematurely in response to an appropriate molecule that has diffused out from prey that is still too far away for the nematocyst's tube to reach. Axons have been found to synapse with cnidocytes. Perhaps it is through these synapses that the threshold of activation is controlled in the starved and well-fed individuals referred to earlier. Lowering of the threshold of activation by the nervous system could serve to bring about firing of an entire battery of nematocysts rather than only a few of them.

The mechanism of nematocyst discharge is not well understood. The two currently favored theories postulate that the pressure within the capsule increases following stimulation of the nematocyst, causing the tube to be forced outward. According to one theory, the increased pressure is produced by contraction of the capsule itself or fibrils associated with it. The second theory suggests that the increased pressure is due to entry of water into the capsule. Mechanisms comparable to those devised to explain the discharge of trichocysts (see Chapter 2) have been applied to nematocysts in order to explain how water would enter the capsule. One suggestion is that the nematocyst operates after stimulation as a simple osmometer to effect water inflow. A second possibility is that hydration and the consequent swelling of a protein in the nematocyst in response to a change in pH are responsible for the inflow of water. It is, of course, possible that mechanisms fitting both theories operate in different species. At discharge, the cap opens and the tube turns inside out as it is shot forth. Nematocysts can be used only once. The cnidocyte is cast off after its nematocyst has discharged, and is replaced by a new one, which develops from an interstitial cell. *Interstitial cells* are unspecialized cells found in both the epidermis and gastrodermis and are capable of developing into other cell types as needed.

After the food has been captured, the tentacles direct it toward the mouth. W. F. Loomis has shown that the feeding response in hydras, evidenced by opening of the mouth, is activated by a tripeptide, reduced glutathione, which is probably released from the prey after the tube of a penetrant has entered it. The food is eventually swallowed, entering the gastrovascular cavity where enzymes that reduce the prey to small particles are released from enzymatic-gland cells which are flagellated. This phase of digestion is extracellular, but digestion is not completed within the gastrovascular cavity. The food particles are engulfed instead, through the use of pseudopodia, by the usually flagellated gastrodermal nutritive-muscle cells, and digestion is completed in food vacuoles in these cells. The beating of the flagella helps to mix the food with the enzymes. The products of digestion then pass to the other cells by diffusion. Undigested matter is finally eliminated through the mouth when the body contracts. The gastrovascular cavity in scyphozoans and anthozoans, unlike that in hydrozoans, is divided into compartments by radiating membranes *(septa).*

RESPIRATION Cnidarians have no special respiratory structures. These animals have a low oxygen requirement and receive an adequate supply of oxygen by diffusion through the epidermis. Carbon dioxide also is lost by diffusion. The siphonoglyph, when present, directs a stream of water into the gastrovascular cavity; this process would appear to provide a chance for the gastrodermis to participate directly in gas exchange.

EXCRETION Marine cnidarians are isosmotic with sea water. Freshwater cnidarians, such as the hydras, are hyperosmotic to their environment. They do not have contractile vacuoles to eliminate the excess water gained by osmosis as do freshwater protozoans and sponges. *Hydra littoralis* makes use of its gastrovascular cavity to get rid of its excess water. This hydra helps maintain itself hyperosmotic in relation to its freshwater environment by expelling from inside its gastrovascular cavity, through its mouth, a fluid it produces which is hyposmotic to its cells, but not as dilute as its environment. Production of this gastrovascular-cavity fluid may be a two-step process: a fluid that is isosmotic with the cells is secreted, then solute is re-absorbed; thus the fluid is formed, which is hyposmotic to the cells but still more concentrated than the environment. Periodically, the body column contracts, forcing the fluid out through the mouth. In an experimental situation, when the solute concentration of the external medium is altered but the hydra is still hyperosmotic to the several test solutions, it has been observed that the rate of column contraction is inversely proportional to the concentration of the external medium. The more dilute the medium, the faster the gastrovascular cavity fills with the fluid that is hyposmotic to the cells (but hyperosmotic to the medium), and, consequently, the faster the cavity empties. The hydra, as a whole, functions in a manner that could be likened to a protozoan contractile vacuole. Hence, in addition to its functions in digestion and as a hydrostatic skeleton, the gastrovascular cavity of a hydra is also involved in osmotic regulation. Freshwater cnidarians are also capable of absorbing sodium and potassium ions directly from the environment to help maintain their hyperosmotic condition. Nitrogenous wastes, primarily in the form of ammonia, are eliminated by diffusion through the body surface.

NERVOUS SYSTEM A primitive nervous system, which sponges apparently lack, is present in cnidarians. The rhythmical swimming movements of a jellyfish are but one example of the type of coordinated muscular activity that cnidarians can perform. The cnidarian nervous system consists of two *nerve nets* (plexuses or networks of neurons), a main one at the base of the epidermis, next to the mesogloea, and a less highly developed one in a comparable position at the base of the gastrodermis. The two nets are connected at various points by transverse fibers. Cnidarians have nothing that could even remotely be considered a brain (or even a central nervous system). Excitation at any point in a typical nerve net spreads freely throughout the net in all directions. This physiological observation that excitation in the nerve net spreads through the net from the point of excitation was corroborated by ultrastructural observations that showed that the most common neuroneuronal junctions in cnidarians are symmetrical synapses, with synaptic vesicles on each side; they are presumably nonpolarized, capable of transmitting impulses in both directions. Polarized synapses have been seen much less frequently.

Neurophysiologists have had interest in cnidarians because these organisms provide the opportunity to study a relatively simple nervous system whose basic principles can be applied to the brains of higher organisms. For example, the neuroneuronal junctions of cnidarians very often require facilitation for transmission to occur; that is, two or more incoming stimuli sufficiently close together in time are required for one neuron to excite another. The requirement for facilitation provides a safety factor against the discharge of efferent neurons in response to very slight afferent stimulation. Many of the neuroneuronal junctions in higher animals also require more than one incoming impulse to initiate an outgoing impulse.

REPRODUCTION Hydras, which are probably the cnidarians most commonly studied in classrooms, are in fact atypical hydrozoans; their life cycle has no medusoid form. In hydras, the polyp gives rise either sexually or asexually to another polyp, whereas the usual hydrozoan pattern is that polyps are the dominant form and give rise asexually to medusae, which then reproduce sexually, giving rise to polyps again. Usually, hydras reproduce asexually by budding in the spring and summer and sexually in the fall. Other atypical life cycles of hydrozoans are seen in a few species where there is only a medusoid form, one medusa generation giving rise to the next medusa generation; or where a medusa is the dominant form and the polyp is small and solitary.

Obelia has a typical hydrozoan life cycle in which the polyp is dominant over the medusa (Fig. 4-2). As mentioned earlier, a colony of *Obelia* consists of gastrozooids and gonangia. Each gonangium contains a *blastostyle,* or central stalk, from which craspedote medusae form asexually by budding. The polypoid generation of *Obelia* never reproduces sexually. Then, the male and female medusae escape from the gonangia into the sea and become sexually mature. There they shed their eggs and sperm into the water, where fertilization occurs. The *zygote,* or fertilized egg, develops into a larva called a *planula,* which is elongated, ciliated, radially symmetrical, and a solid mass of cells, lacking a mouth and a gastrovascular cavity. A *larva* is a developmental stage between the embryo and the adult; it differs strongly in appearance from its parents. After a free-swimming existence, the planula settles down, becomes attached to the substrate, and develops into a colony of polyps. This alternation of polypoid (asexual) and medusoid (sexual) forms is called *metagenesis* or alternation of generations. It should be pointed out, however, that cnidarian metagenesis is not related to the alternation of haploid and diploid generations of plants (see Delevoryas' *Plant Diversification* in this series). In cnidarians, both the medusa and polyp are diploid, and only the gametes are haploid. Medusae always reproduce sexually, whereas polyps can reproduce sexually or asexually.

In all scyphozoans, an acraspedote medusa is the dominant, conspicuous form. The vast majority of scyphozoans have a larval polypoid form in the life cycle. In very few of them does the planula, which develops from an egg produced by a medusa, develop directly into another medusa. Practically all scyphozoans are *dioecious* (the sexes are separate). The jellyfish *Aurelia* is a typical scyphozoan (Fig. 4-6). Its eggs, following their release from the ovary, become caught in the frills of the four oral arms (lobes of tissue that surround the mouth of many scyphozoan medusae), where fertilization occurs and a typical planula develops. The planula then becomes free-swimming for a while and eventually settles to the bottom and becomes attached. The attached planula then develops into a polypoid larva, the *scyphistoma,* which looks like a hydra and produces more scyphistomae asexually by budding. Each scyphistoma ultimately undergoes a process called strobili-

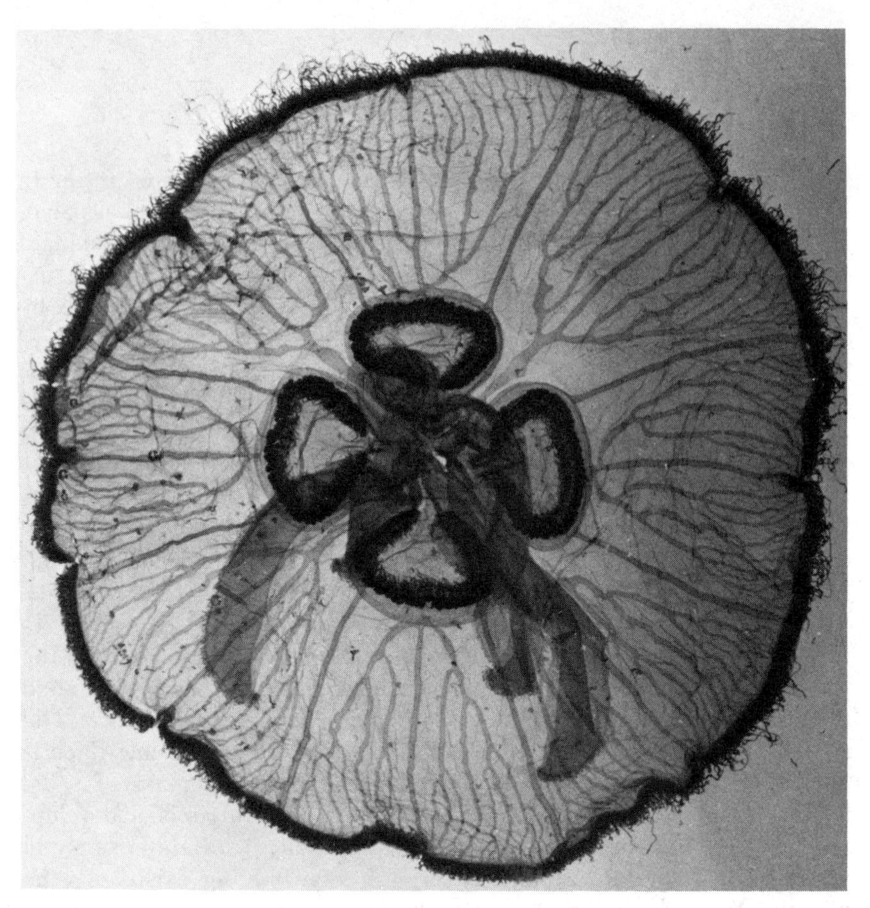

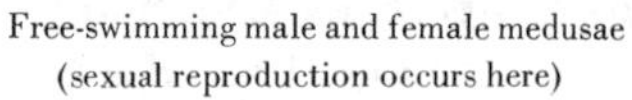

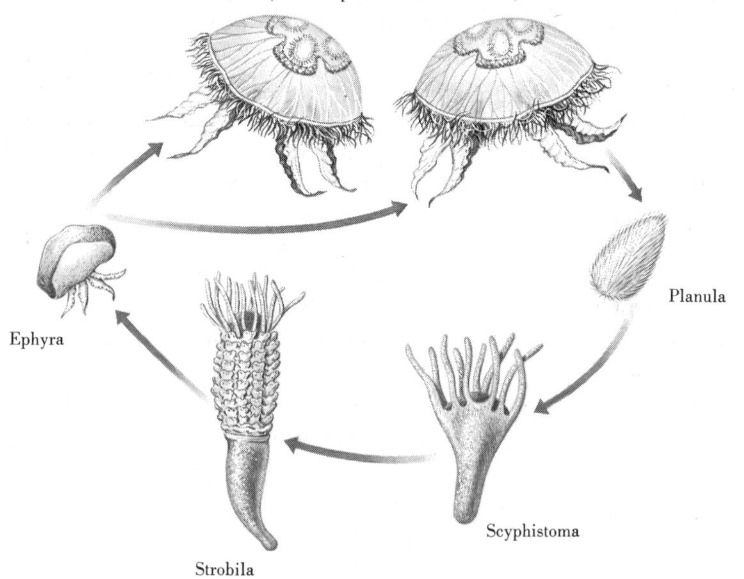

Fig. 4-6 Aurelia aurita. *(A) Preserved specimen; (B) life cycle. [(A) Courtesy of Ward's Natural Science Establishment, Inc.; (B) Modified from Cockrum and MacCauley,* Zoology. *W. B. Saunders Company, 1965.*

zation whereby a stack of immature medusae, called *ephyrae*, is formed at its oral end, by means of transverse fission. The scyphistoma with its attached stack of "saucers" is often called a *strobila*. Each ephyra ultimately breaks away and develops into a mature medusa, and the life cycle then repeats itself.

As stated in the classification section at the beginning of this chapter, the life cycles of anthozoans have no medusa stage. As far as sexual reproduction is concerned, anthozoans may be monoecious or dioecious. A planula develops from the fertilized egg. From the planula, a young polyp of a sea anemone or coral develops and ultimately matures. In colonial corals, the original single polyp attaches to the substrate and then produces the entire colony by budding. Sea anemones, too, can reproduce asexually, not only by budding but by other means also. As a sea anemone crawls, it may leave behind part of the base, which ultimately regenerates into a complete sea anemone. This process is known as *pedal laceration* . Transverse and longitudinal fission of sea anemones have also been reported.

There is evidence in hydras of a hormone that promotes growth (somatic cell proliferation) but at the same time inhibits the development of gametes from the interstitial cells of the epidermis. Gamete production appears to commence only in the absence of this somatic growth hormone. It is apparently a secretion product of some of the neurons.

PHYLOGENY A hollow, spherical colony of flagellates, different from any that might have evolved into the sponges, quite likely became transformed into an organism that was ancestral to all the other metazoans. This transformation involved the conversion of the hollow, spherical colony into a solid, ovoid, multicellular organism with radial symmetry, closely resembling the planula as we know it today. Because of its similarity to the planula, this hypothetical, multicellular organism is known as the *planuloid ancestor*. Cnidarians are most likely the descendants of an organism that branched off the ancestral line leading from this planuloid ancestor to all the higher phyla (see Fig. 1-3). Such a planuloid ancestor readily accounts for the occurrence of a planula larva throughout the phylum Cnidaria, as well as for the radial symmetry in this phylum. It is not difficult to visualize how this planuloid ancestor could have been transformed into an individual with the basic body plan of a cnidarian. Hollowing of the ancestral organism would have formed the gastrovascular cavity. Formation of an opening into the cavity would have provided the mouth, and a circlet of tentacles could have formed as extensions of the body wall around the mouth to assist in feeding. The planuloid ancestor has an important place in evolutionary theory; not only could it have given rise to the first cnidarian but also it was the kind of organism that could have been modified to produce even higher phyla, as will be described in the next chapter.

An intriguing evolutionary question is whether the ancestral cnidarian was a polyp or medusa. Opinion seems to be fairly evenly divided. Those who support the medusa as ancestral take the stand that the ancestral cnidarian indeed was medusoid, but the evolutionary tendency has been to suppress and, finally, as in the hydras, to eliminate the medusa. Furthermore, according to this view, the polyp evolved originally as a larval stage of a cnidarian that had a medusa as its adult form. But with the passage of time, the polyp evolved into one of the two adult forms, along with the medusa, of cnidarians. Retention of the polyp in the life cycle could have allowed cnidarians to make use of new food supplies, the polyp generally being attached to the substrate while the medusa is almost always free-swimming. The tendency to emphasize the polyp reached its climax among the anthozoans, where the medusa is absent from the entire class. On the other hand, those who favor a polyp as the ancestral cnidarian suggest that the more mobile medusa evolved later as a means of wider species-dispersal. No matter whether the polyp or medusa was ancestral, practically all biologists agree that the hydrozoans are the most primitive cnidarians, and that scyphozoans and anthozoans both evolved from hydrozoans.

CTENOPHORA The phylum Ctenophora [Gr. *ktenos*, comb + *phoros*, bearing] consists of about 90 species of marine animals closely related to the cnidarians. Ctenophores are never colonial. Although the number of species is not large, ctenophores are widely distributed, and some species occur in abundance, particularly in warm seas.

Ctenophores are generally considered to have evolved from a medusoid cnidarian. Ctenophores are biradially symmetrical animals. They are commonly called sea walnuts and comb jellies. The body is usually transparent with eight ciliated bands called *comb rows* (Fig. 4-7). Each band consists of a series of transverse plates of fused cilia, which provide the propulsive force for locomotion. The body is somewhat medusalike, with a gelatinous middle layer that contains amoebocytes. The ctenophores have a gastrovascular cavity like that of the cnidarians, but it is much more elaborate, having several elongated branches attached to the main portion.

All ctenophores are hermaphrodites. The gametes are usually shed through the mouth with fertilization taking place in the sea. Their life cycles have no polyplike stage. Only one species, *Euchlora rubra*, possesses cnidocytes on its tentacles. Other ctenophores have *colloblasts* (adhesive cells) on their tentacles for capturing food. *Euchlora*

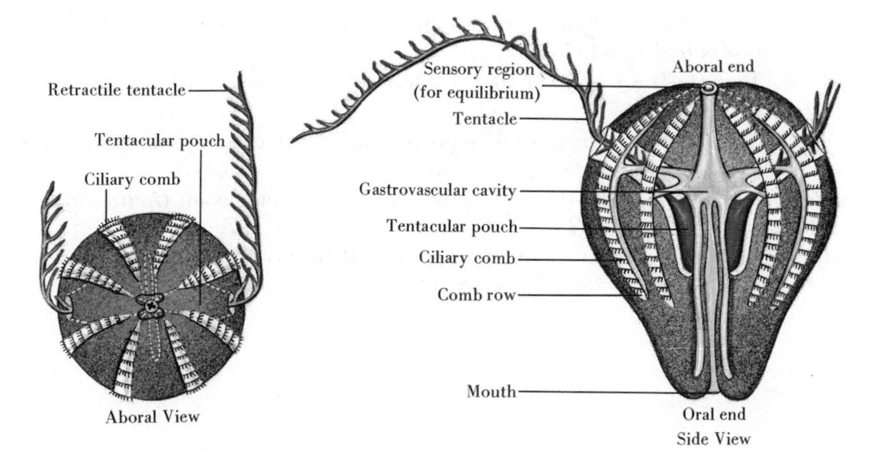

Fig. 4-7 *A typical ctenophore,* Pleurobrachia *sp. (From Stiles, Hegner, and Boolootian,* College Zoology, *8th ed. Macmillan, 1969.)*

rubra lacks colloblasts. Its cnidocytes are considered further evidence of a cnidarian origin of this phylum.

FURTHER READING

Benos, D. J., and R. D. Prusch, "Osmoregulation in *Hydra:* Column Contraction as a Function of External Osmolality," *Comparative Biochemistry and Physiology,* vol. 44A (1973), p. 1397.

Burnett, A. L., ed., *Biology of Hydra.* New York: Academic Press, 1973.

Delevoryas, T., *Plant Diversification.* New York: Holt, Rinehart and Winston, 1966.

Gierer, A.,"Hydra as a Model for the Development of Biological Form," *Scientific American,* vol. 231:6 (1974), p. 44.

Hand, C., "On the Origin and Phylogeny of the Coelenterates," *Systematic Zoology,* vol. 8 (1959), p. 191.

Jones, O. A., and R. Endean, *Biology and Geology of Coral Reefs, Vol. IV, Geology 2.* New York: Academic Press, 1977.

Lenhoff, H. M., and W. F. Loomis, eds., *The Biology of Hydra and of Some Other Coelenterates.* Coral Gables: University of Miami Press, 1961.

———————, L. Muscatine, and L. V. Davis, eds., *Experimental Coelenterate Biology.* Honolulu: University of Hawaii Press, 1971.

Lentz, T. L., *The Cell Biology of Hydra.* Amsterdam: North-Holland Publishing Co., 1966.

Loomis, W. F., "Glutathione Control of the Specific Feeding Reactions of *Hydra,*" *Annals of the New York Academy of Sciences,* vol. 62 (1955), p. 209.

Muscatine, L., and H. M. Lenhoff, eds., *Coelenterate Biology: Reviews and New Perspectives.* New York: Academic Press, 1974.

Pantin, C. F. A., "Capabilities of the Coelenterate Behavior Machine," *American Zoologist*, vol. 5 (1965), p. 581.

Rees, W. J., ed., *The Cnidaria and Their Evolution*. New York: Academic Press, 1966.

Tardent, P., "Gametogenesis in the Genus *Hydra*," *American Zoologist*, vol. 14 (1974), p. 447.

Westfall, J. A., "The Nematocyte Complex in a Hydromedusan, *Gonionemus vertens*," *Zeitschrift für Zellforschung*, vol. 110 (1970), p. 457.

Westfall, J. A., "Ultrastructure of Synapses in a Primitive Coelenterate," *Journal of Ultrastructure Research*, vol. 32 (1970), p. 237.

chapter **5**

Platyhelminthes and Nemertinea

PLATYHELMINTHES

The phylum Platyhelminthes [Gr. *platy*, flat + *helminthes*, worms], consisting of animals commonly called flatworms, seems to be the one in which bilateral symmetry first appeared as the primary symmetry of an entire phylum. Both larval and adult flatworms are bilaterally symmetrical. The phyla that will be discussed in the following chapters are all characterized by a primary bilateral symmetry (although this is supplanted by radial symmetry of secondary origin in most adult echinoderms). The evolution of bilateral symmetry was an important advance; by the very nature of this symmetry, organisms acquired anterior and posterior ends, as well as dorsal and ventral surfaces. These modifications of the body form prepared the way for the appearance of highly mobile organisms.

The ventral surface was applied to the substrate and then became involved in the locomotor process, as is obvious in crawling organisms. With the evolution of a leading or anterior end, sense organs became concentrated anteriorly and the nerve cells associated with them also began to accumulate there to form a brain, resulting in the ultimate evolution of a head. In a crawling or swimming organism, the head is typically the first portion of the body to encounter a new environment. Concentration of nervous and sensory elements at the head end is known as *cephalization.* A result of this process is that the connections between nerves and sensors are short, an obvious advantage in minimizing the time required, for example, for an escape reaction from unfavorable circumstances.

Bilaterally symmetrical organisms are anatomically at the organ grade of construction rather than the cell or tissue grade of the previously described phyla. The presence of organs and organ systems to handle the various functions vastly increases the overall efficiency of the organism. The cell specialization that multicellularity made possible reached a significant plateau with the evolution of organs and organ systems.

In addition to being bilaterally symmetrical, flatworms are, as their common name implies, dorsoventrally flattened. The flatworm is acoelomate, and, if it has a digestive tract it also has a mouth, but there is never, as stated in Chapter 1, an anus in a member of this phylum. Mesenchyme, in which the cells are fairly closely packed, and muscle cells fill the space between the *integument*, or body covering, and the internal organs. Flatworms range in size from microscopic, free-living species to tapeworms about 12 meters long.

CLASSIFICATION The phylum Platyhelminthes consists of three classes, the Turbellaria, Trematoda, and Cestoda.

The class Turbellaria consists of the free-living flatworms. They are found in fresh and salt waters and in moist terrestrial habitats. The epidermis of turbellarians is at least partially ciliated. All turbellarians have mouths, and all except the exclusively marine order Acoela have intestines. These flatworms are unsegmented. The planarian, *Dugesia tigrina* (Figs. 1-7A, 5-1), is a freshwater turbellarian that is commonly studied in student laboratories.

The class Trematoda consists of the flukes (Fig. 5-2), all of which are parasites. Flukes are unsegmented. A fluke has a mouth and an intestine, which almost always has two branches, and one or more suckers for attachment to the host. The Oriental liver fluke, *Opisthorchis sinensis* (Fig. 1-7B), is a parasite of man. Its life cycle is described later in this chapter.

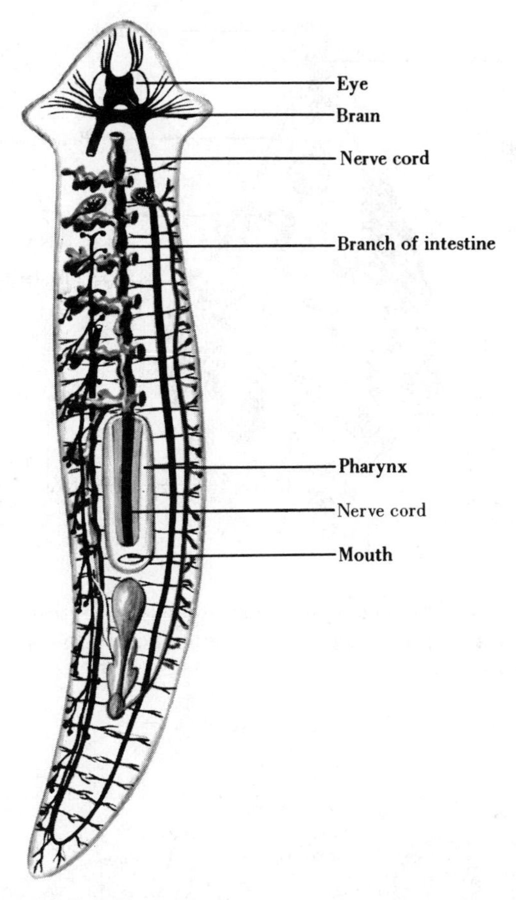

Fig 5-1 *Diagrammatic representation of the planarian,* Dugesia tigrina. *(Modified from Villee and Dethier,* Biological Principles and Processes. *W. B. Saunders Company, 1973.)*

The class Cestoda consists of the tapeworms. All are parasites, and almost all are segmented. A cestode never has a mouth or digestive tract. The pork tapeworm, *Taenia solium* (Fig. 5-3), whose life cycle is described below, is a parasite of man.

MORPHOLOGY AND PHYSIOLOGY

FLATWORM STRUCTURE

The turbellarians and trematodes resemble each other more than either resembles the cestodes. The bodies of turbellarians and flukes are always undivided. In contrast, the bodies of all but a few species of tapeworms are divided into segments called *proglottids* (Fig. 5-3). The body of a typical

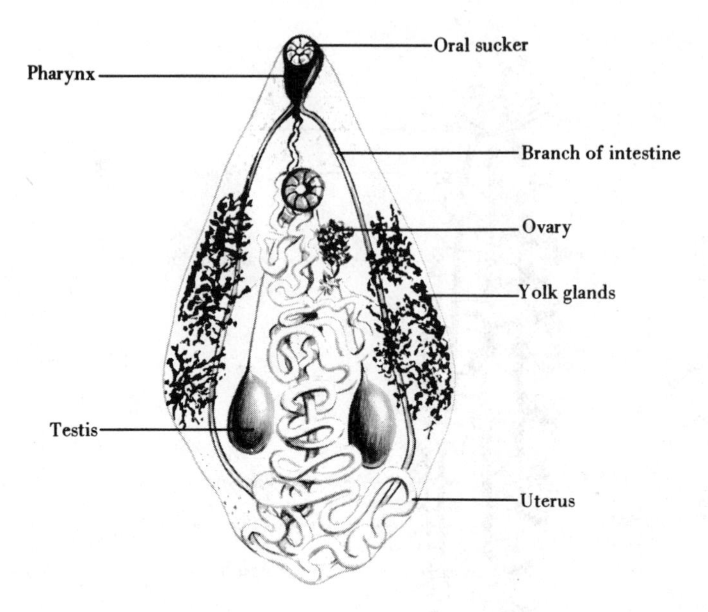

Fig. 5-2 *A fluke,* Prosthogonimus macrorchis, *that parasitizes the oviduct of the domestic hen. (Modified from Clark,* Contemporary Biology. *W. B. Saunders Company, 1973.)*

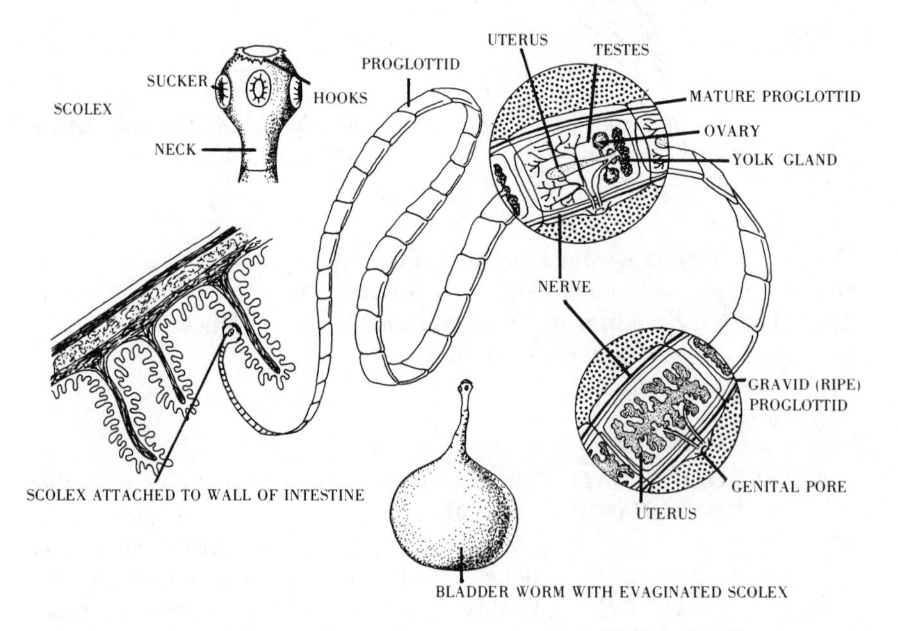

Fig. 5-3 *The pork tapeworm,* Taenia solium. *(After Villee,* Biology, *6th ed. W. B. Saunders Company, 1972.)*

tapeworm consists of a head or *scolex* that bears either suckers alone or suckers and hooks, followed by a relatively short, undivided *neck* (the region where new proglottids are formed), which is followed by the chain of proglottids. The suckers and hooks are used to attach the tapeworm to the host. The oldest proglottid is at the posterior end of the tapeworm. The proglottids show a progressive increase in size and maturity from the newly formed ones at the neck to the older ones at the posterior end of the tapeworm. The chain of proglottids is called a *strobila*, the same term that is often used for the scyphozoan scyphistoma when it has a stack of ephyrae still attached to it. One of the adaptations of the flukes and tapeworms for their parasitic existence is these attachment organs. Other adaptations involve mainly the digestive and reproductive systems and larvae that pass from one host to another, as will be described later in this chapter.

The integument of all flatworms is completely protoplasmic; that is, it is a living integument. Until the integument of the parasitic flatworms was studied by electron microscopy, the exposed surface was thought to be covered by a *cuticle*, a nonliving, protective layer secreted by the underlying cells. However, observations with the electron microscope revealed that parasitic flatworms lack cuticles and have instead unciliated protoplasmic (hence living) integuments consisting of a complex syncytium (Fig. 5-4). Therefore, the term "cuticle" is no longer

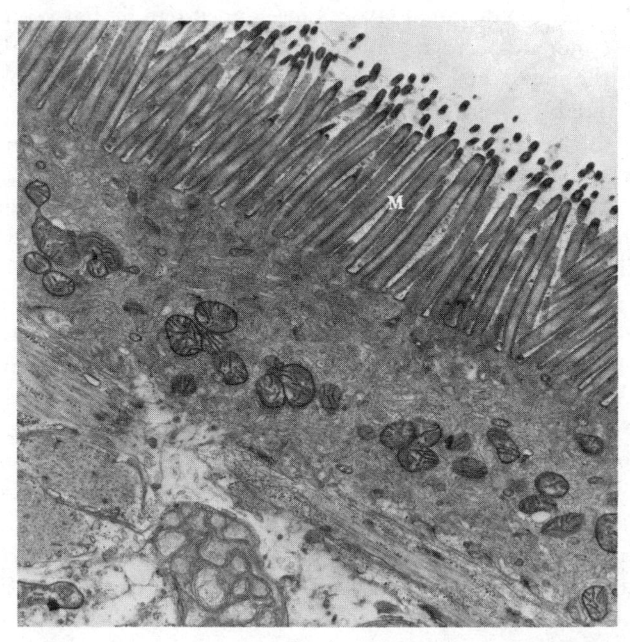

Fig. 5-4 *Electron micrograph of the integument of the tapeworm,* Nybelinia *sp., showing the numerous small projections (microvilli), labeled M, which increase the surface area. No cuticle is present. (Courtesy of R. D. Lumsden, Tulane University.)*

applicable. The integument of the parasitic flatworm is now more specifically referred to as the *tegument.* On the other hand, the protoplasmic nature of the turbellarian integument was apparent even under light microscopy. Among turbellarians, the integument consists of discrete epidermal cells or is syncytial. The epidermis of turbellarians (except in the acoels and primitive members of the other turbellarian orders) contains *rhabdoids.* These are rod-shaped bodies, which, according to J. B. Jennings, can be discharged from the epidermis and then disintegrate to assist in the formation of a protective slimy sheath around the worm. The entire body surface in most turbellarians, including all of the acoels, is ciliated; however, in some of the larger turbellarians, the cilia are restricted to the ventral surface.

LOCOMOTION An evolutionary innovation in the flatworms is subepidermal muscle layers, composed of circularly, longitudinally, and frequently diagonally arranged layers of cells, that, along with the muscle cells that lie among the mesenchymal cells, provide greater locomotor ability than is possible in cnidarians. The outermost muscular layer consists always of circularly arranged cells, but when the diagonal layer is present, it can lie between the other two or be innermost. Turbellarians move from place to place by crawling along the substrate. Smaller turbellarians, such as the acoels, are probably capable of crawling by use of their cilia alone. Mucous glands, found on the ventral surface of all turbellarians, aid in locomotion by secreting a viscous slime against which the cilia beat.

However, among larger turbellarians, although the ventral cilia may have some small role in locomotion, crawling is accomplished largely, if not wholly, by muscular waves that pass backward from the anterior end. Alternating transverse waves of contraction of the circular and longitudinal muscles pass from the anterior to the posterior ends in conjunction with alternate attachment and release of ventral adhesive areas to the substrate. Contraction of the circular muscles in a part of the body, with the longitudinal muscles relaxed, will cause that part of the body to become long and thin; contraction of the longitudinal muscles in a part of a worm, with the circular muscles relaxed, will cause that part to become short and thick. As the wave of contraction of the circular muscles passes along the body from anterior to posterior, it causes successive portions of the worm to be raised off the substrate and moved slightly ahead of where they had been. The subsequent wave of contraction of the longitudinal muscles causes the elevated portions of the worm, after they have advanced, to be set down on the substrate, where they form temporary points of attachment to the substrate. Thus, progression along the substrate occurs by a series of short steps. Adhesive glands that open onto the ventral surface secrete a sticky substance that enables formation of these temporary points of attachment where the longitudinal muscles are more contracted. Inasmuch as the tegument of trematodes and cestodes is not ciliated, their movements depend entirely on muscle cells.

FEEDING AND DIGESTION The digestive system of turbellarians has a wide variety of form; the structure of the digestive system is one of the taxonomic bases on which this class is divided into orders. No member of the order Acoela has an intestine. Among the other turbellarians, intestines are present, ranging from a simple sac to a highly branched structure. All turbellarians have mouths but, as stated previously, no flatworm has an anus. The primitive location of the turbellarian mouth is near the center of the ventral surface, but during the evolution of this class, the mouth has migrated so that now, depending upon the species, it can be found anywhere along the midventral line from the anterior to the posterior end. The mouth of a turbellarian usually leads into a tubular structure, the pharynx, but in some of the acoels there is no pharynx. If an acoel has a pharynx, it opens directly into the mesenchyme, but if it lacks a pharynx, the mouth itself opens directly into the mesenchyme. In higher turbellarians, the pharynx opens into the intestine, often through a short intervening esophagus. The esophagus is narrower than the pharynx and is poorly muscularized. The pharynx, which assists in ingestion of food, ranges in structure in turbellarians from a simple ciliated tube in some acoels to a muscular tube that can be protruded through the mouth during feeding. The protrusible type of pharynx presumably developed through folding of the wall of a simple tubular pharynx. When the pharynx is protruded, food is ingested through the opening at its tip. The majority of turbellarians are carnivorous, feeding mainly on small living animals and even feeding on larger dead ones. Because of the lack of an intestine in acoels, food goes directly into the internal mass of mesenchyme cells. Small bits of food ingested by an acoel are simply engulfed by cells of the mesenchyme and digested intracellularly. Digestion of larger food by an acoel is at first extracellular, enzymes being released by gland cells in the pharynx and by cells of the mesenchyme; larger-sized food is partially digested to a smaller size, it is engulfed by the nutritive cells, and digestion is completed intracellularly. Similarly, in a turbellarian with an intestine, digestion of larger food items is at first extracellular, initiated by enzymes released from gland cells of the pharynx and intestine, followed by intracellular digestion of the resulting smaller particles in phagocytic cells of the intestine. Small bits of food may be engulfed at once by these phagocytic cells. Presumably, diffusion provides for the passage of nutrients from the phagocytic cells to the rest of the body. There is no circulatory system in flatworms.

A trematode, like a turbellarian, has a mouth, but that of trematodes is almost always either at or near the anterior end. The mouth leads into a muscular nonprotrusible pharynx that forces the food into the esophagus, which is like that in a turbellarian, and then on into the intestine, which is joined directly to the esophagus. In most trematodes, the intestine has two simple, cylindrical branches which end blindly. In some trematodes, the Oriental liver fluke for example, the mouth is encircled by a sucker. Depending upon where a trematode lives in the host, its food consists of blood, tissue cells, tissue exudates, or food particles in the host's intestine. Digestion is, in large measure, extracellular. Some nutrients are probably absorbed directly through the tegument also.

Adult cestodes are parasites of vertebrates, nearly always in the intestine, where they have a readily available food supply. A cestode has no mouth or digestive tract. It obtains all of its nutrients by absorbing them through the tegument. C. P. Read and his associates investigated the nutrition of cestodes by use of isotopically labeled molecules. They found, for example, that cestodes use an active transport mechanism to obtain for themselves an adequate quantity of amino acids from the host (diffusion alone would not suffice), and have been able to determine the rate at which the amino acids are incorporated into proteins. Tapeworms also actively take up glucose, certain fatty acids, purines, and pyrimidines. Cestodes presumably lost the digestive tract after they gained the ability to live solely on material they absorbed from the host. The tapeworm tegument bears numerous finger-like projections, called microvilli, that increase the surface area available for the absorption of nutrients (Fig. 5-4).

RESPIRATION Because of dorsoventral flattening, the dimensions of flatworms are such that respiration can be adequately accomplished by diffusion alone.

SENSE ORGANS Sensory mechanisms are well developed in the turbellarians and greatly reduced in the parasitic classes. Eyes occur in all turbellarians and some trematodes, being poorly developed in the latter, but are absent from cestodes. For an internal parasite living in the darkness, there is no selective advantage to having eyes. One pair of eyes is the usual number, but sometimes there are more. The eyes nearly always consist of a pigmented cup, within which lies a light-sensitive layer of neurons whose axons form the optic nerves and travel to the brain. These eyes are not image-forming; they are capable only of distinguishing light from dark. Chemoreceptors (cells sensitive to chemical stimuli) and touch-sensitive cells aid flatworms in locating food and hosts. Both types of cells are more concentrated anteriorly than posteriorly, and in parasitic species the attachment organs are also well endowed with touch receptors.

EXCRETION A flatworm, except an acoel, has an organized excretory system. It consists basically of a series of branched tubules having closed inner ends. The blind end of each tubule consists of a *flame cell*, which is a cup-shaped cell containing a tuft of cilia (Fig. 5-5). The beating of the ciliary tuft is reminiscent of a candle flame in a draft, which is the reason for the name "flame cell." This ciliary activity drives the fluid along the tubules. *Nephridium* is the name given to a unit of the excretory system in inver-

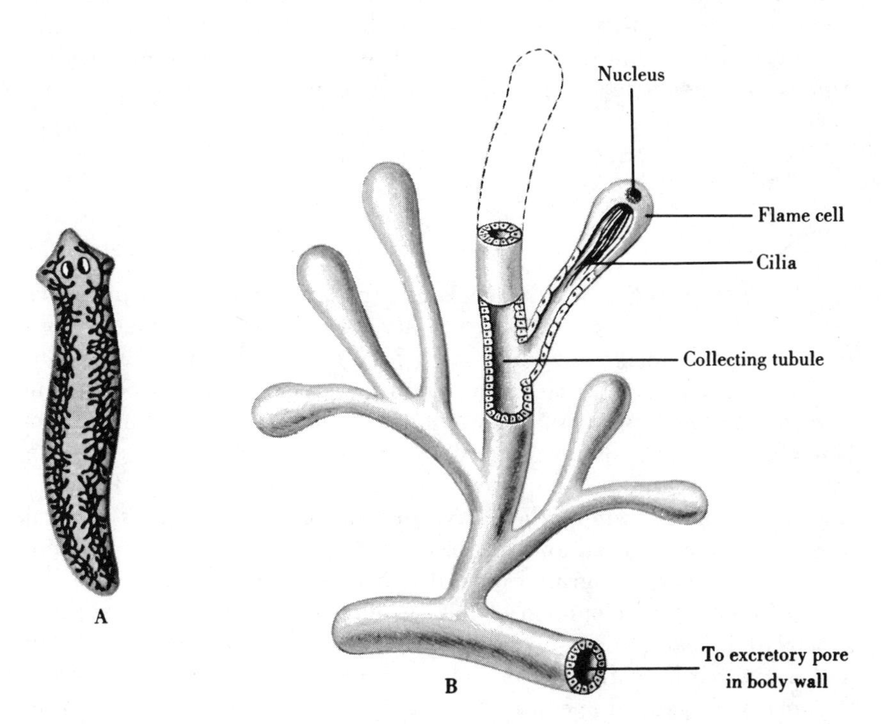

Fig. 5-5 *The protonephridial excretory system of flatworms. (A) Diagrammatic representation of the distribution of the protonephridia; (B) enlarged view of one of the protonephridia shown in A. Each protonephridial tubule of a flatworm has a flame cell at its inner end. (Modified from Griffin,* Animal Structure and Function. *Holt, Rinehart and Winston, 1962.)*

tebrates. One that has a closed inner end, opens (through a nephridiopore) onto a body surface that is in direct contact with the environment, and has a current-producing mechanism such as a flame cell is known as a *protonephridium*. The protonephridial system, highly developed in freshwater turbellarians, evidently serves to eliminate excess water from these hyperosmotic organisms which have the constant problem of water influx. However, there is no direct evidence that their protonephridia are indeed involved in osmoregulation. Acoels which, as stated earlier, are strictly marine, have no excretory organs, and the excretory system is only poorly developed in the rest of the marine turbellarians. The other marine turbellarians, like the acoels, are isosmotic with seawater; and their protonephridia would have no osmoregulatory problem to overcome if, as seems likely, getting rid of excess water is their main role. The protonephridial system of trematodes and cestodes is also reduced in comparison with that of freshwater turbellarians, presumably because there is little osmoregulatory work for this system to perform in these parasites. Whether these protonephridia excrete nitrogenous wastes has not yet been determined, but because of the size and shape of flatworms, these wastes could be eliminated by diffusion alone. Future experiments with flatworms will probably show that their nitrogenous wastes are eliminated by both diffusion and the protonephridia. Ammonia appears to be the major nitrogenous waste product of flatworms.

NERVOUS SYSTEM If acoels as a group form the most primitive order of flatworms, as some biologists contend, then the acoel nervous system may reveal what was the evolutionary pathway of the flatworm nervous system. In some acoels, the nervous system consists of a nerve net, similar to that found in cnidarians, between the epidermis and the subepidermal musculature. In other acoels, the main components of the nervous system are found in the mesenchyme, where they have become organized into a central nervous system with an anterior brain and typically five pairs of nerve cords. This system migrated inward beneath the muscle layers during the evolutionary development of this system, presuming, of course, that acoels are indeed the most primitive flatworms. This arrangement, brain and nerve cords in the mesenchyme, is also found in the rest of the flatworms. Advanced species in all classes of flatworms, however, have a reduced number of cords, as few as one pair. The cords have cross connections, and for this reason the flatworm nervous system is sometimes referred to as "ladder-type." The tapeworm brain is in the scolex. Nerve cords consist of the cell bodies and fibers of neurons, whereas nerves consist only of the fibers of neurons.

Some biologists, however, are of the opinion that the order Macrostomida, not the Acoela, consists of the most primitive flatworms. The macrostomid intestine is a simple saclike structure, reminiscent of the cnidarian gastrovascular cavity. This fact supporters of the macrostomids as the most primitive flatworms use as an argument in their favor. If macrostomids are, in fact, the most primitive of all flatworms, then the nerve-net type of nervous system in some of the acoels (as well as the lack of an intestine in the entire order) may be a degenerate condition, not the primitive archetypal one. The macrostomid nervous system does indeed consist of a brain and nerve cords.

REPRODUCTION The majority of flatworms are hermaphroditic. However, each class has at least one dioecious representative, the most commonly cited example probably being the blood fluke, *Schistosoma*, which numbers man among its hosts. Many adult turbellarians also reproduce asexually either by transverse fission, each portion then regenerating the missing parts, or by fragmenting into many portions, each of which reorganizes itself to form a complete, small worm.

The life cycles of the parasitic flatworms are complicated by the introduction of *intermediate hosts*, organisms parasitized by the larval stages rather than the adults. The *definitive host* is the organism in which the adult parasite lives. A parasite that requires intermediate hosts runs all the greater risk of not being able to complete its life cycle because its larvae may not be able to find the proper organisms. Parasites, however, typically compensate for this risk by producing exceedingly large numbers of eggs and sperm. Segmentation in tapeworms can best be explained as a reproductive adaptation; the continuous differentiation of new segments, each segment having a complete reproductive system, increases the reproductive potential of the tapeworm.

The Oriental liver fluke, *Opisthorchis sinensis*, will be used to illustrate just how complex the life cycle of a trematode can be (Fig. 5-6). It has three hosts. As an adult, it lives in bile passages in the liver of a human, where it reaches a size about 15 mm long. Fertilized eggs pass into the intestine with the bile and are eliminated in the feces. Meanwhile a larva, the *miracidium*, is developing within the egg shell. The developing eggs, having been either deposited directly into a pond or river or washed there by a rainfall, are then ingested by an appropriate species of snail. Within the snail, the miracidium hatches, burrows into the snail's tissues, and transforms into another larval form, a saclike *sporocyst* that reproduces asexually. The sporocyst's offspring are still other larvae, *rediae;* these in turn reproduce asexually and form still more larvae, *cercariae*, which then escape from the snail into the water, where they have

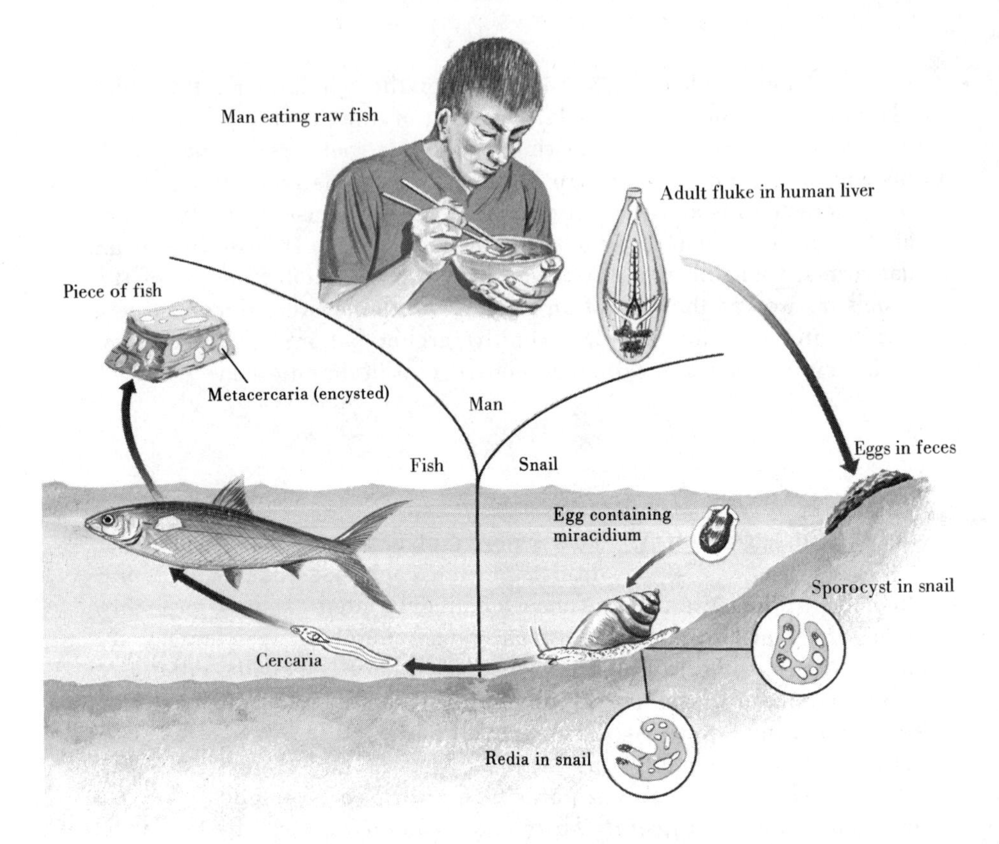

Fig. 5-6 *The life cycle of* Opisthorchis sinensis, *the Oriental liver fluke. (From Johnson, Laubengayer, DeLanney, and Cole,* Biology, *3rd ed. Holt, Rinehart, and Winston, 1966.)*

a free-swimming existence until they encounter one of several species of freshwater fishes. Upon coming in contact with the fish, the cercaria attaches to it, burrows under the scales, encysts in skin or muscles, and finally transforms into a *metacercaria,* which is a juvenile fluke. If the cercaria does not encounter a fish of the proper species within about 24 hours after having left the snail, it settles to the bottom and dies. After raw or insufficiently cooked infected fish is ingested by a human, the metacercaria becomes liberated in the small intestine and ultimately makes its way up the common bile duct to the liver, where it attains sexual maturity. Because of the asexual reproduction of these larval forms, the number of adult worms that can develop from a single egg is extremely large. Infected individuals suffer mainly from liver dysfunction. Continued infection of the population with this parasite is a public health problem involving sanitation as well as education in proper cooking of fish to kill the metacercariae. One difficulty in eradicating this disease is the custom in the Far East of eating raw fish.

Even more widespread than opisthorchiasis, the disease caused by *Opisthorchis*, is schistosomiasis. Schistosomiasis is sometimes also called bilharziasis because it was Theodore Bilharz who, in 1851, first discovered the worm, *Schistosoma*, that is responsible for the disease. *Schistosoma* affects many millions of people in areas such as Brazil, Africa, and the Far East. Three species of *Schistosoma* infect humans. The life cycle of *Schistosoma* is simpler than that of *Opisthorchis*, involving only two hosts. Adult *Schistosoma*, about 8 mm long, inhabit abdominal veins. Depending upon the species, the fertilized eggs pass to the outside either through the urinary bladder or intestine. Once the eggs are discharged into water, the miracidia that developed while within the eggshell hatch and enter an appropriate snail, where two generations of sporocysts are produced, and finally cercariae develop. Unlike *Opisthorchis*, *Schistosoma* does not form rediae. The eggs get into the bladder and intestine through ruptures in the walls of the blood vessels in which the eggs were deposited. These ruptures are probably caused by substances that are secreted by the developing miracidia inside the eggs and ooze out through minute pores in the eggshell. The infection is obtained when the cercaria, having left its intermediate snail host, burrows directly through the skin of a human who might be wading, working, or washing in an infected stream. It can even be obtained by drinking the water, thereby allowing the cercaria to penetrate into the body proper from, for example, the throat. Schistosomiasis is characterized by the presence of a rash, an enlarged liver, and a general weakness of the host that leaves him or her susceptible to other diseases.

The life cycles of cestodes are, in general, not as complicated as those of trematodes because they usually do not involve multiplication by asexually reproducing larval stages. Typically, each egg ultimately produces no more than one adult. The pork tapeworm, *Taenia solium*, is an example (Fig. 5-3). This worm lives in the small intestine of humans and attains a length of about 7 meters. The scolex, about 1 millimeter in diameter, is armed with hooks and suckers for attachment. Gravid proglottids, containing thousands of fertilized eggs, break off from the tapeworm and reach the outside through the host's feces. The proglottids then burst open, releasing the eggs, which the intermediate host, a pig, obtains by ingestion of contaminated food or water. As the uterus of an older proglottid is becoming crowded with eggs, it is developing lateral branches, increasing its volume. An embryo, the *oncosphere*, which develops while still within the eggshell, is then liberated, burrows through the intestinal wall of the pig, and makes its way to one of several possible sites, most commonly a muscle. Here it encysts and develops into a bladder worm *(cysticercus)*, which is a small, whitish larva with an invaginated scolex. When a human eats insufficiently cooked, infected pork, the cysticercus is released into the small intestine, and the scolex evaginates, attaches, and grows into an adult tapeworm in about 10 weeks. Proper cooking is essential to ensure death of the cysticerci. The worm will cause a mild nutritional drain on the host, ordinarily not causing any serious pathological problems.

PHYLOGENY The role of the planuloid ancestor in the evolution of new phyla does not appear to have ended with the cnidarians and ctenophores. The theory known as the planuloid-acoeloid theory seems to explain the origin of the flatworms better than any other. This theory proposes that the radially symmetrical planuloid ancestor not only gave rise to the cnidarians and through them the ctenophores, but also that this planuloid ancestor then went on to become a bilaterally symmetrical form, called the *acoeloid ancestor*, which in turn gave rise to all the flatworms and ultimately to the rest of the higher phyla as well. Acoels, which, as noted above, may be the most primitive living flatworms, can be linked without difficulty to a planuloid form that assumed a creeping mode of existence, and which then became dorsoventrally flattened and acquired a ventral mouth. Differentiation of dorsal and ventral surfaces produced the change from radial to bilateral symmetry. If acoels are indeed the most primitive flatworms, then the nerve net not only in cnidarians but also in those acoels that have it was most likely inherited directly from the planuloid ancestor. Other planuloid characteristics of the acoels are the completely ciliated epidermis and the internal mass of cells in place of an intestine. On the other hand, if the macrostomids are indeed the most primitive flatworms, then it must be assumed that the acoeloid ancestor developed a digestive cavity and a brain with nerve cords before giving rise to the earliest flatworms, and that acoels, which have no digestive cavity (and some now have only a nerve-net type of nervous system) are the products of degenerate changes.

Some recent proponents of the syncytial theory for the origin of all metazoans other than sponges from ciliated protozoans (J. Hadži for one) do not accept the role of the planuloid ancestor described in this chapter and the previous one. Instead, they point to flatworms as the group that most likely evolved directly from ciliated protozoans, and consider the cnidarians as having been derived from flatworms that had become sessile. Through natural selection cnidarians would then have attained their fundamental radial symmetry. (The advantage of radial symmetry to a sessile animal was discussed earlier.) Flatworms are clearly more highly evolved than cnidarians. According to this theory, called the ciliate-flatworm theory, cnidarians are essentially degenerate flatworms. Also, the anthozoans, according to this ciliate-flatworm theory, were the first cnidarians because many adult anthozoans are bilaterally symmetrical, as were their supposed flatworm ancestors, the hydrozoans and scyphozoans evolving from anthozoans. Anatomically, however, anthozoans actually are the most complicated members of their phylum, and consequently, if this theory is correct, the rest of the cnidarians must have evolved from anthozoans by a retrogressive process. This ciliate-flatworm theory has not been well received.

With respect to the phylum Platyhelminthes itself, it is generally agreed that both the flukes and the tapeworms evolved directly from the turbellarians. The use of intermediate hosts by the parasitic flatworms raises the question of how such complicated life cycles could have evolved. The most likely answer is that the life cycles recapitulate the phylogeny of the species. What are now the intermediate hosts were probably in the past each in turn the definitive hosts and, as higher forms evolved, new definitive hosts were added to the life cycles.

To summarize, the major innovations of the flatworms were bilateral symmetry, cephalization, and organ systems. These three provided a firm morphological base on which future phyla could be built. It would be difficult to overemphasize the contribution that flatworms made to the evolution of higher forms.

NEMERTINEA

The phylum Nemertinea [Gr. *Nemertes*, a sea nymph] is also often referred to as the phylum Rhynchocoela [Gr. *rhynchos*, snout + *koilos*, hollow]. About 600 species are known. Nemertines are bilaterally symmetrical acoelomates with the same solid type of body construction as flatworms have (mesenchyme and muscle cells filling the space between the integument and organs). They are unsegmented, slender, usually elongated, frequently dorsoventrally flattened worms (Fig. 5-7). The majority of nemertines are

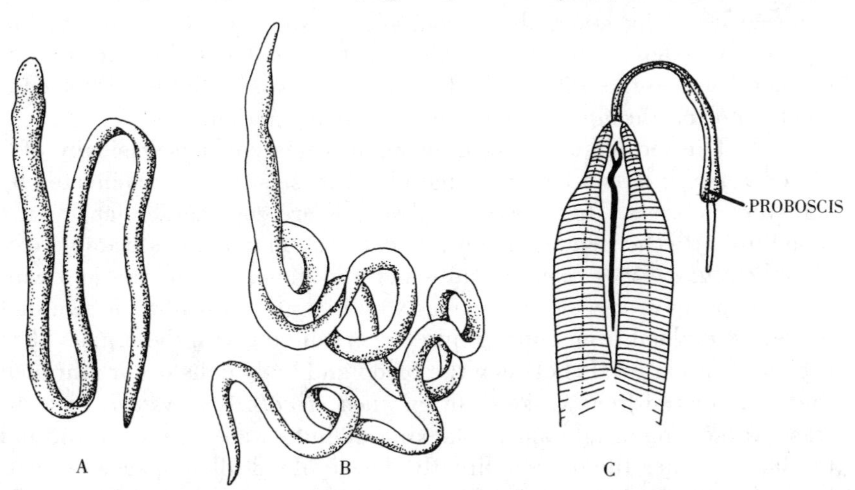

Fig. 5-7 *Nemertines. (A)* Lineus ruber; *(B)* Procephalothrix spiralis; *(C) the anterior portion of* Plotonemertes aurantiaca, *with its proboscis everted. (From Hyman after Coe.* The Invertebrates, *Vol. II. Copyright © 1951, McGraw-Hill Book Company. Used by permission.)*

less than 20 centimeters long, but *Lineus longissimus* from the North Sea is reported to be as much as 30 meters long. Most nemertines are bottom-dwelling marine forms, but one genus has invaded freshwater, where its members are also bottom-dwelling, while still another genus is terrestrial, occupying humid tropical and subtropical habitats. At the anterior end, a nemertine has a long eversible proboscis used for capturing prey, and consequently one of its common names is proboscis worm. Because of their elongated and frequently flattened bodies, nemertines are also commonly called ribbon worms. The proboscis, which lies in a special chamber, the *rhynchocoel*, located dorsal to the digestive tract, is shot forth through a pore at the anterior end of the worm, everting in the process, and coils around the prey. Nemertines have a flame-cell type of excretory system, a nervous system similar to that found in higher flatworms, and a ciliated epidermis. Nemertines show two significant evolutionary advances over the flatworms, a digestive tract with two openings (mouth and anus) and a circulatory system. Nevertheless, the fact that nemertines share so many characteristics with the free-living flatworms has convinced most biologists that nemertines evolved from turbellarian stock. However, at least one biologist has suggested that nemertines evolved from annelids by a series of degenerative changes, both groups having circulatory systems and two openings (mouth and anus) to their digestive tracts, with the nemertines having lost, among other annelid features, segmentation and a coelom.

The evolution of an elongated digestive tract with two openings (as in nemertines) was highly significant because structural and functional specializations along the length of the tract were then possible. Ingestion, digestion, absorption, and egestion could then proceed in an orderly fashion. The mouth of a nemertine is located on the ventral surface close to the anterior end, and the anus is at the posterior end. Food is moved through the intestine by ciliary action.

The circulatory system of nemertines consists basically of a closed network of two lateral, longitudinal vessels that are connected by transverse vessels. These two lateral vessels are contractile, causing the blood to flow. The nemertine circulatory system is a good example of a *closed circulatory system*, one in which the blood is confined in a continuous circuit of vessels and spaces with distinct walls of their own that prevent the blood from coming into direct contact with the organs. The exchange of materials between the blood and body cells occurs through the walls of the finer vessels. In an *open circulatory system*, by contrast, vessels open into spaces lacking definite walls of their own, and the blood bathes the organs directly. These blood-filled spaces are collectively called a hemocoel, which was discussed earlier, in Chapter 1. Circulatory systems are adaptations that provide a more efficient mechanism for the transport of oxygen and nutrients from the source of supply to the tissues than is possible by diffusion alone in larger animals. Efficient circulatory systems were essential for the evolution of higher, more complex, animals.

FURTHER READING

Erasmus, D. A., *The Biology of Trematodes*. New York: Crane, Russak and Co., 1972.

Greenberg, M. J., "Ancestors, Embryos, and Symmetry," *Systematic Zoology*, vol. 8 (1959), p. 212.

Hadži, J., "An Attempt to Reconstruct the System of Animal Classification," *Systematic Zoology*, vol. 2 (1953), p. 145.

Hanson, E. D., "On the Origin of the Eumetazoa," *Systematic Zoology*, vol. 7 (1958), p. 16.

Hyman, L. H., *The Invertebrates: Platyhelminthes and Rhynchocoela*, Vol. II. New York: McGraw-Hill, 1951.

Jennings, J. B., "Further Studies on Feeding and Digestion in Triclad Turbellaria," *Biological Bulletin*, vol. 123 (1962), p. 571.

Jones, H. D., "Observations on the Locomotion of Two British Terrestrial Planarians (Platyhelminthes, Tricladida)," *Journal of Zoology*, vol. 186 (1978); p. 407.

Lumsden, R. D., "Surface Ultrastructure and Cytochemistry of Parasitic Helminths," *Experimental Parasitology*, vol. 37 (1975), p. 267.

Mueller, J. F., "Helminth Life Cycles," *American Zoologist*, vol. 5 (1965), p. 131.

Read, C. P., and J. E. Simmons, Jr., "Biochemistry and Physiology of Tapeworms," *Physiological Reviews*, vol. 43 (1963), p. 263.

Riser, N. W., and M. P. Morse, *Biology of the Turbellaria*. New York: McGraw-Hill, 1974.

Senft, A. W., D. E. Philpott, and A. H. Pelofsky, "Electron Microscope Observations of the Integument, Flame Cells, and Gut of *Schistosoma mansoni*," *Journal of Parasitology*, vol. 47 (1961), p. 217.

Wilson, R. A., and L. A. Webster, "Protonephridia," *Biological Reviews*, vol. 49 (1974), p. 127.

chapter **6**

Pseudocoelomates: Nematoda, Acanthocephala, and Rotifera

NEMATODA

The phylum Nematoda [Gr. *nematos*, thread] consists of bilaterally symmetrical, pseudocoelomate, unsegmented organisms (Fig. 1-8). The body is circular in cross section, which accounts for nematodes' common name, roundworms, and in most species the body is tapered at both ends. Some biologists consider the nematodes merely one of five classes of pseudocoelomates in a phylum named the Aschelminthes [Gr. *ascos*, cavity + *helminthes*, worms]. Aschelminths, in addition to being pseudocoelomate, also share other characteristics, such as having a mouth and anus, having an intestine that has no musculature, and being unsegmented. However, most biologists hold the view, as herein, that each one of these five groups that are sometimes lumped together as one phylum deserves phylum rank. In spite of similarities, these five groups are, in other aspects, very diverse and lack any major distinctive, unifying characteristics, such as the unicellularity of protozoans or the pores of sponges, that would justify lumping them into one phylum. The phylum Nematoda is not divided into classes. Nematodes are a fairly uniform group of organisms; the differences among the orders are not sufficient to warrant their elevation to classes.

MORPHOLOGY AND PHYSIOLOGY

NEMATODE STRUCTURE

Nematodes are geographically very widespread animals, being found either free-living in soil, fresh water, and sea water, or living as parasites. They are common parasites of man. In total number of individuals, nematodes probably rank second only to insects. Free-living nematodes are usually 0.1–1.0 millimeters long, but some parasitic species may attain a length of about 2 meters.

The traditional view of the nematode body wall has been that it is covered by an acellular cuticle that is wholly external to an underlying epidermis that secretes it. However, recent work with the electron microscope has revealed that the cuticular material in the body wall of the nematode *Nippostronglylus brasiliensis* is covered by a trilaminate membrane, approximately 100 Å thick, that appears to be the original cell membrane of the outer edge of the epidermal cells that secrete the cuticle (Fig. 6-1). During the formation of the cuticle, a new cell membrane appears to be formed beneath the original outer one, and the cuticular material is laid down between these membranes. It seems, therefore, that the cuticular material is limited externally by a cell membrane and is, consequently, an intrinsic part of the epidermis, derived internally from it, rather than simply an extracellular secretion of it. The epidermis of nematodes can be syncytial, but it is more commonly cellular. The innermost layer of the body wall is a layer of muscle that consists of longitudinally arranged cells only. These muscle cells consist of three distinct parts, a saclike belly containing the nucleus and a store of glycogen, an obliquely striated, long contractile portion, and an arm that extends toward the nerve cord to make contact with a neuron (Fig. 6-2). The muscle cells of most other animals do not have such an arm; the neurons send a process to the muscle cells instead.

DIGESTION, CIRCULATION, AND RESPIRATION

The mouth, at the anterior end, leads into a cuticular buccal capsule, a short tube that joins the muscular pharynx. The pharynx is used for swallowing food; pharyngeal mus-

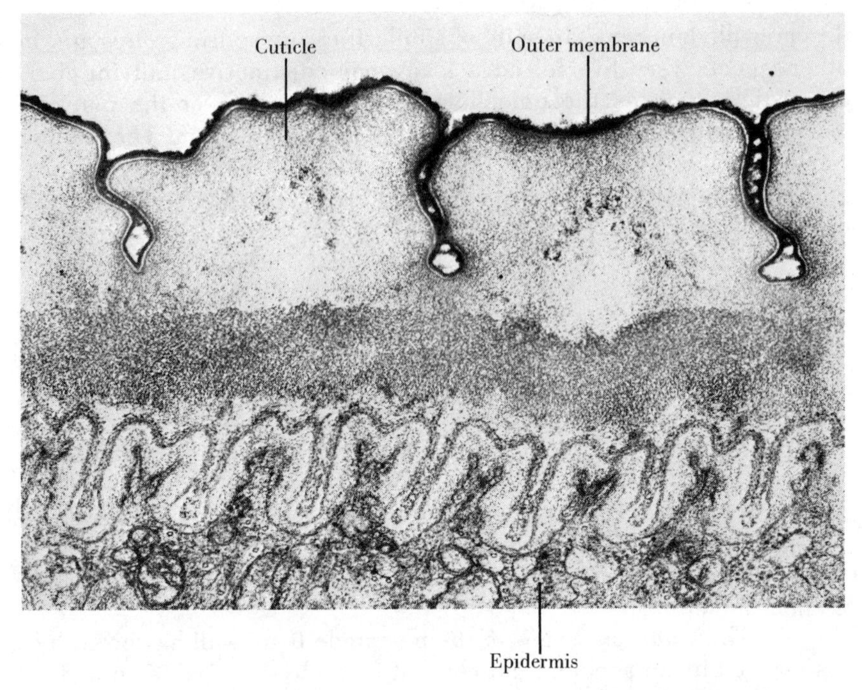

Fig. 6-1 *The cuticle of the nematode,* Nippostrongylus brasiliensis. *Note the trilaminate membrane at the outer edge of the body wall and the cuticle just below this membrane. (From Lee,* Tissue and Cell, *vol. 2, 1970.)*

cles force the food directly into the intestine. There is no esophagus. The intestine is a straight tube, its cells being a single layer of cuboidal or columnar epithelium. In large nematodes such as *Ascaris lumbricoides,* the basal portion of each of these intestinal cells often is fixed to a homogeneous layer of slightly basophilic material, which is probably secreted by the intestinal cells as a supporting structure for them. Because of the absence of both a ciliated lining in the intestine and a muscular layer around the intestine, the propulsive force for moving food through the intestine is most likely generated by the muscular pharynx. As recently ingested food is pumped by the pharynx into the intestine, the new meal forces previously eaten food ahead of it along the intestine. The anus is located on the ventral surface near the posterior end of the body.

The diet of nematodes is highly varied. Free-living species are herbivorous or carnivorous. Depending upon the species, parasitic nematodes survive on material in the host's intestine, disintegrated host tissues, or blood. A nematode feeds very frequently, presumably because the digestive tract lacks a region where food eaten in bulk could be stored. Digestion is probably all extracellular, but some intracellular digestion is possible. The end-products of protein, fat, and carbohydrate digestion are passed into the fluid of the pseudocoel, where they are available to the body cells. No special organs are present to handle circulation and respiration. Oxygen is presumably obtained by diffusion

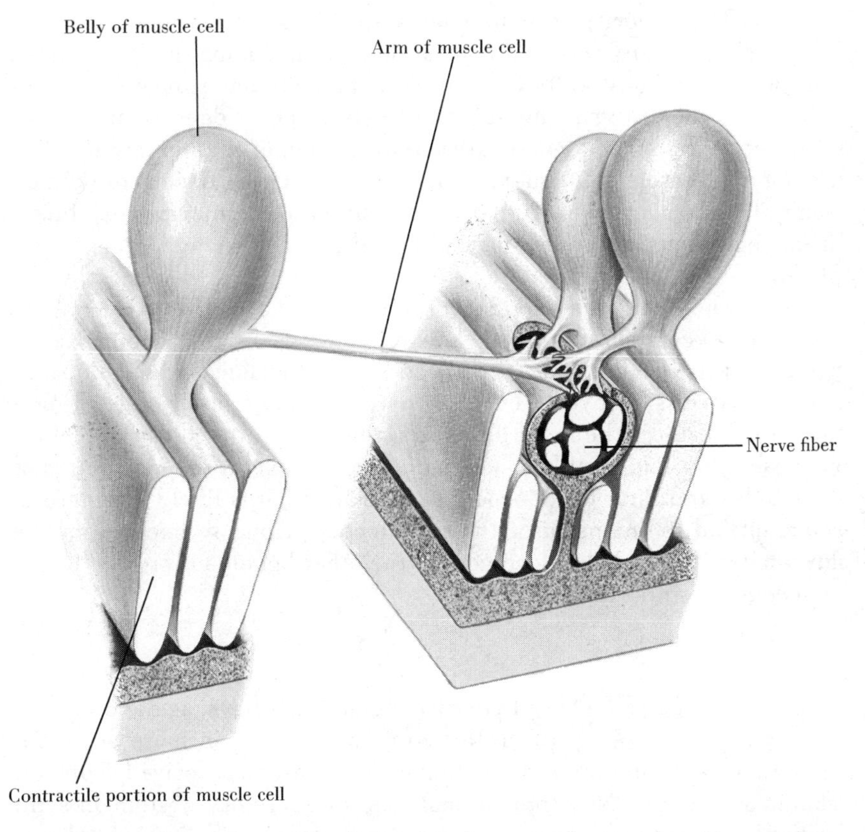

Fig. 6-2 *The anatomical relationship between the muscle cells and nerve fibers of the nematode,* Ascaris lumbricoides. *(After DeBell,* Quarterly Review of Biology, *vol. 40, 1965.)*

through the body wall. Diffusion and churning of the fluid in the pseudocoel as a result of body movements presumably suffice to move oxygen and nutrients from the locations where they enter the pseudocoelomic fluid to the cells that need them. The substitution of the pseudocoel for the mesenchyme of flatworms and nemertines around the intestine provided a faster means of distributing nutrients to body cells. Absorbed nutrients will move more quickly through the fluid in the pseudocoel than through mesenchyme.

LOCOMOTION Nematode movements are restricted because of the presence of a restraining cuticle and the lack of circular muscles to antagonize the action of the longitudinal muscles and return these longitudinal ones to their resting length. But nematodes are nevertheless capable of moving from place to place.

The majority of nematodes move forward by serpentine undulations produced by waves of contraction passing along the longitudinal muscles in the body wall. The cuticle commonly has ridges which enable nematodes to grip the substrate better. The ridges should, however, not be construed as an external indication of segmentation. They are not a reflection of segmentation in the underlying tissues or organs. None of the structures internal to the cuticle of nematodes exhibits a linear repetition that would suggest the existence of segmentation in this phylum.

The nematode pseudocoel functions as a hydrostatic skeleton and helps to compensate for the lack of circular muscles. The high fluid pressure produced in the pseudocoel when the longitudinal muscles contract is resisted by the cuticle, thereby automatically generating a force that restores the longitudinal muscle cells to their resting length. Because the pseudocoel has such an important role in the locomotion of nematodes and their pseudocoelomate relatives, it is likely that natural selection led to retention of the pseudocoel primarily because of the advantages it provided in locomotion; the other benefits it provided were of secondary importance.

EXCRETION Primitive nematodes have a glandular excretory system, whereas in more advanced species it is a tubular system that was apparently derived from the glandular system. No other animal has an excretory system like the glandular or tubular types seen in nematodes, not even their closest pseudocoelomate relatives. Nematodes have neither flame cells nor any other current-producing mechanism in the excretory system. The types of excretory system seen in nematodes were apparently not derived directly from the excretory system of any other group, but evolved independently among the nematodes.

By definition, a protonephridium must have a current-producing mechanism. Consequently, the types of excretory systems found in nematodes are not considered protonephridial. Nematodes most likely evolved from an ancestral form whose excretory system had a current-producing mechanism, but the nematodes lost it during their evolution. Examination of various nematodes provides evidence that there was indeed a gradual evolution from *renette cells* (glandular excretory cells that are found only in nematodes) to a system of tubules derived from them (Fig. 6-3). In free-living, marine species, which are the most primitive nematodes, one or two (usually one) renette cells located in the pseudocoel open to the outside *via* the excretory pore located on the ventral surface near the anterior end. More advanced members of the phylum have an H-shaped system of tubules associated with renette cells, the tubules appearing to have developed as outgrowths from the renette cells. In even more advanced nematodes, the H-shaped system of tu-

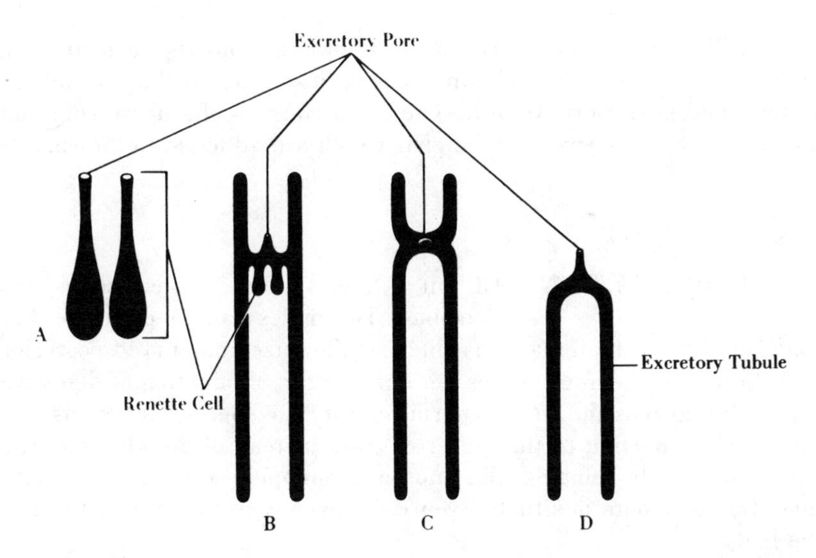

Fig. 6-3 *The presumed sequence of events in the evolution of the nematode excretory system. (A) Renette cells alone; (B) renette cells combined with an H-shaped system of excretory tubules; (C) H-shaped system of excretory tubules without renette cells; (D) the most advanced type, which lacks the anterior tubules and renette cells.*

bules is present, but the renette cells have been lost. Finally, in the most advanced condition, the anterior tubules have also been lost, thereby resulting in a tubular system that has the shape of an inverted U. The main roles of these excretory structures appear to be osmoregulation and ion regulation, with elimination of nitrogenous wastes occurring mainly by other routes. Ion regulation is the maintenance of different concentrations of specific ions between the external and internal environments of an animal, these differences being required for the normal functioning of cells and tissues. Osmoregulation, on the other hand, is the maintenance of a constancy of the osmotic concentration of the internal environment, the osmotic concentration depending upon the total number of dissolved particles, regardless of the kind. There is experimental evidence that in freshwater nematodes these excretory structures serve to eliminate excess water that has entered by osmosis. On the other hand, in those nematodes which are isosmotic with their environments, these structures have little or no osmoregulatory role to perform, and ion regulation appears to be their primary role, producing a urine that is isosmotic with the fluid in the pseudocoel but which differs from this fluid with respect to the concentrations of particular ions, potassium for example. Ammonia is the principal nitrogenous waste of nematodes and seems to have two main routes of removal that are not associated with these excretory structures: through the body wall by diffusion and *via* the digestive tract along with the solid waste products.

NERVOUS SYSTEM The nervous system consists basically of a brain and a series of nerves that extend anteriorly and posteriorly from it. The brain consists of a nerve ring that surrounds the pharynx, and ganglia, which are attached to the ring.

REPRODUCTION All but a few species of nematodes are dioecious. The males can generally be distinguished from the females by their smaller size and curved posterior end. In males, the anus serves not only for the elimination of digestive wastes but also as the site of sperm release; the sperm duct opens into the terminal portion of the digestive tract instead of directly onto the body surface. In females, the anus and gonopore are separate openings; this gonopore is situated ventrally, usually in the middle third of the body.

Ascaris lumbricoides and *Trichinella spiralis* are two of the commonly encountered nematode parasites of man. The strain of *Ascaris* that parasitizes humans was probably originally one that man obtained from pigs when he began to domesticate them. The human and pig strains are structurally indistinguishable, but the eggs of the human strain will not ordinarily develop to maturity in pigs and vice versa. *Ascaris lumbricoides* is a long nematode; males measure up to 31 cm long, females up to 49 cm. A female lays approximately 200,000 eggs per day. The adults live in man's small intestine, where they feed on partially digested food, although they may occasionally suck blood. After an egg is passed to the outside with the feces, it incubates in the soil, a juvenile worm developing within the shell. Juveniles, unlike larvae, closely resemble their parents in appearance, differing from their parents primarily in size and reproductive maturity. When an egg happens to be swallowed, for example, by a human eating unwashed vegetables that were grown in contaminated soil, the juvenile worm emerges in the small intestine. No intermediate host is required. However, the juvenile worm does not then remain in the small intestine. Instead, it penetrates the wall of the intestine and enters a blood vessel, ultimately getting to one of the lungs, where it burrows into the air spaces. The juvenile then moves out of the lung and up to the throat, then down the esophagus again, and back to the small intestine, where it finally matures into an adult. At the start of the migration, the worm is 0.25 mm long; it is 10 times longer when it returns to the intestine. If there is a heavy infection in the lungs, ascaroid pneumonia may result. Masses of adults in the small intestine can become entangled, producing a blockage.

The adult trichina worm *(Trichinella spiralis)* also inhabits the small intestine, but this worm has a life cycle that is quite different from that of *Ascaris lumbricoides*. The more severe pathological effects of infection by *Trichinella* are due not to the generation of the worm that first gained entrance to the body but to its offspring. The worms arrive in the human intestinal tract as juveniles encysted in poorly cooked pork, and they are freed from the cysts in the small intestine, where they mature and mate (Fig. 6-4). Adult females are 3–4 mm long, twice as long as adult males. The juvenile worms hatch from the fertilized eggs within the female's uterus. When juveniles are born, they are 0.1 mm long. They are deposited in a lymph vessel in the intestine by the female, which has burrowed into this vessel. The juveniles eventually invade and become lodged in skeletal muscles, of course damaging them in the process, where they roll themselves into a coil and eventually become enclosed in cysts. The inflammation produced by these juveniles in the muscles results in the formation by the host of the cyst capsules around the worms. The encysted juveniles are 1 mm long. The juveniles cannot mature unless the meat containing them is eaten by a susceptible mammal. The cysts become calcified within 6 to 9 months. Ingestion of 30 grams of heavily infected pork can result in at least 75 million worms distributing themselves throughout an unsuspecting individual. A piece of pork of this size can contain more than 100,000 juvenile worms.

Encystment is the last developmental stage a trichina worm attains in a human host. A hog usually becomes infected by eating contaminated scraps of pork in garbage. Symptoms of trichinosis, the disease caused by *Trichinella*, include fever, muscle pains, and difficulty in breathing and swallowing. A heavy infection can be fatal, particularly if many juveniles invade the heart musculature.

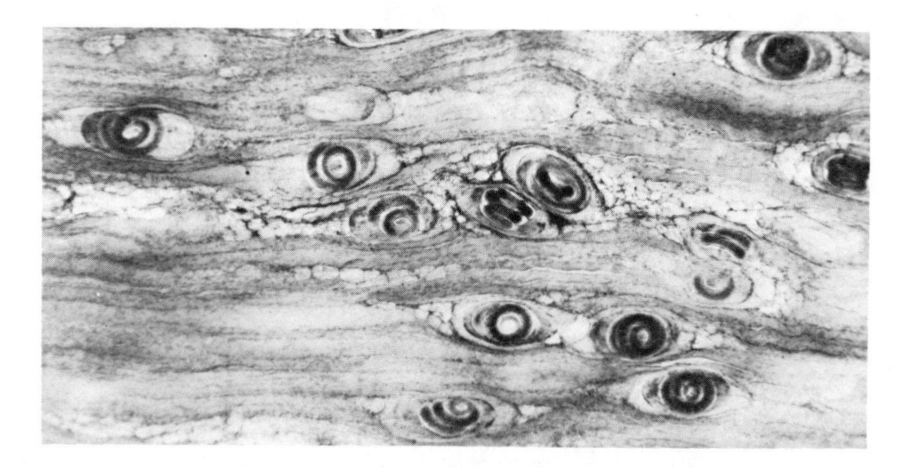

Fig. 6-4 *Juveniles of* Trichinella spiralis *encysted in the diaphragm of a rat. (Courtesy of Ward's Natural Science Establishment, Inc.)*

PHYLOGENY The ancestral nematodes were most likely free-living, marine species. The phylogenetic origin of the pseudocoelomates is not at all clear. One view is that they branched from the main line leading from the acoeloid ancestor to the coelomate ancestor, as shown in Fig. 1-3, perhaps arising from a free-living flatworm. Another theory that has been proposed, but has not been well received, is that pseudocoelomates evolved from an annelid (which, as stated earlier, is a coelomate) that underwent degenerative changes, a theory comparable to the one not highly regarded, described in Chapter 5, that suggested that nemertines are degenerate annelids.

It is generally agreed that pseudocoelomates were another evolutionary dead end; and even if they evolved from flatworms, it is highly unlikely that they were a direct link between the acoelomates and coelomates. The coelom could not have been phylogenetically derived from the pseudocoel because these body cavities have different modes of development. It will be recalled that the coelom does not develop from the blastocoel, whereas the pseudocoel does.

ACANTHOCEPHALA The phylum Acanthocephala [Gr. *acanthos*, spine ´*kephale*, head] consists of about 500 species of pseudocoelomate, bilaterally symmetrical worms, which as adults are parasites in the intestines of vertebrates, mainly fishes, birds, and mammals (Fig. 6-5). The juveniles are parasites in crusta-

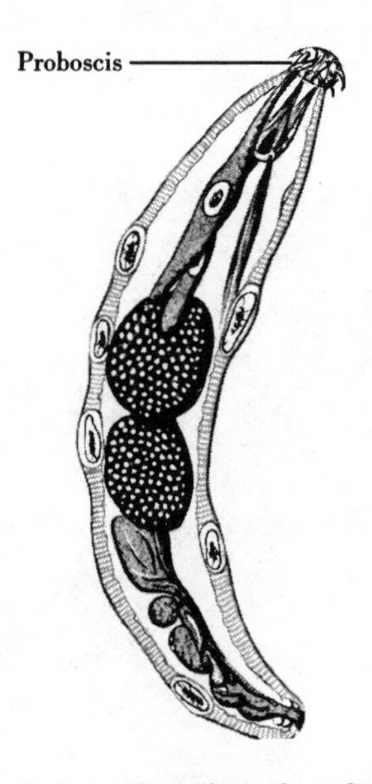

Fig. 6-5 *An acanthocephalan,* Neoechinorhynchus *sp. (After Borradaile and Potts,* The Invertebrates, *3rd ed. Cambridge University Press, 1959.)*

ceans and insects. A majority of the species are less than 25 mm long. A proboscis armed with hooks for attachment to the host is the diagnostic feature of the phylum and provides the basis for the common name of these animals, the spiny-headed worms. Acanthocephalans lack a digestive system, nutrients being obtained by absorption through the integument. The body wall is covered with a cuticle. They are unsegmented and dioecious. The fertilized eggs develop in the pseudocoel of the female to a larval stage that becomes enclosed by a shell. The shelled larva, known as an *acanthor,* is freed from the pseudocoel *via* the female's reproductive tract, and ultimately leaves the vertebrate host with feces. If an intermediate host eats the encased larva, it leaves its shell and enters the blood of the intermediate host, where it develops into a juvenile worm. When the intermediate host is eaten by a definitive host, the juvenile attaches to the intestinal wall and completes its development, becoming sexually mature.

ROTIFERA The members of the phylum Rotifera [L. *rota,* wheel + *fere,* to bear] are microscopic pseudocoelomates characterized by a "wheel organ" *(corona)* at the anterior end (Fig. 6-6). Cilia on the corona beat in a whirling motion, which gives the illusion of a rotating wheel. Food particles are brought to the mouth in the water current produced by the beating of the coronal cilia. The body, which is divided into three regions (head, trunk, and foot), is covered with a cuticle. They are unsegmented. Rotifers move about by creeping or swimming. The coronal cilia provide the propulsive force for swimming; but for crawling, rotifers use the foot and trunk muscles. Most rotifers are the size of ciliated protozoans, being among the smallest of the metazoans. The majority of the 2,000 species inhabit fresh water, where they occur in abundance. There are also a few marine species and even some that are able to survive living among mosses and lichens.

Rotifers are closely related to the nematodes, being one of the groups some authors combine with nematodes and three others (Gastrotricha, Kinorhyncha, and Nematomorpha) into the phylum Aschelminthes, which was referred to earlier in this chapter. One of the interesting features of rotifers and of the rest of the "Aschelminthes" is that many of their adult organs have a characteristic number of cells that is constant for the species. For example, in newly hatched individuals as well as in adults, the rotifer *Epiphanes senta* always has 183 brain cells and 15 esophageal cells. This constant makes these animals useful for analyzing developmental processes.

The females of one of the three classes of rotifers, the Monogononta, interestingly can produce different kinds of eggs, apparently in response to various environmental conditions, such as the arrival of

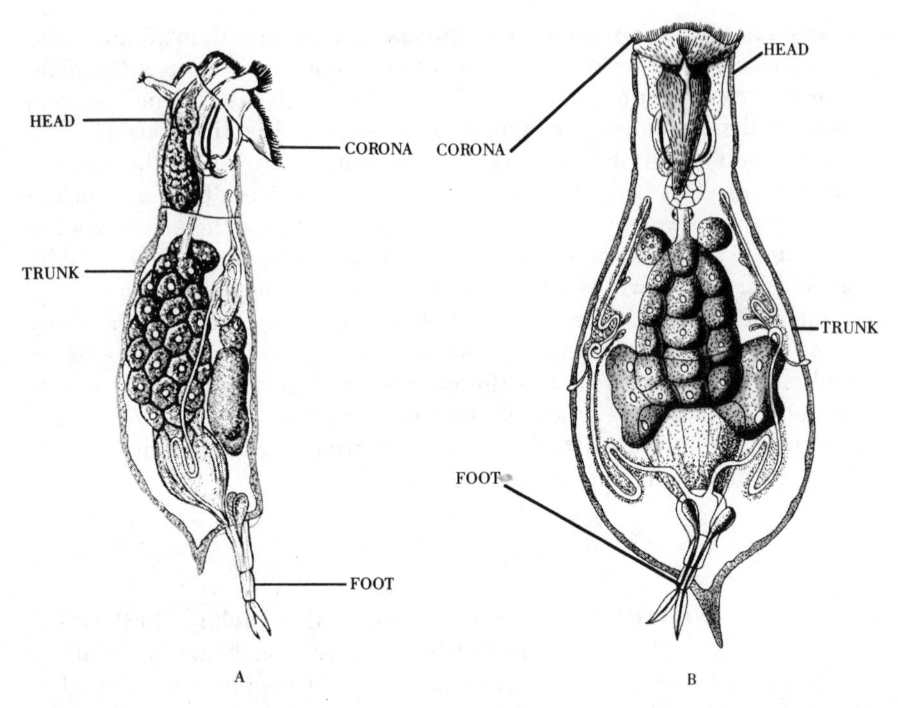

Fig. 6-6 *A rotifer,* Notommata copeus. *(A) Lateral view; (B) ventral view. (After Hyman,* The Invertebrates, *Vol. III. Copyright © 1955, McGraw-Hill Book Company. Used by permission.)*

spring or winter, or the drying up of their habitat. One kind of egg, called *amictic*, is thin-shelled, diploid, and develops parthenogenetically (without being fertilized) into a diploid female that will produce either amictic eggs or another type, called *mictic* eggs. A *diploid* cell has both chromosomes of every pair, whereas a *haploid* cell has only one member of every pair. Fusion of a haploid egg with a haploid sperm will produce a diploid zygote. A mictic egg is a thin-shelled, haploid egg which, if not fertilized, develops parthenogenetically into a haploid male. But if a mictic egg is fertilized, it accumulates additional yolk and secretes a thick shell. Such a thick-shelled diploid mictic egg is even more specifically called a *dormant* egg. A dormant egg always hatches into a diploid female that produces amictic eggs only. Dormant eggs are more capable of withstanding unfavorable environmental conditions, such as desiccation, than are the thin-shelled kinds. Furthermore, dormant eggs may not hatch for several months, as over the winter, whereas the thin-shelled ones usually hatch in 7–10 days. Dormant eggs will hatch in the spring into females that, as noted above, produce the rapidly hatching amictic eggs which develop parthenogenetically. Thus, in the spring, the population size can increase rapidly.

FURTHER READING

Bird, A. F., *The Structure of Nematodes.* New York: Academic Press, 1971.

Chapman, G., "The Hydrostatic Skeleton in the Invertebrates," *Biological Reviews,* vol. 33 (1958), p. 338.

Clark, R. B., *Dynamics in Metazoan Evolution.* New York: Oxford University Press, 1964.

Croll, N. A., *The Behaviour of Nematodes.* London: Edward Arnold, Ltd., 1970.

DeBell, J. T., "A Long Look at Neuromuscular Junctions in Nematodes," *Quarterly Review of Biology,* vol. 40 (1965), p. 233.

Deppe, U., E. Schierenberg, T. Cole, C. Krieg, D. Schmitt, B. Yoder, and G. von Ehrenstein, "Cell Lineages of the Embryo of the Nematode *Caenorhabditis elegans,*" *Proceedings of the National Academy of Sciences,* vol. 75 (1978), p. 376.

Gray, J., and H. W. Lissmann, "The Locomotion of Nematodes," *Journal of Experimental Biology,* vol. 41 (1964), p. 135.

Hyman, L. H., *The Invertebrates: Acanthocephala, Aschelminthes, and Entoprocta,* Vol. III. New York: McGraw-Hill, 1951.

Jarman, M., "Electrical Activity in the Muscle Cells of *Ascaris lumbricoides,*" *Nature,* vol. 184 (1959), p. 1244.

Kessel, R. G., J. J. Prestage, S. S. Sekhon, R. L. Smalley, and H. W. Beams, "Cytological Studies on the Intestinal Epithelial Cells of *Ascaris lumbricoides suum,*" *Transactions of the American Microscopical Society,* vol. 80 (1961), p. 103.

Lee, D. L., and H. J. Atkinson. *Physiology of Nematodes,* 2nd ed. New York: Columbia University Press, 1977.

Read, C. P., "Some Aspects of Nutrition in Parasites," *American Zoologist,* vol. 8 (1968), p. 139.

Thorne, G., *Principles of Nematology.* New York: McGraw-Hill, 1961.

chapter 7

Annelida

The most obvious feature of the members of the phylum Annelida [L. *annelus*, little ring] is segmentation. This segmentation is, in fact, the basis for the phylum name. The body appears to be divided externally into a linear series of rings. This division of the body into a linear series of similar parts or segments is clearly seen not only externally, but also internally in the serial repetition of the nerves, ganglia, muscles of the body wall, and blood vessels. Segmentation among the annelids evolved independently of that in tapeworms, and, as noted in Chapter 1, there is a basic difference between tapeworms and annelids with respect to the region of formation of new segments; in tapeworms, the new segments differentiate near the anterior end, whereas in annelids, they form near the posterior end. However, as noted in Chapter 5, the only segmented flatworms are tapeworms (and even some of the tapeworms are not segmented), but segmentation is characteristic of all annelids.

Annelids have the largest average size among the wormlike animals. Giant earthworms of Australia attain a length greater than 3 meters and a diameter of 2.5 centimeters. Annelids are schizocoelomates. In most annelids, the coelom is spacious and compartmentalized by transverse intersegmental septa formed from two layers of peritoneum, one layer from the segment ahead of the septum and the second from the segment behind it. However, in leeches (except for one primitive species), the coelom is greatly reduced, and the septa have disappeared. Biologists have tried for many years to explain adequately the phylogenetic origin and adaptive significance of segmentation in coelomates. One explanation that has been well accepted is the theory that segmentation arose in coelomates (and was retained by natural selection) because an animal with an elongated body would be able to move from place to place more effectively if its body musculature were segmentally arranged. Support for this theory is the fact that the locomotory capabilities of annelids are clearly greater than those of any other worms. Even in unsegmented, bilaterally symmetrical animals there is a tendency for the organs to be distributed along the length of the body rather than be bunched up. Once the organs had become dispersed along the trunk of an ancestral form, the body musculature could have become segmented, and afterwards, the rest of the organ systems. In all likelihood, the ancestral annelid was a free-moving marine organism that had a head followed by a trunk composed of a series of identical segments. These segments probably resembled the segments of some annelids living today, and each segment would have borne a pair of lateral, paddlelike, fleshy appendages known as *parapodia*, from which bristles *(setae)* projected (Fig. 7-1). Parapodia are supported internally by chitinous rods *(acicula)*. *Chitin* is a horny substance consisting of a long-chain polymer of D-glucosamine units, in which the amino groups are acetylated. How this hypothetical ancestral

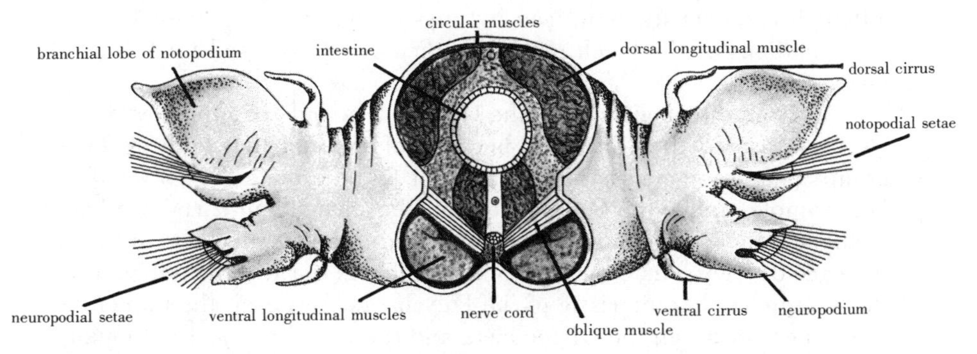

Fig. 7-1 *Cross section of the polychaete,* Neanthes *sp. (Redrawn from Burnett and Eisner,* Animal Adaptation. *Holt, Rinehart, and Winston, 1964.)*

annelid could have given rise to all of the modern annelids will be discussed below.

CLASSIFICATION The phylum is usually separated into three classes.

The class Polychaeta (Fig. 1-9A and B) consists of annelids having parapodia and setae. The class has been divided into two major groups—the errant and sedentary forms. In general, errant polychaetes are free-swimming, crawling, or actively burrowing forms, whereas sedentary polychaetes mostly dwell permanently in tubes they have manufactured about themselves, but some live in permanent burrows. These two groups of polychaetes show morphological differences that reflect their adaptation to either the more or the less active way of life. For example, the active forms have more highly developed parapodia than do the sedentary forms. However, the division between these two groups is not a sharp one. Practically all polychaetes are marine, although a few inhabit fresh water. Reproductive individuals *(epitokes)* of a polychaete, the Samoan palolo worm, *Eunice viridis,* are considered a delicacy by the natives of Samoa.

The class Oligochaeta consists of the earthworms and some of the freshwater annelids. They have lost their parapodia but have retained a few setae (Fig. 1-9C). The earthworm, *Lumbricus terrestris*, provides a service to farmers because its burrowing activities allow air and water to penetrate the soil more easily.

The class Hirudinea consists of the leeches. They occupy marine, freshwater, and terrestrial habitats. Of the three classes, the Hirudinea exhibit the least diversity of form among the members of its class. Leeches have lost their setae as well as their parapodia (Fig. 1-9D). Early in the nineteenth century when bloodletting was most widely, although erroneously, practiced by physicians, as a treatment for disease, millions of the leech *Hirudo medicinalis* were used for this purpose.

Some biologists treat the Oligochaeta and Hirudinea as a single class, called the Clitellata, because of the fact that both of these groups possess some characteristics not found among the Polychaeta. For example, unlike the Polychaeta, the Oligochaeta and Hirudinea have a clitellum (hence the name "Clitellata") that secretes a cocoon in which the eggs develop. Also, neither the Oligochaeta nor the Hirudinea have the parapodia characteristic of the Polychaeta. However, the numerous differences between the Oligochaeta and the Hirudinea justify to many biologists, as herein, dividing these two groups into separate classes. Where the class Clitellata is used, the Oligochaeta and Hirudinea are treated as subclasses.

MORPHOLOGY AND PHYSIOLOGY

ANNELID STRUCTURE

In general, the errant polychaetes (Fig. 1-9A) resemble the ancient fossil annelids more closely than any other annelid does; they have retained a body form closer to the hypothetical ancestral type than have the other annelids. Many of the modifications from the supposed ancestral type that are seen in the sedentary polychaetes (Fig. 1-9B) can be attributed to their mode of existence. Usually no more than their heads ever protrude from their tubes or burrows. Their heads are modified (often, for example, by the presence of tentacles) to provide an effective mechanism for obtaining food; such a means of food-getting contrasts with the more active modes of feeding, such as predation, used by the errant polychaetes. Errant polychaetes usually have strong, chitinous jaws that enable them to capture prey; sedentary polychaetes lack them. Furthermore, the fact that the parapodia are reduced in sedentary polychaetes is a reflection of their sedentary mode of life. The parapodia of free-swimming errant polychaetes, by contrast, are the most highly developed of any annelid's, both in size and strength, being important in providing a propulsive force for locomotion. Even the parapodia of burrowing errant polychaetes, although not as large as those of free-swimming individuals, are, nevertheless, larger than those of the sedentary forms.

Oligochaetes (Fig. 7-2) have lost the parapodia of the ancestral annelid, and the oligochaete head is greatly reduced compared with that of the polychaete. These modifications of oligochaetes are probably a reflection of the burrowing mode of existence adopted by almost all oligochaetes; a few species of aquatic oligochaetes live on submerged vegetation. Oligochaetes exhibit less diversity of external form than polychaetes, and leeches the least of all.

The head of a modern polychaete is made up of a well-developed *prostomium* and a *peristomium,* and in some polychaetes, some modified segments just behind the peristomium are also part of the head; the oligochaete head consists of a small prostomium and the peristomium; and the leech head consists of a small prostomium plus the first six segments of the body. The prostomium is a fleshy preoral lobe, which projects forward, overhanging the mouth. It is not considered a true segment. The first segment is the peristomium, which surrounds the mouth. It is likely that the head of the ancestral annelid consisted of the prostomium alone with the peristomium being the first trunk segment. As cephalization progressed, however, the peristomium, and in some instances even additional segments, became incorporated into the head. In the process, the peristomium and additional trunk segments became modified so that they no longer looked exactly like the rest of the trunk segments.

The bodies of all leeches are usually dorsoventrally flattened, and always consists of 34 segments; polychaetes and oligochaetes by

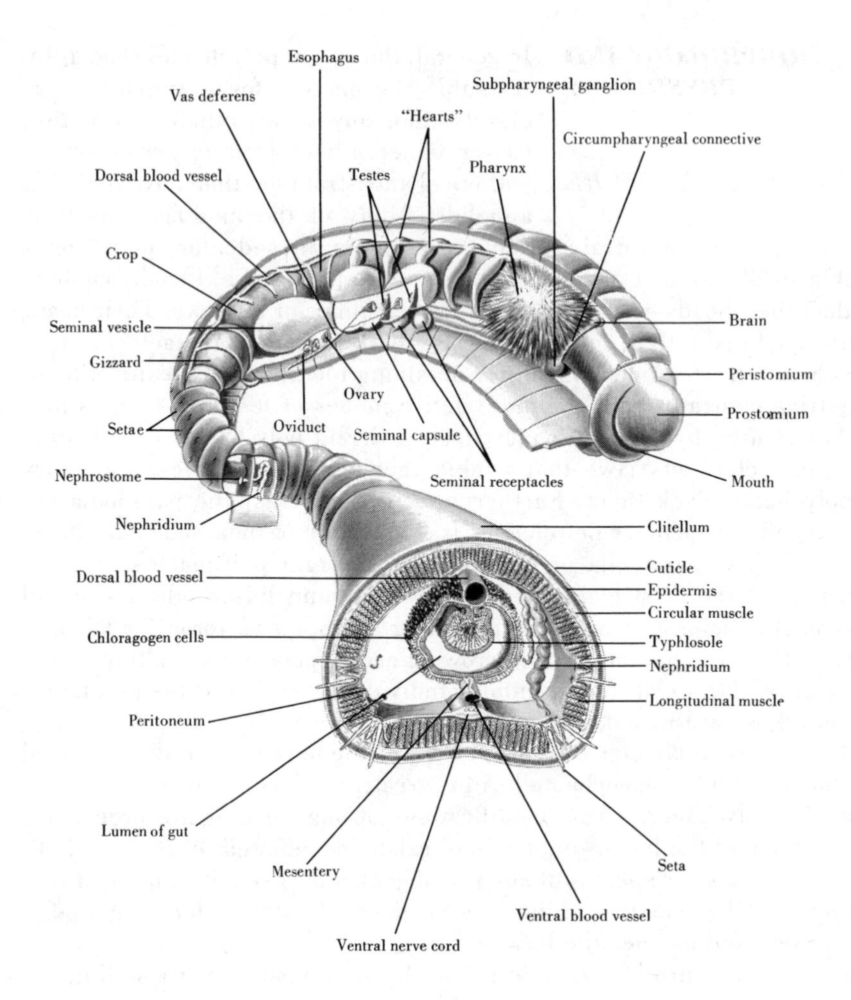

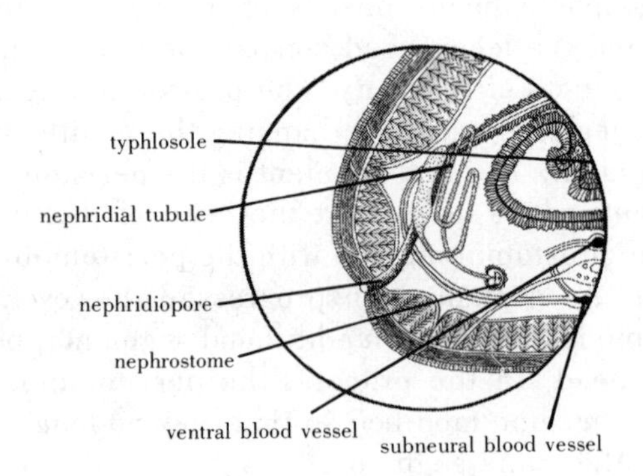

Fig. 7-2 *The structure of an earthworm,* Lumbricus terrestris. *(After Johnson, Laubengayer, DeLanney, and Cole,* Biology, *3rd ed. Holt, Rinehart, and Winston, 1966.)*

contrast, show no constancy in the number of segments from one species to another. However, it is difficult to count the segments in leeches because their segments are externally subdivided. A leech has an anterior sucker, usually surrounding the mouth. There is also a posterior sucker that is usually larger than the anterior one. Both suckers are formed from modified segments. In the anterior half of mature oligochaetes and leeches, there is a glandular group of adjacent segments that compose the *clitellum.* The clitellum is a reproductive structure that secretes a cocoon into which the eggs are deposited and where they develop into young worms. The clitellar segments may become permanently swollen at the onset of sexual maturity through the development of the glandular cells in these segments, or the segments may become conspicuous only during the breeding season, as a result of activation of these glandular cells. There is no uniformity among oligochaetes in regard either to the number of segments the clitellum occupies or to its location with respect to the genital pores, but in leeches it always occupies segments 9, 10, and 11.

The structure of the body wall is essentially the same in all annelids. The outermost layer is a thin, pliant cuticle, below which is a single layer of columnar epidermal cells. Within the epidermis is a thin layer of connective tissue, then a layer of circular muscle cells, followed by a layer of longitudinal muscle cells, and, finally, innermost is the peritoneum.

LOCOMOTION Errant polychaetes have highly developed parapodia, which are used very effectively for locomotion, but in the sedentary polychaetes, as noted above, the parapodia are reduced. Sir James Gray has analyzed the mechanism of locomotion in an errant polychaete by means of slow-motion photography. Slow crawling is accomplished by a backward, effective stroke of the parapodia while they are in contact with the substrate. During the effective stroke, the parapodia are extended and lowered toward the substrate. The reverse occurs during the forward, recovery stroke. Fast crawling and swimming are accomplished by a paddlelike action of the parapodia, plus undulatory movements of the body that are produced by alternate contractions of the longitudinal muscles on the left and right sides of the body wall.

Although oligochaetes have lost their parapodia, they have retained setae, four groups per trunk segment and usually two per group, which are used in locomotion. Crawling in oligochaetes, sedentary tubiculous polychaetes, and burrowing polychaetes is not undulatory, but depends instead on peristaltic waves of contraction of the circular and longitudinal muscles that affect both sides of the body simultaneously, along with use of the setae as an anchoring mechanism. Forward progression commences with the initiation of a wave of contraction of the circular muscles that sweeps from the anterior to the posterior end of the animal. This wave of contraction results in elongation of the worm, because the circular muscles contract against the incompressible coelomic fluid trapped between the septa, causing the segments to become elongated as they become narrower. The fluid-filled coelom functions in locomotion as a hydrostatic skeleton. However, until the wave of contraction of the circular muscles has extended to about midway along the worm, the posterior end remains anchored in place because of protruding setae. Consequently, as the worm becomes elongated while the posterior end is anchored, the anterior end is propelled forward. The setae become protruded when the longitudinal muscles contract. After the wave of contraction of the circular muscles has reached the posterior half of the worm, a wave of contraction of the longitudinal muscles commences at the head end and passes to the posterior end. The result of this wave is not only the anchoring of the anterior end but also the pulling forward of the posterior end because contraction of the longitudinal muscles results in shortening and widening of the segments. The wave of longitudinal muscle contraction is followed by another wave of contraction of the circular muscles, again commencing at the head end, and so on.

Crawling in leeches is similar to the pattern in earthworms except that instead of anchoring by means of setae, leeches have anterior and posterior suckers for attachment. After attachment of the posterior sucker, the body elongates because of a wave of contraction of the circular muscles, the anterior sucker then attaches, the posterior sucker releases its hold, and finally a wave of contraction of the longitudinal muscles causes the posterior end to be pulled forward. The cycle then is repeated.

FEEDING AND DIGESTION

Annelids show wide diversity both in the kinds of food they use as energy sources and in the mechanisms for obtaining this food. Free-swimming and crawling errant polychaetes are mostly carnivorous predators, capturing small invertebrates; but some are omnivores—algae are included in their diets. The burrowing errant polychaetes subsist on organic material in mud that they swallow. Many sedentary polychaetes use tentacles which extend from the head to collect food particles that have settled onto the substrate. They are not filter feeders; they neither create a water current nor collect suspended food particles. For example, *Amphitrite ornata* (Fig. 1-9B), a tube-dweller, spreads its tentacles over the substrate, and any food particles on the substrate that adhere to mucus secreted onto the tentacle surfaces are passed to the mouth by the action of the cilia on the tentacles. Other sedentary polychaetes are, however, indeed filter feeders, usually having special ciliated feeding appendages protruding from the head that create a water current by the beating of the cilia and collect suspended food particles. However, the sedentary polychaete, *Chaetopterus*, has an interesting mucoid filter feeding mechanism that is very different from the ciliary mechanism of the other filter-feeding polychaetes. This worm lives in a U-shaped tube, the openings of which extend above the surface of the substrate. Through the action of certain fanlike parapodia, a current is set up that brings water containing suspended food particles into the tube, where it passes through a baglike mucous sheet, secreted by other parapodia, that collects the food particles. The mucous sheet is continuously being rolled into a ball. When the ball reaches a certain size, it is cut loose and transported by ciliary activity to the mouth, where it is swallowed.

Earthworms mainly consume decomposing organic material, primarily vegetation at the surface of the ground; however, they also obtain some organic matter from soil they swallow while burrowing. Aquatic oligochaetes subsist not only on decomposing organic matter, but also include algae in their diet.

About three-fourths of the leech species are bloodsucking. The rest are predacious, feeding on such invertebrates as snails and insect larvae. The bloodsucking leeches feed on a wide variety of animal blood, including that of snails, insects, and all types of vertebrates.

The digestive tract of a typical polychaete is a straight tube, which runs from the anterior mouth to the posterior anus and is differentiated into a pharynx, esophagus, stomach, and intestine. The pharynx is an ingestive organ. The jaws of those errant polychaetes that have them, and most do, are attached to the pharynx to aid in capturing prey or biting off pieces of algae. In such species, the pharynx can be everted through muscular action, bringing the jaws to the anterior tip of the body from their position inside the pharynx. The esophagus in polychaetes is typically a short, straight tube that connects the pharynx to the posterior portions of the digestive tract. Digestion is extracellular; the enzymes are released from the epithelial lining of the stomach (a dilated portion of the digestive tract) into the cavity. Absorption of the nutrients occurs in the intestine. The wall of the digestive tract contains muscle cells that provide a means both of mixing the food with the digestive enzymes by churning movements, and forcing it along the tube by peristalsis. The intestinal epithelium is typically ciliated; these cilia assist the musculature in mixing and moving the intestinal contents, particularly in species in which intestinal muscles are not well developed. This musculature provides a more effective means of moving food from one end of the intestine to the other than is possible without such an intestinal musculature. In polychaetes and oligochaetes, the intestine is surrounded by a layer of yellowish, greenish, and brownish *chloragogen cells*, the main center for synthesis and storage of fat and glycogen. Deamination of amino acids and formation of the nitrogenous waste products, ammonia and urea, also occur here.

The oligochaete digestive tract is also a straight tube. As in polychaetes, the pharynx functions as the ingestive organ in oligochaetes. No oligochaete has pharyngeal teeth. In aquatic oligochaetes, the pharynx can be everted to pick up particles, and in earthworms it functions as a pump to draw in food. The pharynx leads into the esophagus. In earthworms, part of the esophagus is modified to form a crop and a gizzard posterior to the crop. The crop is thin-walled and used for food storage, whereas the gizzard is highly muscular and used to grind food particles. Neither of these organs produces digestive enzymes. The location of the crop and gizzard along the esophagus is variable, being at the anterior end in some oligochaetes and at the posterior end in others. The wall of the oligochaete esophagus contains a cluster of glands, called calciferous glands, whose function is excretory rather than digestive, despite their association with the esophagus. They excrete excess calcium ions from the blood as calcite, which is eliminated with the feces. It appears, from electron microscopy of calciferous glands, that the mitochondria are intimately involved in the production of the calcite; crystalline structures have been found in the mitochondria. Few oligochaetes have stomachs, in the absence of which both digestion and absorption occur in the intestine. In most earthworms, the dorsal surface of the intestine is folded into the intestinal cavity. This fold *(typhlosole)* provides a greater surface area inside the intestine, thereby increasing the intestine's absorptive efficiency.

Leeches have two types of ingestive organs. A predacious leech has a muscular proboscis, which can be protruded out of the mouth, whereas bloodsucking leeches have a muscular, sucking pharynx. The bloodsucking medicinal leech, *Hirudo medicinalis*, has three jaws just within its oral cavity (its mouth is surrounded by the oral sucker), which are used to cut through the skin of the host. Unicellular salivary glands in the pharyngeal wall of bloodsucking leeches secrete an anticoagulant, *hirudin*, that prevents the blood meal from clotting as it is being ingested. Posterior to the pharynx is a short esophagus. It joins a stomach that occupies approximately the middle third of the body and quite often has pouches that increase its volume. The intestine, which may also have pouches, comprises about the last third of the digestive tract. The anus of a leech, formed by fusion of the last eight segments of the body, is on the dorsal surface just anterior to the posterior sucker, but the anus is terminal in polychaetes and oligochaetes.

CIRCULATION In annelids, as a whole, a closed circulatory system is usually considered characteristic. A few annelids, however, do not have the completely closed circulatory system that the rest of the annelids have. In a few polychaetes, the blood passing through the walls of the stomach and intestine is contained simply in a cavity in the walls of these organs and not within blood vessels or spaces having definite walls of their own. A similar situation occurs in the intestinal wall of the aquatic oligochaete, *Aeolosoma*. Consequently, these few annelids could be said to have an open circulatory system, but it is not a typical open circulatory system because, except for the blood-filled space in the digestive tract, everywhere else in the body the blood is contained in vessels with definite walls of their own. In contrast, animals such as arthropods, which have the typical open circulatory system, have blood-filled spaces throughout their bodies, not only in one or two organs. Large blood-containing spaces without definite walls of their own, such as in these few polychaetes, are called *sinuses*, small ones *lacunae*. These lacunae and sinuses together comprise the hemocoel of the animal. As noted in Chapter 1, a hemocoel develops from the primary body cavity (blastocoel). A hemocoel is neither part of nor derived from the coelom. An animal that has an open circulatory system must then also have a hemocoelas part of this system.

With the appearance of larger-sized animals, diffusion and churning became very inefficient means of transporting nutrients and oxygen from one part of the body to another. Large, active animals use oxygen at a faster rate than can be supplied simply by diffusion and churning. Natural selection favored a well-developed circulatory system.

Blood flow in annelids is accomplished by means of peristaltic waves of contraction moving along blood vessels. The dorsal blood vessel is the most important contractile vessel of all, but others may also contribute to the maintenance of the blood flow. In fact, the earthworm, *Lumbricus terrestris*, has five pairs of esophageal vessels connecting the dorsal and ventral blood vessels; these pairs of vessels are so conspicuously contractile that they are sometimes referred to as "hearts."

The basic design of the annelid circulatory system, as seen in all annelids except those that have a sinus in the digestive tract and some leeches, is such that blood flows anteriorly in the dorsal blood vessel, then ventrally in lateral vessels around both sides of the anterior portion of the digestive tract, and finally posteriorly in a vessel ventral to the intestine. In each segment, vessels branch from the ventral blood vessel and carry blood to the structures in that segment, and corresponding vessels are received by the dorsal blood vessel. These segmental vessels that carry blood from the ventral blood vessel to the structures in each segment and then from these structures to the dorsal blood vessel are connected to each other by a thin-walled network of smaller vessels (capillaries) that readily allows for the exchange of materials between the blood and tissues. However, among those annelids that have a sinus in the digestive tract, a portion of the dorsal blood vessel is absent from the affected section of the digestive tract that has the sinus. Also missing are the network of capillaries and segmental vessels that occur in the corresponding regions of the digestive tract of annelids with a completely closed circulatory system. The dorsal and ventral blood vessels and the smaller vessels that are associated with the digestive tract could have evolved from such a sinus. Possession of such a sinus is probably a primitive annelid trait; the organ is, presumably, derived from the circulatory system of the hypothetical ancestral annelid discussed earlier in this chapter.

Some leeches have retained the typical annelid closed type of circulatory system, but in others the original circulatory system has disappeared. Instead, through reduction of the coelom, a network of coelomic vessels has developed that functions as a closed circulatory system. The lateral longitudinal coelomic vessels are contractile and thereby are able to provide the propulsive force for circulation of the blood.

The blood of most annelids contains a respiratory or transport pigment that increases its oxygen-carrying capacity. Hemoglobin, an iron-containing, red transport pigment, is the most common one; but chlorocruorin, an iron-containing, green transport pigment, is found in a few families of polychaetes. When hemoglobin is present in the blood of annelids it is dissolved in the plasma. Chlorocruorin, which occurs only in the blood of annelids, is likewise dissolved in the plasma, not enclosed in blood cells. In addition, the blood cells of the polychaete *Magelona* have an iron-containing, violet transport pigment hemerythrin, but there is no transport pigment in its plasma. Oligochaetes and leeches have only hemoglobin.

RESPIRATION Oligochaetes and leeches have no special organs of respiration; gaseous exchange occurs across the body wall by diffusion, oxygen being taken up by both the blood and coelomic fluid. Among polychaetes, in addition to the body wall, the parapodia serve a respiratory role. Being thin structures, parapodia can perform effectively in the uptake of oxygen. Also, as an aid to respiration, in many polychaetes parts of the parapodia became enlarged, forming gills, thus providing even more body surface area for use in oxygen uptake. Any structure which is directed outward from the body surface and is concerned primarily with gas exchange is called a gill. Gills are almost always richly supplied with blood vessels. Tubiculous polychaetes pump water through their tubes by means of peristaltic contractions or undulatory movements of the body wall, or by use of cilia, thereby obtaining oxygen for themselves. The filter-feeding tube-dwellers make use of this respiratory current to obtain food also.

SENSE ORGANS Photoreceptors, useful in locating food, are particularly well developed in errant polychaetes. On the other hand, the photoreceptors of oligochaetes are poorly developed, which is not surprising in view of the burrowing mode of existence all but a few of them have adopted. Similarly, the photoreceptors of practically all sedentary polychaetes are as poorly developed as those of oligochaetes, in keeping with a sedentary existence. Leeches appear to have photoreceptors that are more highly developed than those in oligochaetes and sedentary polychaetes, but not at the level of those in errant polychaetes. The photoreceptors of errant polychaetes, a very few sedentary polychaetes, and leeches are localized structures (true eyes) located at the head end. They consist essentially of clusters of photosensitive cells, pigment cells, and supporting cells. In contrast, oligochaetes and most sedentary polychaetes do not have true eyes. Instead, they are provided with individual photoreceptor cells, which are scattered along the body wall but are more concentrated at the anterior end than elsewhere. In addition to their light-sensitivity, annelids are highly sensitive to chemical stimuli and touch. The chemoreceptors are present in the epidermis along with free nerve endings, which are responsible for the sense of touch. Predacious annelids are aided by their chemoreceptors in finding prey. A well-developed sense of touch is particularly important to a burrower, living in darkness.

EXCRETION Some polychaetes have a protonephridial type of excretory system, whereas all other annelids have another type of nephridium, a *metanephridium* that occurs only in coelomates. A metanephridium is an excretory tubule that is open at both ends. Its inner end always opens into the coelom—as opposed to the more primitive protonephridium, which has a closed inner end. A typical metanephridium consists of a *nephrostome* (open, ciliated funnel) in the coelom and a ciliated *nephridial tubule* opening at the *nephridiopore* on the body surface. The cilia cause the fluid to flow through the metanephridia. The cavities of the metanephridia and gonadal ducts are apparently extensions of the coelom. Instead of flame cells with a tuft of cilia, as in flatworms, the protonephridium of annelids has at its closed coelomic end one or more *solenocytes*, long, tubular cells, with a single flagellum beating inside the tubular portion of each cell. Fluid apparently enters a solenocyte through the wall of the tubular portion of the cell and is then driven into the nephridial tubule by the flagellum. As the fluid moves along either type of nephridium it is converted into urine.

Oligochaetes typically have one pair of metanephridia per body segment, except at the extreme anterior and posterior ends, where there are none. The oligochaete peristomium lacks not only nephridia and nephridiopores but also setae. Presumably, all of these were lost when the peristomium ceased to be a trunk segment and became part of the head. Polychaetes have either one pair of nephridia per segment or simply one pair for the whole organism. Leeches have 10 to 17 pairs of nephridia in the middle portion of the body, one pair per segment. In spite of the reduction of the coelom in leeches, their metanephridia still maintain their association with the coelom, opening into the coelomic vessels. The nephridium typically opens to the surface in the segment behind the one containing its inner end, the nephrostome being in one segment and the rest of the metanephridium in the adjoining posterior segment. Although the urine is produced predominantly from the coelomic fluid, some waste products go directly into the nephridia from the blood that passes through vessels surrounding the nephridial tubules. During the passage of the fluid being processed into urine along the nephridial tubule, nutrients, such as glucose and amino acids, which might have been present in the fluid that entered the tubules, are reabsorbed, but the wastes are excreted.

Freshwater annelids are hyperosmotic to their environment. Their nephridia are adapted for conservation of salt and elimination of the excess water that enters by osmosis. In addition, they can actively take in sodium and chloride ions across the body wall in order to compensate for loss in the urine. Earthworms are adapted for living in moist soil, where the osmotic stress is intermediate between that faced by a freshwater annelid and life in air, where desiccation would be a large problem. As in the case of freshwater annelids, the nephridia of terrestrial annelids minimize loss of salt by reabsorbing it from the fluid in the nephridial tubules, producing a urine that is hyposmotic to the coelomic fluid and blood. Marine annelids are isosmotic with sea water, and the work their nephridia must perform is consequently minimal. As would be expected, nephridia are more highly developed in those annelids that are under constant osmotic stress.

Such animals as polychaetes and leeches, whose major nitrogenous waste product from protein catabolism is ammonia, are called *ammonotelic*. But in earthworms there is a tendency for urea to become their principal nitrogenous waste product. Animals whose main nitrogenous waste is urea are called *ureotelic*. Earthworms exhibit what is referred to as a mixed ammonotelism–ureotelism. For example, when the earthworm *Pheretima posthuma* is immersed in water it is ammonotelic, but when kept in moist air it is ureotelic, that is, more nitrogen is excreted as urea than ammonia.

A shift from ammonotelism to ureotelism, as in *Pheretima posthuma*, is an adaptation for survival in a terrestrial habitat. When this worm is immersed in water, ammonia can diffuse through the body surfaces (no special organ is needed to excrete ammonia) into the surrounding water and away from the worm. In a terrestrial situation, the ammonia would instead accumulate in the animal to a lethal level because the ammonia could not be eliminated fast enough to keep its concentration in the animal at a safe level. Urea is much less toxic than ammonia; thus, animals can tolerate a much higher concentration of urea than ammonia in their bodies. In general, ureotelic animals have neither a greatly restricted nor an unrestricted water supply, whereas ammonotelic animals must live in an abundance of water for the efficient excretion of ammonia. Ureotelic animals prevent the build-up of ammonia in their bodies by converting it to urea. Apparently, with the emergence of some oligochaetes from water onto land, a change in nitrogen metabolism occurred through natural selection, which resulted in the better adaptation of this group of annelids for a terrestrial existence. The problem of nitrogen excretion was just one of many that had to be solved when animals were converting from aquatic to terrestrial life. Moreover, for annelids to survive on land, they had to avoid desiccation. Selection of a damp habitat by earthworms is a behavioral mechanism that helps avoid some of the stress. The cuticle is also important because it reduces evaporative water loss through the body wall.

NERVOUS SYSTEM The annelid nervous system consists essentially of an anterior dorsal brain, a pair of connectives (one on each side of the digestive tract anteriorly) that join the brain to one or two solid ventral nerve cords, which extend beneath the digestive tract to the posterior end of the worm. The polychaete brain is in the prostomium, but among the oligochaetes and leeches the brain has tended to shift posteriorly. The brain of the earthworm, *Lumbricus*, for example, is in the third segment, while the brain of a leech occupies the fifth and sixth segments. A pair of ventral nerve cords, each with a ganglion in every segment, is the primitive condition; but medial fusion of these nerve cords has been the evolutionary trend. The originally paired, segmental ganglia have had a greater tendency toward fusion than have the intersegmental nerve trunks that run from one ganglion to the next; these trunks often are still paired, whereas the ganglia are fused. Where the two ganglia in the same segment have not fused, there is a transverse connection between them. Primitive polychaetes and primitive oligochaetes have nerve cords that are completely unfused. The rest of the annelids exhibit varying degrees of medial fusion of their nerve cords, fusion being complete in only a few polychaetes, a few leeches, and nearly all oligochaetes. The ganglia in each segment supply nerves to the structures in that segment, for example, to the muscle cells in the body wall.

The brain is the principal center of sensory input, and the subpharyngeal ganglion, the first ganglion of the ventral nerve cord, is the major center controlling the onset of motor activity. However, the subpharyngeal ganglion is regulated by the brain, which has inhibitory or restraining control over it, as evidenced by the fact that annelids exhibit increased motor activity after the brain has been removed.

REPRODUCTION Most polychaetes are dioecious, whereas all oligochaetes and leeches are hermaphroditic. The gonads (organs in which the gametes are produced) of polychaetes are not permanent structures; they are present only during the breeding season and arise from the peritoneum lining the coelom simply as masses of gametes. Their gametes are released into the coelomic fluid (usually in an immature form and ripen in the coelom), and from there they make their way to the outside. The gonads of oligochaetes and leeches, on the other hand, are permanent structures associated with special ducts for the gametes.

A few polychaetes have ducts that function only to conduct gametes to the exterior. However, in most polychaetes, the gametes pass out either through the metanephridia or through ducts that lead into the protonephridia.

In a few species of polychaetes, shedding occurs by rupture of the body wall of the adults. Fertilization of polychaete eggs is almost always external. When copulation occurs, nephridia and modified parapodia serve as sperm receptacles.

Three families of polychaetes (nereids, syllids, and eunicids) exhibit epitoky; that is, during the breeding season, reproductive individuals *(epitokes)* develop from the nonreproductive ones *(atokes)* either by direct transformation (nereids, eunicids) or by budding (syllids). Epitokes differ markedly in external appearance from atokes. The epitokes, for example, have larger eyes that are much more sensitive to light, and often the segments containing the gametes are much larger than the rest of the trunk segments. Also, epitokes are better swimmers than atokes because the parapodia of epitokes are larger, making them more effective swimming paddles.

The earthworm, *Lumbricus terrestris,* will serve as an example to describe the oligochaete type of reproductive system. This worm has one pair of ovaries and two pairs of testes. The testes are located in a pair of seminal capsules. Three seminal vesicles are attached to each capsule. The sperm are released from the testes as spermatogonia. They complete their maturation in the seminal vesicles and remain there until copulation occurs. The ovaries, in contrast, release mature eggs into the coelom. A pair of oviducts, originating in the coelom in the segment that contains the ovaries, and a pair of sperm ducts *(vasa deferentia)* from the seminal capsules, lead to the body surface, where they terminate at the genital pores (gonopores). There are two male and two female gonopores. The female reproductive organs also include two pairs of seminal receptacles, in which sperm received during copulation are stored. These seminal receptacles are not connected to the oviducts; they have their own openings to the body surface. Copulation involves the mutual exchange of sperm. A few days after copulation, the clitellum secretes around the worm's body a tubular cocoon, into which eggs are discharged *via* the female gonopores, and sperm are released from the seminal receptacles as the worm backs out of the cocoon. Fertilization is external, occurring within the cocoon.

Leeches have one pair of ovaries and four to ten pairs of testes. Fertilization in leeches is internal. Some leeches have a penis and are able to deposit sperm directly into the female reproductive tract. In the case of leeches that lack a penis, packets of sperm, called *spermatophores,* are driven, by muscular contraction of the wall of the chamber in which the spermatophores are stored, directly through the integument of the mate (into its body). There the sperm are discharged from the spermatophores and make their way to the ovaries. Unlike oligochaetes, leeches have only one male gonopore and one female gonopore. One pair of ducts, carrying sperm, fuses to produce a single opening for the exit of sperm and another pair of ducts (the oviducts) join together to form a single opening for the release of the eggs. The clitellum products a cocoon, which receives the fertilized eggs when they are laid.

A few polychaetes and many aquatic oligochaetes have the ability to reproduce asexually by budding or by transverse division of the body into two or more parts. Each part then regenerates its missing parts. No leech can reproduce asexually.

Evidence has been gathered showing that reproduction in annelids is under hormonal control. The development of epitokes, for example, during the nonbreeding part of the year is inhibited in nereid

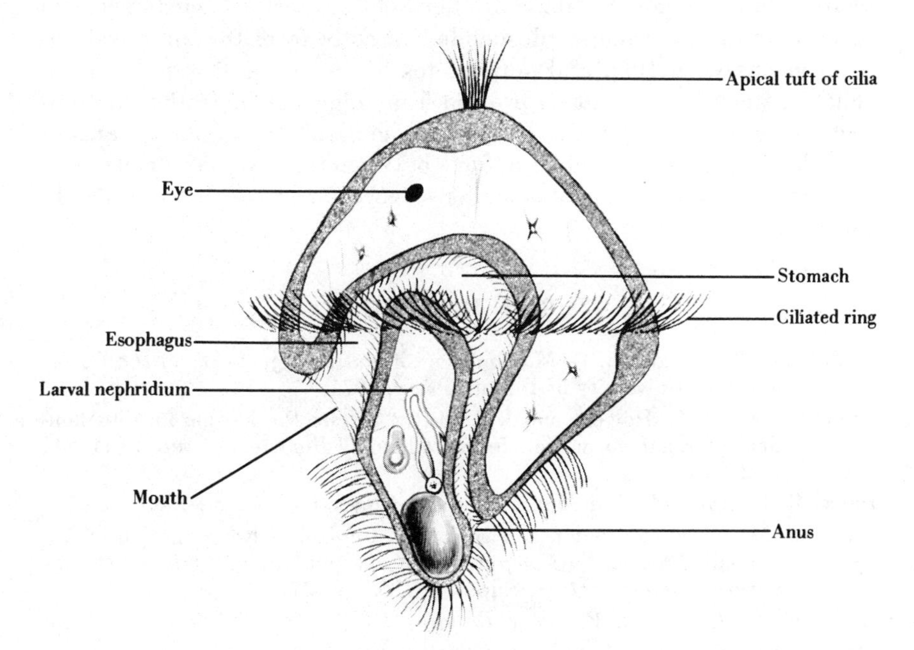

Fig. 7-3 *A trochophore larva of the polychaete,* Eupomatus *sp. (Modified from Stiles, Hegner, and Boolootian,* College Zoology, *8th ed. Macmillan, 1969.)*

polychaetes by a substance produced in the brain. During the breeding season, production of this substance declines. In contrast, in oligochaetes, a substance from the brain is necessary for maturation of the gonads.

The polychaete zygote develops into a free-swimming larva called a *trochophore larva* (Fig. 7-3), which later metamorphoses into a young worm. A ring of ciliated cells is one of the characteristics of this larva. Mollusks also have a trochophore larva, and similar larvae occur in other protostome phyla. Oligochaetes and leeches have no larvae; juvenile worms develop directly from the fertilized eggs, then emerge from the cocoons.

PHYLOGENY The evolutionary origin of the annelids is not clear. However, it is likely that these coelomates were derived from an ancestral form that was also coelomate (the coelomate ancestor in Fig. 1-3) and that evolved from an acoelomate. This acoelomate was quite likely the acoeloid ancestor described in Chapter 5, or a turbellarian that was closely related to it. The first coelomate was apparently schizocoelous (and a protostome).

As was discussed in Chapter 6, it is improbable that pseudocoelomates were ancestral to coelomates. With respect to phylogenetic relationships among the annelids themselves, most biologists seem to agree that the polychaetes descended directly from the ancestral annelid discussed earlier, and polychaetes in turn gave rise to the oligochaetes. Leeches most likely evolved from oligochaetes (rather than directly from the polychaetes) because, in general, the organ systems of oligochaetes are intermediate in form between those of polychaetes and leeches. The aquatic oligochaetes are more primitive than the earthworms.

FURTHER READING

Brinkhurst, R. O., and B. G. M. Jamieson, *Aquatic Oligochaeta of the World.* Toronto: University of Toronto Press, 1971.

Crang, R. E., R. C. Holsen, and J. B. Hitt, "Calcite Production in Mitochondria of Earthworm Calciferous Glands," *BioScience*, vol. 18 (1968), p. 299.

Dales, R. P., *Annelids.* London: Hutchinson University Library, 1963.

Durchon, M., "Neurosecretion and Hormonal Control of Reproduction in Annelida," *Progress in Comparative Endocrinology, General and Comparative Endocrinology,* Suppl. 1 (1962), p. 277.

Edwards, C. A., and J. R. Lofty, *Biology of Earthworms*, 2nd ed. London: Chapman and Hall, 1977.

Fingerman, M., "Lunar Rhythmicity in Marine Organisms," *American Naturalist*, vol. 91 (1957), p. 167.

Fingerman, M., "Endocrine Mechanisms in Marine Invertebrates," *Federation Proceedings*, vol. 32 (1973), p. 2195.

Giese, A. C., and J. S. Pearse, eds., *Reproduction of Marine Invertebrates, Vol. 3, Annelids and Echiurans*. New York: Academic Press, 1975.

Goodrich, E. S., "The Study of Nephridia and Genital Ducts Since 1895," *Quarterly Journal of Microscopical Science*, vol. 86 (1945), p. 113.

Gray, J., "Studies in Animal Locomotion. VIII. The Kinetics of Locomotion of *Nereis diversicolor*," *Journal of Experimental Biology*, vol. 16 (1939), p. 9.

Laverack, M. S., *The Physiology of Earthworms*. New York: Pergamon-Macmillan, 1963.

Mann, K. H., *Leeches (Hirudinea), Their Structure, Physiology, Ecology, and Embryology*. New York: Pergamon, 1962.

Mill, P. J., ed., *Physiology of Annelids*. New York: Academic Press, 1978.

Nicholls, J. G., and D. Van Essen, "The Nervous System of the Leech," *Scientific American*, vol. 230:1 (1974), p. 38.

Stephenson, J., *The Oligochaeta*. New York: Oxford University Press, 1930.

chapter 8

Mollusca

The approximately 100,000 described, living species in the phylum Mollusca [L. *mollis*, soft] are exceeded in number only by the species in the phylum Arthropoda. The phylum name refers to the soft body of these animals, which is usually protected by a calcareous shell. As stated in Chapter 1, the members of this phylum include the chitons, snails, clams, squids, and octopuses. The worldwide popularity of shell-collecting has led to the discovery of a number of previously undescribed species and has contributed significantly to our knowledge of the geographic distribution of mollusks.

The primary symmetry of mollusks is bilateral, but in many snails, coiling of the animal has brought about secondary alteration to an asymmetrical condition. There is

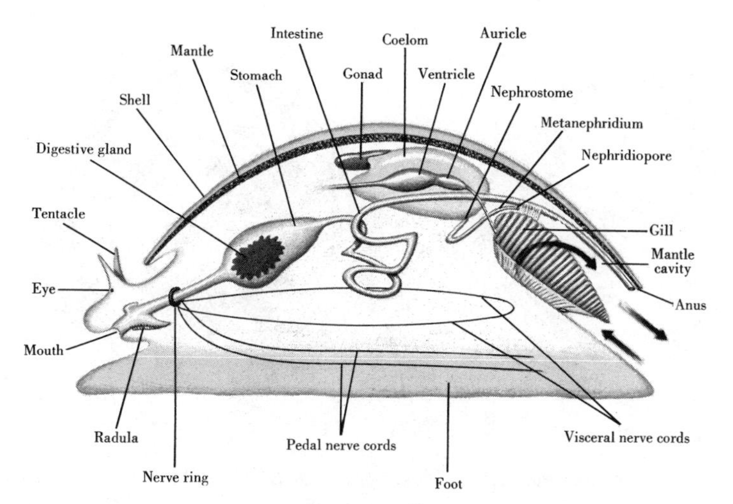

Fig. 8-1 *Hypothetical ancestral mollusk. The arrows at the posterior end show the path of the water current through the mantle cavity. (After Yonge, Pelseneer, and others, from Barnes,* Invertebrate Zoology, *W. B. Saunders Company, 1963.)*

no one structure that is unique to mollusks and found among all of them. Rather, it is a combination of characteristics that distinguishes this phylum of schizocoelomates from other animals. This combination has classically been used in the construction of a hypothetical ancestral mollusk (Fig. 8-1). Its anatomy will be described in the section on mollusk structure. Seemingly, each of the mollusk classes could have been derived relatively easily from the hypothetical form. The hypothetical ancestor embodies the traits of living primitive mollusks and can, therefore, also be regarded as a generalized mollusk; this animal may not have ever existed, but something like it might have existed.

The size variation among members of this phylum is great. Some adult snails have a normally developed shell only 1 millimeter high, whereas the giant squid of the North Atlantic *(Architeuthis princeps)* is reported to be over 18 meters long, body and arms combined, the largest invertebrate known.

CLASSIFICATION The phylum Mollusca has been divided into the following seven classes.

The class Aplacophora consists of only about 130 species of marine wormlike mollusks commonly called solenogasters (Fig. 1-10A). They are usually less than 5 centimeters long. Some solenogasters live at great ocean depths. *Pachymenia abyssorum* has been found off California at a depth of 4,000 meters.

The class Polyplacophora consists of the chitons (Fig. 1-10B), which are elliptical mollusks whose shell consists of eight plates. A

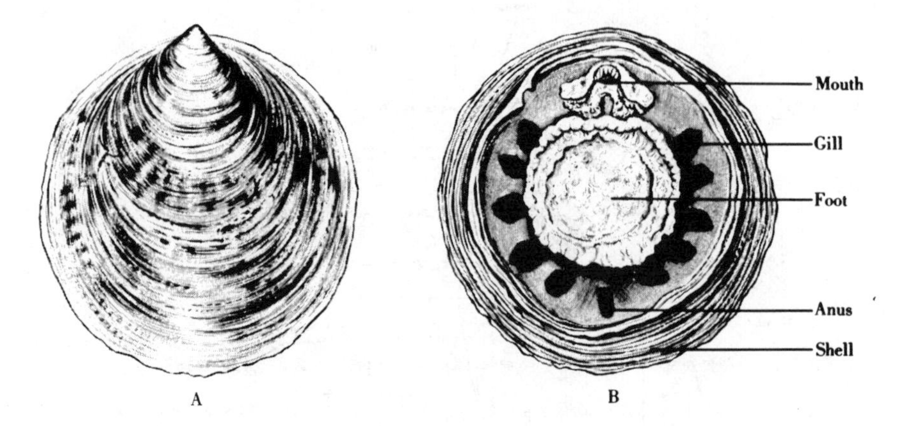

Fig. 8-2 Neopilina galatheae, *(A) dorsal and (B) ventral views. (From Lemche,* Nature, *vol. 179, 1957).*

chiton's head is much reduced, and its body is dorsoventrally flattened. The class is strictly marine. *Chiton tuberculatus* is a polyplacophoran that is eaten by humans in the West Indies. Its foot is called "sea beef."

The class Monoplacophora consists of the genus *Neopilina* alone (Fig. 8-2). Monoplacophorans have a shell of one piece, are bilaterally symmetrical, and are the only mollusks whose organs show any sign of segmentation. Until the discovery, in 1952, of living *Neopilina galatheae,* monoplacophorans had been considered an extinct group.

The class Gastropoda (Fig. 1-10C) has remained closest to what is considered to have been the ancestral mollusk. Gastropods, or snails, comprise the largest class of mollusks in number of species. Most gastropods have remained aquatic, but some have become terrestrial—the only mollusks to do so. The class consists of species that have a single undivided shell, or species in which the shell is either reduced and internal or absent. *Parafossarulus striatulus* is one of several snails that serve as an intermediate host for the Oriental liver fluke.

The class Scaphopoda is composed of mollusks that have tusk-shaped shells (Fig. 1-10D). Consequently, scaphopods are commonly called tusk or tooth shells. The body is enclosed by a single tubular shell that is open at both ends. Scaphopods are strictly marine. They are burrowers, lying buried with the large anterior end down and the smaller end extending slightly above the substrate. The head is small. *Dentalium pretiosum* is a scaphopod from the Pacific Coast of North America, occurring from Sitka, Alaska, to lower California. Its shell was used by Indians there for wampum.

The class Bivalvia (Fig. 1-10E) consists of the bivalve mollusks. The shell consists of two parts, left and right, or *valves* that are hinged dorsally and usually similar in size and shape. Examples are the scallops, clams, and oysters. The head is greatly reduced, the gills are enlarged, and the entire body is laterally compressed. The class is mainly marine, but some freshwater bivalves also exist. The oyster, *Crassostrea virginica*, which grows on the east coast of the United States and in the Gulf of Mexico, is one of the commercially important bivalves.

The class Cephalopoda (Fig. 1-10F) is composed of strictly marine mollusks which have a well-developed head that is encircled by 8 (octopus), 10 (squid), or about 90 (nautilus) arms. The arms are homologous with the anterior part of the foot in other mollusks. The shell is external (nautilus), internal (squid), or absent (octopus). When present, the shell is always a single piece. *Loligo pealei* is a squid commonly collected along the eastern coast of North America from Maine to South Carolina.

MORPHOLOGY AND PHYSIOLOGY

MOLLUSK STRUCTURE

The close affinity between annelids and mollusks is revealed by comparison of embryos and larvae from both phyla. Because of this closeness, most biologists agree that annelids and mollusks had a common ancestry. However, biologists have debated for many years whether mollusks had a segmented ancestor or not. In other words, did segmentation appear in the protostome line before or after mollusks became a distinct group? Presently, the debate is more heated than it was before because of one of the major biological finds of the twentieth century. In 1952, 10 specimens of a group of mollusks, the monoplacophorans, known until then only as fossils from the Paleozoic Era and presumed extinct for 400 million years, were dredged up from where they had been living at a depth of 3,570 meters off the west coast of Costa Rica. The specimens did not survive being pulled up from such a depth. The largest was 37 mm long. The anatomy of these mollusks, named *Neopilina galatheae*, was described by H. Lemche. Since then, at least six additional species of living monoplacophorans have been found. Furthermore, monoplacophorans from shallower water (348–402 meters) have not only been brought to the surface alive, but have been kept alive for up to 25 days after having been collected. *Neopilina* has an undivided, bilaterally symmetrical shell (Fig. 8-2). Its head is reduced and lacks eyes and tentacles. The underside of *Neopilina galatheae* reveals five pairs of metamerically arranged gills. Internally, five pairs of typical metanephridia and eight pairs of foot retractor muscles originating on the underside of the shell also reveal that *Neopilina* is segmented. It is dioecious, with two pairs of gonads near the middle of the body. Some biologists consider *Neopilina* evidence that mollusks had a segmented ancestor, but the absence of segmentation in all other living mollusks (either in the larval or adult stages) has led more investigators to the conclusion that the segmentation of *Neopilina* is a secondarily acquired repetition of parts and is not a primary characteristic derived directly from a segmented ancestor.

There is so much diversity among the numerous mollusks that a list of phylum characteristics must be very broad to include all of the members. The easiest approach to a study of this large phylum is to begin with a description of what the ancestor of present-day unsegmented mollusks, and quite likely of *Neopilina* also, probably looked like (Fig. 8-1). Although at first glance a scallop and an octopus may appear to have little in common, they do share a basic structural plan.

The hypothetical ancestral mollusk was probably bilaterally symmetrical and about the size of *Neopilina*. Presumably, it possessed a definite head with tentacles and eyes and a ventral *foot*, which was a flattened, muscular structure used for locomotion. The sole of the foot was most likely ciliated. Above the foot would lie the *visceral mass*, the portion of the body that contained most of the organs. The ancestor was probably covered by an undivided shell secreted by the underlying *mantle*. The mantle develops in all mollusks from a fold or a pair of folds of the body wall and encloses a space, the *mantle cavity*, between the main portion of the body and the mantle itself. The digestive tract would have had both a mouth and an anus, the mouth opening into the oral cavity that contained a *radula* (a membranous belt on the surface of which were teeth used for scraping and grinding and for transporting food back to the stomach as along a conveyor belt). The nervous system most likely consisted basically of a nerve ring around the esophagus and bilaterally arranged nerve cords. There were probably no ganglia, the neurons simply being scattered throughout the nervous system. One pair of gills, a pair of metanephridia, and two pairs of gonads would also have been present. Gametes would have been released into the coelom and transported to the body surface through the metanephridia. This ancestor was probably dioecious, because the most primitive living members of each class are dioecious. The coelom enclosed the heart and a portion of the intestine. The cavities of the gonads and metanephridia are also part of the coelom. Its circulatory system would have been an open one with a three-chambered heart consisting of one ventricle and two auricles.

For the major groups of modern mollusks to have evolved from this ancestral form, several changes would have had to occur. Dorsoventral flattening of the visceral mass, restriction of the mantle cavity to a groove along each side of the foot, and division of the shell into eight plates would have provided the basic chiton form. Monoplacophorans would also have undergone a dorsoventral flattening of the visceral mass, but unlike the rest of the mollusks, they apparently evolved to become secondarily segmented. The basic bivalve form could have been achieved by lateral compression of the body, division of the shell into two equal

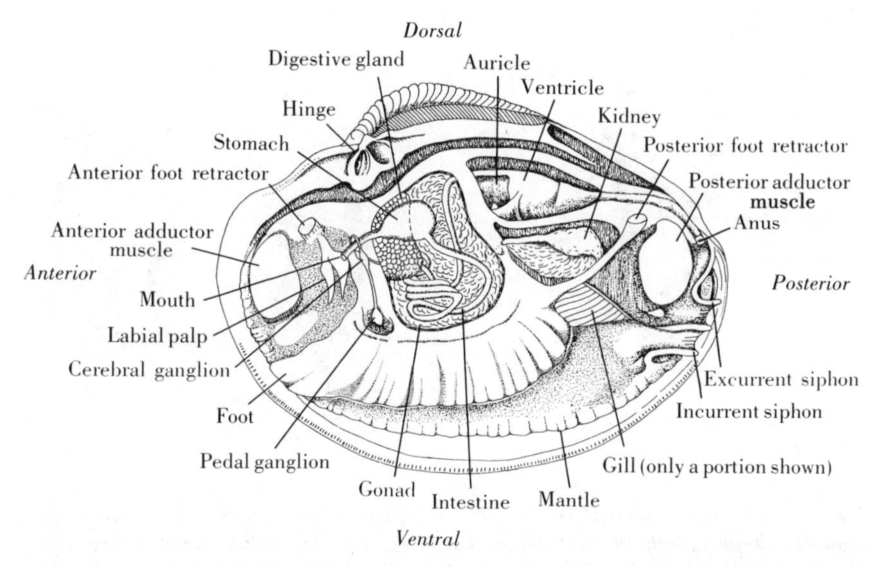

Fig. 8-3 *The bivalve,* Mercenaria mercenaria, *with the left valve removed. (After Schechter,* Invertebrate Zoology. *Prentice-Hall, 1959. Reprinted by permission.)*

lateral parts, and enlargement of the mantle hanging down on each side of the body (Fig. 8-3). The enlarged mantle would have provided for an expanded mantle cavity in which there would be enlarged gills. The foot would have been extended ventrally as a bladelike adaptation for burrowing. Scaphopods could have been formed simply by sealing the mantle and shell ventrally, forming a tube. Cephalopods could have formed by extending the visceral mass upward and forming a circle of arms from the anterior part of the foot, the arms coming to surround the head. The resulting lengthened body is coiled in the nautilus but not in other cephalopods such as the squid. Gastropods retained a form closest to the ancestral mollusk. In them, as in the cephalopods, the visceral mass also extended upward, and tended to be coiled, but whereas in the cephalopods the head and foot have merged, the head and foot in the gastropods have remained distinct from each other. Solenogasters are aberrant mollusks; they have no distinct head, nor do they have any longer a mantle, foot, or shell; solenogasters are, nevertheless, unmistakable mollusks.

The anatomy of most adult gastropods is complicated by *torsion*, twisting of the visceral mass about 180° counterclockwise, relative to the head and foot, during development (Fig. 8-4). Torsion brings the gills, anus, and excretory pores to an anterior position behind and above the head. This also involves the nervous system, causing the nerve cord (called the pleuroparietal connective) joining the left pleural ganglion to the original left parietal ganglion, and that joining the right pleural ganglion to the original right parietal ganglion, to become twisted. The

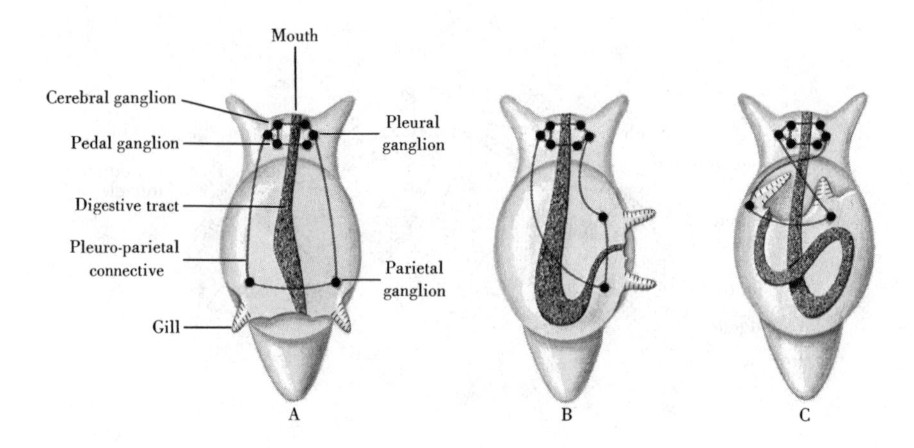

Fig. 8-4 *Diagrammatic representation of the torsion process. (A) Before torsion; (B) displacement 90° counterclockwise; (C) torsion complete. (Modified from Parker and Haswell after Korschelt and Heider,* A Text-Book of Zoology, *Vol. 1, 6th ed. Macmillan, 1949.)*

result is that the nervous system has the configuration of an "8," with the former left parietal ganglion now on the right and the former right parietal ganglion now on the left. Torsion imposes a secondary asymmetrical body form of the untorted larva. The spiral coiling of the shell was a separate evolutionary event that also resulted in disturbance of the primary bilateral symmetry of gastropods. It is also of interest that the members of a group of gastropods, the opisthobranchs, which evolved from snails that underwent torsion during development, have experienced various degrees of detorsion during their evolution with the mantle cavity and the structures in it having shifted back to the right side, some species even becoming so detorted that as adults they are bilaterally symmetrical. Opisthobranchs are all marine, with a strong tendency toward having a reduced shell or, in some species, no shell at all. A sea slug (nudibranch) is an example of an opisthobranch that has reacquired bilateral symmetry; it lacks a shell.

A number of theories have been proposed to explain the adaptive significance of torsion in gastropods, but the arguments used by most of the authors are weak, being based on premises that do not now seem valid. However, the theory put forth by M. T. Ghiselin appears to be quite adequate. According to this theory, torsion had selective advantage because it helps the larval snail to balance its shell, both in settling to the bottom after its free-swimming existence and then in crawling about while still a larva. Without torsion, the heavy part of the shell would project forward over the head and disturb the animal's balance. Torsion puts the heaviest part of the shell over the posterior part of the larva's body; consequently, the center of gravity is more advantageously placed. However, with respect to adults, L. H. Hyman stated: "Torsion must be disadvantageous to adult snails as many of them have undergone detorsion processes."[1] Because the anus is over the head after torsion, fouling of the gills is possible.

Presumably the secondary detorsion that occurred among opisthobranchs is related to the tendency in them for the shell to be reduced or eliminated, thereby doing away with the balancing problem to which M. T. Ghiselin referred. Coiling was probably retained by natural selection, according to Ghiselin, primarily because it also made locomotion easier for larval gastropods. The apparent advantage of a larva having a coiled shell is that an uncoiled one would be tall and unstable because the geometric center of the shell would be well above the center of gravity of the animal. Because of this instability, after the larva settled it would have difficulty balancing the shell when crawling about. There would be a tendency for the shell to fall over, hampering locomotion. Coiling of the shell not only lowers the gastropod's center of gravity and the geometric center of the shell but also reduces the energy output required for the animal to move forward, by reducing the surface area of the shell surface to decrease fluid drag.

Molluscan shells such as those of *Neopilina*, gastropods, and bivalves typically consist of three layers. The thin outer layer *(periostracum)* is composed of a horny material called *conchiolin* or *conchin*. The hinge ligament that joins the two valves of bivalves is elastic and consists of conchiolin also. When the adductor muscles (usually two, but some bivalves have only one) that pull the valves together relax, the hinge ligament causes the valves to gape. This is so because when the adductor muscles pull the valves together, the outer (that is, dorsal) part of the hinge ligament is stretched whereas the inner (that is, ventral) part is compressed, but when the adductor muscles relax, the hinge ligament because of its inherent elasticity causes the valves to gape. In some species of burrowing mollusks, however, the periostracum becomes worn off by the abrasive action of the mud and sand in which they burrow. The middle *(prismatic)* layer is formed of minute vertical crystals of calcium carbonate, and the inner *(nacreous)* layer consists of thin sheets of calcium carbonate deposited parallel to the shell surface. Depending upon the arrangement of these thin calcareous plates, the nacreous layer appears either porcelainlike or iridescent and pearly. The prismatic layer is the thickest.

The gastropod shell consists of a single chamber no matter how large the animal. The shell of the nautilus is chambered, because as the nautilus grows, it moves its body forward periodically and secretes a perforated partition posteriorly. However, in each nautilus, there is a cord of tissue (called the *siphuncle*) which extends from the visceral mass through the perforations and back through each chamber.

[1] L. H. Hyman, *The Invertebrates:* Mollusca I, Vol. VI (New York: McGraw-Hill, 1967, p.10).

The nautilus shell is a bilaterally symmetrical planospiral and is used as a buoyancy structure. The chambers, when initially formed, are full of fluid which is soon replaced by gas, mainly nitrogen, thus providing near neutral buoyancy and enabling the animal to swim more easily. The gas is secreted by the siphuncle.

LOCOMOTION Squids are highly mobile, active swimmers, whereas oysters are sessile. The vast majority of mollusks, however, move about slowly. Chitons, *Neopilina*, and most gastropods have retained the ancestral type of foot, but in all other mollusks it has become modified, adapting to changes in both habitat and feeding behavior. The large, flat foot of the chitons can be used for creeping and adhesion to smooth surfaces by a mechanism similar to that of a suction cup. Chitons can move about by muscular waves that pass anteroposteriorly along the foot. The sole of the foot is ciliated, but chitons move about by using these muscular waves alone. Nevertheless, chitons generally exhibit very little locomotor activity, tending to remain in one spot for long periods of time. Scaphopods have a conical foot, adapted for burrowing and situated at the head end, where the large opening of the shell is located.

Most gastropods, and apparently *Neopilina* also, use the foot as a means of slowly gliding from place to place, across the substrate, as in the ancestral mollusk. However, in some gastropods the foot is adapted for burrowing, the anterior portion functioning as a plow, and in free-swimming gastropods (such as the opisthobranch sea butterflies) the foot has been transformed, by the development of lateral expansions, into a finlike structure. In many shelled gastropods the foot has a horny disk *(operculum)* which seals the opening when the snail withdraws into the shell. The sole of the gastropod foot is typically ciliated. The sole of *Neopilina* is ciliated also.

Small gastropods move about by use of the cilia on the sole of the foot, ciliary waves proceeding from anterior to posterior. But the propulsive force is most commonly provided by muscular waves. In most species, these waves progress from posterior to anterior (in some species the wave moves in the reverse direction) along the sole of the foot. It is also typical of snails that each propulsive wave extends across the entire width of the foot. However, some snails have evolved a different method of locomotion, in which the foot is functionally halved. For example, H. W. Lissmann, who has analyzed the locomotion of gastropods in detail, found that *Pomatias elegans* uses a type of movement that is comparable to the locomotion of a bipedal vertebrate: The snail lifts one side of the foot heel first, puts it down heel first, and then repeats the performance with the other side—always alternating sides.

Most bivalves live buried in sand or mud. The foot of a bivalve is well adapted for burrowing. It has become compressed and wedge-shaped as part of the general lateral compression of the entire body. The foot is put forth as an anchor (the distal end of the foot becoming dilated through engorgement with blood), and the body is then dragged forward. Projection of the foot from the shell is the result of the action of protractor muscles; retractor muscles pull the foot back. Scallops, however, do not burrow. The scallop foot is much reduced in size, now being too small to assist in locomotion. A scallop swims by a type of jet propulsion. Rapid closure of the shell shoots streams of water out of the mantle cavity on either side of the hinge line, resulting in movement of the scallop in the opposite direction. Some bivalves, on the other hand, are sessile. Oysters, for example, are attached to the substrate by one valve, and the foot has been lost. All scaphopods are burrowers and use the foot for burrowing in the same way as do burrowing bivalves. Some solenogasters appear to move about by use of cilia on the body surface, while others make use of peristaltic waves that move along the body.

In cephalopods, as stated earlier, the anterior part of the foot has become divided into a series of arms that have come to surround the head; the remainder (posterior end) of the foot forms a ventrally located funnel. Swimming is accomplished by rapid expulsion of water from the mantle cavity through the funnel (another example of jet propulsion among the mollusks), and the speed of locomotion depends largely on the force with which the water is expelled. The funnel can be directed either backward or forward, thereby changing the direction of locomotion. Squids, with their streamlined bodies, are the most mobile cephalopods. The nautilus also is a capable swimmer, but octopuses, although able to swim, generally crawl about using the adhesive suction disks on their arms for attachment as they pull themselves along.

FEEDING AND DIGESTION

Mollusks show much diversity in feeding mechanisms; the structure of the digestive tract, however, has remained fairly constant. Once food has been ingested, it passes from the oral cavity to the esophagus and then to the stomach. In herbivores, part of the esophagus is often modified into a crop, or both a crop and an adjacent gizzard. A pair of digestive glands is connected to the stomach. Extracellular digestion occurs in the stomach, but in some mollusks, such as bivalves, there is also intracellular digestion, in the digestive glands, of food particles conveyed there by cilia. The digestive glands are the main site of absorption of digested material. The intestine, which runs from the stomach to the anus, also participates in the absorption of the products of digestion, although the main function of the intestine is the consolidation of feces.

The mouth of chitons and *Neopilina* is located on the ventral surface anterior to the foot. *Neopilina* has a radula and apparently feeds on bottom mud, which is rich in microorganisms. Examination of the intestinal contents of *Neopilina* has revealed the remains of diatoms, radiolarians, and foraminiferans. Chitons feed primarily on algae, using the radula to scrape encrusted algae from rocks and shells.

The type of feeding behavior exhibited by scaphopods is well adapted to their burrowing habit: they feed on microscopic organisms in the surrounding sand and mud, which are caught by the tentacles or *captacula* about the mouth. The scaphopod radula functions mainly to assist in ingestion of the food captured by the tentacles. Most solenogasters have a radula; they feed on other invertebrates such as cnidarians and small crustaceans.

Gastropods have the most diverse feeding habits of all the mollusks. Scavengers, herbivores, carnivores, and filter feeders occur in this class. The filter feeders, through use of the cilia on the gills, cause a current of water to flow into the mantle cavity, bringing food particles with it. (This current also brings in oxygen for respiration.) Near the mouth, the food particles become trapped in mucus, and the radula then pulls the mucus mass into the oral cavity. It is clear that the gills, in addition to their normal respiratory function, have assumed an important role in feeding. The radula is especially well developed in gastropods. Some carnivorous snails that feed on bivalves use it first to bore a hole through the shell and then to tear off pieces of flesh and ingest them.

Many of the filter feeders among the gastropods, and all bivalves except the primitive ones, possess a *crystalline style*, which is a gelatinous rod impregnated with amylase, an enzyme that hydrolyzes starch. The style is produced in a style sac, an evagination of the stomach, and projects into the stomach, where the free end gradually erodes, liberating the enzyme. However, the rest of the digestive enzymes secreted into the stomach are products of the digestive glands.

Filter feeding is an adaptation that reaches a very high level of development among the bivalves, most of them making use of the following feeding mechanism. A water current created by cilia on the gills, which contains suspended food particles, is drawn over the gill surface, where other cilia deliver the food to ciliated food grooves on the gills. The cilia in these grooves then deliver the food to the two *labial palps*, thin appendages on either side of the mouth, where the particles are sorted by size and the smaller ones are directed to the mouth for ingestion while the larger ones are rejected. The few bivalves that are not filter feeders ingest bottom detritus. None of the bivalves has a radula. Presumably, disappearance of the radula among bivalves is related to their more or less sedentary type of existence and the predominance of filter feeding in this class. Filter feeders have no need for a rasping organ.

The water current enters most bivalves posteroventrally, flows through the gills into the excurrent *suprabranchial cavity,* and leaves posteriorly and dorsally, having made a U-turn. This suprabranchial cavity was formed originally when the mantle cavity was divided into lower and upper portions by extension of the gills across it. A major problem for burrowing bivalves is the large amount of silt carried by the incurrent water. Because this silt could clog the gills, the shells are rapidly closed periodically, forcing water and accumulated silt back out the incurrent opening. This system of having the current both enter and leave at the posterior end enables the clam to burrow down until the posterior end is just above the surface, thereby diminishing the silting problem. In addition, the edges of the mantle have tended to seal where openings are not needed. The result of this sealing process has been the evolution of excurrent (dorsal) and incurrent (ventral) siphons at the posterior end. In some species, the siphons are greatly elongated, tubular structures that allow the main portion of the clam to remain well below the surface, with only the tips of the siphons extending to the surface.

Cephalopods are carnivorous, and, in addition to a radula, they possess a beak for tearing food. The arms are used first to capture prey and then to pull it toward the mouth. Adhesive suckers are found on the arms of squids and octopuses, but not on the nautilus.

The feeding mechanisms among mollusks show excellent adaptation to the general mode of existence of the members of each class, from the scraping of algae from rocks by the sluggish chitons to the active capture of prey by the squids. The class with the most active individuals, the cephalopods, has the most efficient means of obtaining large quantities of food rapidly. This is to be expected if we compare the greater metabolic demands of the cells in a squid with the demands of those in a chiton. Those bivalves that assumed a burrowing, filter-feeding way of life became tied to a relatively sedentary existence. On the other hand, the higher degree of development of filter feeding in the bivalves excellently illustrates how animals can ensure their survival by assuming a mode of existence in which they meet little competition. The burrowing bivalves, having adopted a relatively sedentary way of life, enjoy the protection that a burrow provides, while reaping the benefits of being able to feed from the water above by means of the feeding current their cilia draw into the burrow.

CIRCULATION Cephalopods have a closed circulatory system, whereas all other mollusks have an open system—probably the ancestral type. In contrast to the pulsatile blood vessels of the annelids, mollusks (except scaphopods) possess a chambered heart. The type of heart, consisting of one ventricle and two auricles, that was suggested for the hypothetical ancestor is present in many mollusks. However, in most gastropods the right auricle is either vestigial or, most frequently, absent, as a result of the loss of the right gill, which supplied it with blood. This gill apparently disappeared after torsion evolved, because the right side of an ancestral gastropod that underwent torsion would have been pressed against the shell and unable to function properly.

The typical pattern of blood flow in mollusks with an open circulatory system is the following. Blood pumped from the ventricle passes through an anterior aorta and its branches into sinuses and lacunae, directly bathing the internal organs. It is then conveyed through sinuses to the gills (passing through lacunae in the gills), and then passes into veins that lead to the auricles. The auricles are thin-walled and deliver the blood to the thicker-walled, more muscular ventricle. An aorta is a large *artery*, an artery being any blood vessel that carries blood in the direction away from the heart toward the organs; a *vein* is a vessel that carries blood back toward the heart from the organs. Whereas in animals with a closed circulatory system, arteries are connected to the veins by capillaries, in animals with a typical open circulatory system, such as mollusks, there are no capillaries to join the arteries to the veins.

Unlike the situation in any of the other mollusks, the scaphopod circulatory system consists only of sinuses; there is no heart. The closed circulatory system of cephalopods is an extensive set of blood vessels that is able to meet the metabolic demands of these active animals by delivering a supply of oxygen to the cells in a more effective manner than would be possible with an open system, where the blood flow is more sluggish. Cephalopods have a fairly large coelom, but the coelom of mollusks with an open circulatory system is relatively smaller. This smaller size is, presumably, related to the fact that the hemocoel of such mollusks is large, crowding the coelom and reducing the importance of the coelomic fluid as a vehicle for internal transport.

Two oxygen-transport pigments occur in the blood of mollusks. Chitons, cephalopods, and some gastropods possess *hemocyanin*, a copper-containing protein. Hemocyanin is blue when oxygenated and colorless when not. It is always dissolved in the plasma. Molluscan hemocyanin has a molecular weight of several million daltons, about nine million in the snail *Busycon*. The second oxygen-transport pigment occurring in mollusks is hemoglobin. Snails of the family Planorbidae are the only mollusks that have, so far, been shown with certainty to have hemoglobin dissolved in their plasma; and only rarely is hemoglobin present in the blood cells of mollusks, having been demonstrated for certain in the blood cells of a few bivalves only—for example, *Arca*. However, the blood of solenogasters is reddish; the color might be due to hemoglobin. But the pigment in solenogaster blood has not been identified, nor, apparently, has anyone determined whether this pigment is in the blood cells or dissolved in the plasma. The molecular weights of hemoglobins in solution in the plasma of invertebrates are usually greater than one million daltons, whereas hemoglobins from blood cells are much smaller molecules. For example, hemoglobins of vertebrates occur only in blood cells, and have a molecular weight of about 65,000 daltons. Because of the large size of the oxygen-transport pigments in plasma, they tend to be confined to the circulatory system, diffusion into the surrounding tissues being reduced.

RESPIRATION Aquatic mollusks other than the scaphopods depend primarily upon gills for the uptake of oxygen. No doubt some gaseous exchange occurs across exposed body surfaces (such as the mantle) in all mollusks. Scaphopods, however, lack gills; the mantle serves as the main respiratory surface. Through the action of cilia on the mantle, oxygen-laden water is brought into a scaphopod's mantle cavity *via* the opening at the small end of the tubular shell that extends above the substrate. By muscular contraction, the water is then forced out the same opening.

Those gastropods that became terrestrial lost their gills. The mantle cavity became modified into a *lung*, a respiratory surface that is folded into the body. The wall of the lung is highly vascularized for efficient uptake of oxygen.

All aquatic mollusks except the cephalopods and, of course, the scaphopods use gill cilia to create an inhalent current of water for respiration. Chitons have a large number of gills—from 6 to 88 pairs, depending upon the species. Beating cilia draw in water at the anterior end; the water then bathes the gills attached to the roof of the grooves that are in the mantle cavity on both sides of the foot, and leaves at the posterior end. Bivalves have kept the single pair of gills that the ancestral mollusk presumably had, but in many species, each gill is folded, giving the appearance of two gills on each side.

Among the cephalopods, the nautilus has four gills, whereas the squid and octopus each have two. Water for respiration, as for locomotion, is drawn into the mantle cavity as a result of the alternate action of antagonistic circular and radial muscles in the mantle. When the circular muscles relax and the radial muscles contract, the mantle cavity enlarges and water is drawn in, between the head and the mantle. When the reverse—relaxation of the radial muscles and contraction of the circular muscles—occurs, the *collar* (anterior edge of the mantle) is drawn tightly around the head, and the water is forced out the funnel.

SENSE ORGANS The best-developed sense organs in mollusks are the photoreceptors. They are found in all mollusks except *Neopilina*, scaphopods, and solenogasters, and show a wide range of structure. The simplest photoreceptor is an optic cup, similar to that of turbellarians, capable only of distinguishing between light and darkness, whereas the most advanced type (in squids and octopuses) bears a striking resemblance to the vertebrate eye and is capable of image-formation (Fig. 8-5). The eyes of cephalopods consist essentially of a transparent cornea, an iris that regulates the amount of light entering the eye by changing the size of the opening (pupil) in its center, a lens that is held in position by the ciliary muscle, and the light-sensitive retina in the rear of the eye. When the ciliary muscle is relaxed, the eye is focused on near objects. The eye *accommodates* (adjusts itself so it can focus on an object) for distant vision by contraction of the ciliary muscle, which draws the lens backward, closer to the retina. Because of the mode of development of the eyes of cephalopods, their eyes are of the *direct type;* that is, the photosensitive cells point directly toward the source of light. This type contrasts with the *indirect type* of eye, found in vertebrates, where the retina is inverted, with the photosensitive elements pointing away from the light source. The eyes of squids and octopuses are large and located on the sides of the head; those of gastropods are at the base of the cephalic tentacles. The eyes of chitons are on the dorsal surface of that portion of the mantle that extends beyond the edges of the shell plates; eyes in bivalves occur most often along the edge of the mantle, but, in some instances, on the siphons. Burrowing bivalves have eyes that are more poorly developed than those of the more mobile bivalves, such as the scallop, which lives on the substrate.

Gastropods, the nautilus, chitons, and bivalves have sensory patches, called *osphradia,* in the mantle cavity, that sample the water which enters the mantle cavity. More specifically, osphradia monitor the amount of sediment in the water, and also test the incoming water for dissolved chemicals. This chemoreceptive function of the osphradia seems to play a role in detecting the presence of food.

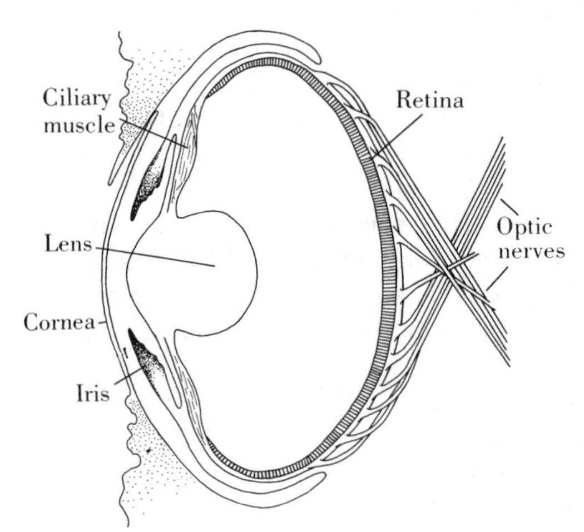

Fig. 8-5 *The eye of an octopus. (After Wells, in* The Physiology of Mollusca, *Vol. II., K. M. Wilbur and C. M. Yonge, eds. Academic Press, 1966.)*

EXCRETION The hypothetical ancestor presumably had one pair of metanephridia. A single pair of metanephridia is now present in all mollusks, except those gastropods in which the right one has been lost, presumably as a result of compression caused by torsion, as in the case of the right gill. Two other exceptions are the nautilus, which has two pairs, and *Neopilina,* which has the five pairs referred to earlier. The typical method of urine-formation is an interesting one. The pericardial cavity connects directly with each metanephridium through a nephrostome. The fluid in the pericardial cavity enters the metanephridia, where it is converted to urine. This pericardial fluid consists of a blood filtrate formed of fluid that passes directly through the wall of the heart plus nitrogenous wastes secreted by glands in the pericardium. Additional nitrogenous waste is eliminated by direct secretion from the blood through the tubule wall into the fluid passing along the metanephridial tubule; materials such as nutrients that are to be retained are reabsorbed. The urine is eliminated into the mantle cavity. In many bivalves, part of the metanephridial tubule is enlarged to form a bladder. Marine species are isosmotic with sea water. In freshwater species, which are, of course, hyperosmotic to their environment, salts are conserved by reabsorption into the blood from the metanephridia during urine-formation; thereby, a urine that is hyposmotic to the blood is produced, the excess water is eliminated, and salts are conserved. There are, of course, variations in this basic plan; a scaphopod, for example, has a pair of metanephridia but no heart, and a solenogaster lacks excretory organs altogether. Diffusion through the body surfaces is apparently a satisfactory way for solenogasters to eliminate their soluble wastes.

The principal nitrogenous excretory product of aquatic mollusks is ammonia, but in terrestrial snails, it is uric acid. The relationship between water availability and nitrogen excretion discussed in the previous chapter can be carried one step further with these mollusks, which excrete uric acid. Uric acid, like urea, is less toxic than ammonia. Furthermore, uric acid is only very sparingly soluble in water, much less so than is either ammonia or urea. Because of uric acid's very low solubility, it can be excreted in a semisolid crystalline form, with a minimal amount of water. But because of urea's high solubility, it will be excreted dissolved in water. Hence, more water is needed to eliminate urea than is needed to eliminate uric acid. Also, because most of the uric acid would be precipitated on account of its low solubility, it is almost completely osmotically inactive, whereas the same quantity of the highly soluble urea dissolved in a very limited volume of water could create unfavorable osmotic conditions, perhaps damaging the excretory organs. Consequently, excreting nitrogen from protein catabolism mainly as uric acid is a very effective mechanism for conserving water among those terrestrial animals that have an even greater need to retain water than do ureotelic animals. Desiccation is a constant threat to terrestrial snails; they continually lose water by evaporation from the body surface, and they are generally restricted to a damp habitat. Animals are called *uricotelic* when nitrogen from proteins is excreted mainly as uric acid. Appropriately, terrestrial gastropods burrow into the soil and become dormant during hot, dry periods of the summer in order to compensate for their poor ability to control water loss through the body surface. This summer dormancy is known as *estivation*.

COLOR CHANGES Cephalopods, other than the nautilus, show strikingly rapid color changes, because they have integumentary structures called chromatophores. A cephalopod chromatophore consists of a central, pigment-containing cell, to which are attached 6 to 20 or more radiating muscle cells. Muscle contraction stretches the pigment cell into a disk having a diameter 15 to 20 times the original, thereby rendering the pigment much more obvious. Electron microscopy has revealed that the pigment cell contains a small elastic sac, within which is the pigment. The sac is attached to the inner surface of the cell membrane of the pigment cell. When the muscle cells relax, the pigment cell is restored to its retracted condition because of the elasticity of this sac. In squids, for example, the chromatophoric pigments are brown, red, and yellow. The ability to change color enables these cephalopods to mimic the color of the background and thereby camouflage themselves—an obvious advantage to them as predators.

INK GLAND Cephalopods, other than the nautilus and some deep-sea species, have an ink gland, whose duct opens into the hind end of the intestine. The ink is a fluid that contains a high concentration of melanin, a brown-black pigment. When the animal is disturbed, the ink is released through the anus, and the animal is able to escape behind the inky screen. An interesting modification of this "smoke screen" mechanism is seen in the small, deep-sea squid, *Heteroteuthis dispar,* which lives at a depth where there is complete, permanent darkness. Near its small ink gland, this squid possesses a large organ that opens into the mantle cavity and contains a luminous secretion. When attacked, this squid releases the luminous material through the siphon. The resulting luminous cloud may draw the attention of the attacker while the squid moves off into the darkness.

NERVOUS SYSTEM The evolutionary trend in the nervous system of mollusks has been toward the formation of ganglia from neurons scattered throughout the nervous system and the subsequent consolidation of these ganglia around the esophagus. In chitons, the nervous system closely approximates the one postulated earlier in this chapter for the hypothetical ancestor, showing little evidence of the formation of ganglia. The nervous systems of the solenogasters and *Neopilina* are similar to that of the chiton, but somewhat more advanced in that they clearly show the beginnings of the aggregation of nerve cell bodies that were scattered throughout the nervous system into discrete ganglia. The rest of the molluscan classes show further advances. Not only do they have obvious ganglia, but these ganglia have tended first to become concentrated around the esophagus and then, to a greater or lesser degree, to fuse together. The basic gastropod nervous system after torsion is depicted in Figure 8-3C.

The cerebral ganglia of gastropods, bivalves, and scaphopods are the main coordinating centers of the nervous system. But they are nevertheless referred to as "cerebral ganglia" rather than a "brain" because the other ganglia are still capable of sufficient independent action to render use of the term "brain" unwarranted. However, the evolution of the molluscan nervous system reaches its climax in the cephalopods, in which the ganglia have become so fused that a true brain surrounding the esophagus has been formed; the individual ganglia have lost their integrity. The supraesophageal portion contains, among others, the primary sensory centers for chemoreception and photoreception and the vertical lobe, an association area that has a role in learning and memory. The subesophageal portion contains centers controlling the eye muscles, chromatophores, mantle, funnel, and fins. The cephalopod brain may be the most highly developed one among invertebrates. J. Z. Young, B. B. Boycott, and M. J. Wells have performed interesting training experiments with octopuses. For example, with the aid of electric shocks as punishment, an octopus was trained not to attack a crab when a white square was placed into the aquarium. When the square was absent, the trained octopus attacked the crab, which is one of the animals this predator normally eats.

REPRODUCTION Most mollusks are dioecious. Where hermaphroditism occurs, it appears to be a specialized, secondarily acquired condition, not a primitive one. As stated earlier, the most primitive members of each class are dioecious. In the hypothetical ancestor, the gametes were supposedly released from the gonads into the coelom and reached the exterior through the metanephridia. In living mollusks this plan has been modified in various ways with respect both to the number of gonads and their relationship to the nephridia. Mollusks now have one, two, or four gonads. Furthermore, in some species (for example, cephalopods and some bivalves), the gonadal ducts open directly into the mantle cavity; the gonads and nephridia are no longer associated with one another.

A trochophore larva, very similar to that of annelids, develops from the eggs of solenogasters, chitons, scaphopods, gastropods, and bivalves. The trochophore larva of scaphopods, gastropods and bivalves next forms another larva called a *veliger.* In chitons, however, the trochophore larva metamorphoses directly into a juvenile chiton. When torsion occurs, it is the veliger that undergoes it. *Neopilina* presumably also has a trochophore larva.

The development of opisthobranch gastropods reveals some of the presumed evolutionary past of this group. As a result of detorsion, in some opisthobranchs the anus opens at the posterior end in contrast to its anterior position in torted gastropods; but in most opisthobranchs detorsion is not complete and the anus opens in a median or near median position on the right side. Furthermore, almost all of the organs of opisthobranchs develop in their definitive sites. That is, the opisthobranch veliger generally does not go through torsion and then detort itself either completely or only partially during its development. But in opisthobranchs in which detorsion is not complete, it is not uncommon that at metamorphosis the anus migrates backward along the right side, further reversing torsion. Also, the veliger of sea slugs, which as adults are shell-less, develops a shell that is subsequently lost during the metamorphosis of this larva into a juvenile. These observations support the conclusion that opisthobranchs evolved from shelled gastropods that underwent torsion.

In two families of freshwater clams, the Unionidae and Mutelidae, development is highly specialized. The veliger (in these families it is called a *glochidium*), because it is so highly modified for a parasitic existence, develops within the gills of the female parent. It develops a shell of two parts and a long adhesive thread which is attached to its foot. Mouth and anus are absent, and the digestive tract is poorly developed. After the glochidium leaves the parent, it settles to the bottom, where it remains until it comes in contact with a fish resting or feeding on the bottom. The larva then quickly attaches to the fish, whose tissues are stimulated to form a cyst around the glochidium. Phagocytic cells from the mantle of the larva then begin to feed on the fish's tissues, development proceeds, and a juvenile clam is formed. The juvenile clam ultimately excysts, falls to the bottom, burrows into the mud, and completes its development into an adult.

With respect to the cephalopod, the sperm leave the single testis through the sperm duct and enter a seminal vesicle, where they are packaged to form spermatophores. The spermatophores are then stored in a reservoir until needed. During mating, the *hectocotylus,* one of the male's arms that is modified for copulation, transfers spermatophores from the reservoir to a sperm receptacle under the female's mouth, to the wall of her mantle cavity near the openings of the oviducts, or into the oviducts themselves. The position depends upon the species. Although there is only one ovary, some cephalopods have paired oviducts. In some octopuses, the tip of the hectocotylus has a cup-shaped cavity in which the spermatophores are carried, and this spermatophore-laden tip becomes detached during mating and remains in the mantle cavity of the female.

Eggs of cephalopods contain much more yolk than those of other mollusks. Because of this abundance, the developmental pattern of cephalopods has become greatly modified. Unlike the embryos of other mollusks, those of cephalopods cannot undergo spiral cleavage because of their large amount of yolk. Instead, they undergo meroblastic cleavage, and all cell division is restricted to the small cytoplasmic disc atop the yolk, resulting in the formation of a germinal disk or cap of cells. These cells develop directly into a juvenile individual. There is no larval stage in cephalopods. In the head of cephalopods is a pair of glands, the optic glands, which secrete a hormone that produces maturation of the gonads in both sexes.

PHYLOGENY The early history of the mollusks is obscure, and very little explicit information is available concerning the phylogeny of this group. Because of the presence of spiral cleavage, a schizocoel, and a trochophore larva in both annelids and mollusks, it is highly likely that annelids and mollusks evolved from a common coelomate stock. The coelomate ancestor discussed in Chapter 7 is a likely candidate here for this role of common ancestor. Annelids evolved toward a segmented body plan, whereas mollusks remained unsegmented except for *Neopilina,* whose segmentation was more likely a secondary acquisition rather than a primitive molluscan characteristic. Among the difficulties in accepting the view that the segmentation of *Neopilina* is a primary characteristic, and not secondarily acquired, are the different numbers of replicated structures (for example, five pairs of gills and eight pairs of foot retractor muscles) and the difficulty in determining what structurally or functionally represents a segment. Furthermore, no larval mollusk that has been studied shows evidence of segmentation, which is certainly hard to understand if segmentation is indeed a primitive characteristic of mollusks. Study of the larval development of *Neopilina* should prove very enlightening. The fact that chitons have a shell consisting of eight plates might be misconstrued as a sign of segmentation. However, study of the chiton trochophore larva reveals that the eight plates are derived from a single, unsegmented shell-secreting portion of the ectoderm, called the shell gland, whereas a truly segmented chiton would be expected to have eight shell glands. Also, adult chitons show no internal sign of segmentation. Therefore, the common annelid–mollusk ancestral stock was probably unsegmented at the time the ancestral forms that gave rise to the mollusks and annelids split off from each other. Only after the split did segmentation evolve in the annelid line.

FURTHER READING

Boycott, B. B., "Learning in the Octopus," *Scientific American*, vol. 212:3 (1965), p. 42.

Cousteau, J. Y., and P. Diolé, *Octopus and Squid, The Soft Intelligence*. Garden City, New York: Doubleday, 1973.

Florey, E. "Ultrastructure and Function of Cephalopod Chroomatophores," *American Zoologist*, vol. 9 (1969), p. 429.

Fretter V., ed., *Studies in the Structure, Physiology and Ecology of Molluscs, Symposia of the Zoological Society of London, No. 22*. New York: Academic Press, 1968.

Galtsoff, P. S., "Physiology of Reproduction in Molluscs," *American Zoologist*, vol. 1 (1961), p. 273.

Ghiselin, M. T., "The Adaptive Significance of Gastropod Torsion," *Evolution*, vol. 20 (1966), p. 337.

Giese, A. C., and J. S. Pearse, eds., *Reproduction of Marine Invertebrates, Vol. 4, Molluscs: Gastropods, and Cephalopods*. New York: Academic Press, 1977.

Hyman, L. H., *The Invertebrates: Mollusca I*, Vol. VI. New York: McGraw-Hill, 1967.

Lemche, H., "A New Living Deep-Sea Mollusc of the Cambro-Devonian Class Monoplacophora," *Nature*, vol. 179 (1957), p. 413.

Linsley, R. M., "Shell Form and the Evolution of Gastropods," *American Scientist*, vol. 66 (1978), p. 432.

Lissmann, H. W., "The Mechanism of Locomotion in Gastropod Molluscs. I. Kinematics," *Journal of Experimental Biology*, vol. 21 (1945), p. 58.

Lowenstam, H. A., "Recovery, Behaviour and Evolutionary Implications of Live Monoplacophora," *Nature*, vol. 273 (1978), p. 231.

Morton, J. E., *Molluscs, An Introduction to Their Form and Functions*. New York: Harper Torchbooks, 1960.

Nixon, M., and J. B. Messenger, *The Biology of Cephalopods, Symposia of the Zoological Society of London No. 38*. New York: Academic Press, 1977.

Purchon, R. D., *The Biology of the Mollusca*, 2nd ed. New York: Pergamon Press, 1977.

Van Weel, P. B., "The Comparative Physiology of Digestion in Molluscs," *American Zoologist*, vol. 1 (1961), p. 245.

Wells, M. J., *Octopus: Physiology and Behavior of an Advanced Invertebrate*. New York: Halsted Press, 1977.

Wilbur, K. M., and C. M. Yonge, eds., *Physiology of Mollusca*, Vol. 1. New York: Academic Press, 1964.

Wilbur, K. M., and C. M. Yonge, eds., *Physiology of Mollusca*, Vol. II. New York: Academic Press, 1966.

Young, J. Z., "Two Memory Stores in One Brain," *Endeavor*, vol. 24:91 (1965), p. 13.

chapter **9**

Arthropoda and Onychophora

ARTHROPODA

The phylum Arthropoda [Gr. *arthros*, joint + *podos*, foot] contains about 80 percent of the described, living species of animals. Furthermore, approximately 90 percent of the species of arthropods are insects. According to the criteria listed in Chapter 1 that are used to decide whether a group of animals is a "successful" one, insects are certainly a very successful class of animals.

Arthropods can be found in the seas, in fresh water, on land, and in the air. They are, in fact, the only invertebrates that colonized land on a large scale, and insects are the only invertebrates that evolved the ability to fly. The phylum shows tremendous diversity of form, inasmuch as it includes animals such as the barnacles (which were classified as mollusks until 1830 because of their "limestone houses"), crabs, lobsters, spiders, ticks, mites, centipedes, millipedes, and insects. The size of living arthropods ranges from microscopic mites to a Japanese species of spider crab that spreads to 4 meters when its legs are extended. In general, the larger arthropods are aquatic; the buoyancy of the water helps to support them.

The most obvious characteristic of arthropods is their external skeleton or *exoskeleton*. The skeletal covering of the body proper is divided into a number of separate, more or less rigid plates, which are joined together by flexible membranes; on the appendages, this covering is divided into tubelike sections with flexible membranes at the joints. An exoskeleton of this sort has built-in advantages; its firmness equips it to provide the muscle attachments necessary for rapid locomotion and ample protection for the internal organs. In addition, its almost impermeable nature is important to terrestrial arthropods, because water loss by evaporation from the body surface is minimized. However, the arthropod exoskeleton presents a major problem to a growing animal because exoskeletons must be shed periodically for growth to occur. After the old exoskeleton has been shed, arthropods are extremely vulnerable to predators until the new one hardens. Additional characteristics of arthropods are bilateral symmetry, primitive segmentation, an open circulatory system, an annelidlike nervous system consisting of a brain located dorsally in the head and a segmentally ganglionated ventral nerve cord, a complete digestive tract, and a schizocoel that is greatly reduced in size.

CLASSIFICATION The living arthropods can be readily separated into nine classes.

The class Merostomata is represented by only five living marine species of horseshoe crabs (Fig. 1-14C). The body consists of a horseshoe-shaped *cephalothorax* (formed by fusion of the head to the thorax) and an abdomen, which ends in a long spikelike structure, the *telson*. Just anterior to the mouth are a pair of clawlike appendages, the *chelicerae*. There are five pairs of walking legs on the ventral side of the cephalothorax. Antennae are absent. The larvae resemble trilobites, an extinct group of marine arthropods (Fig. 9-1). *Limulus polyphemus* is the horseshoe crab of eastern North America.

The class Arachnida (Fig. 1-11D) consists of animals such as the scorpions, spiders, ticks, and mites. The body is composed of a cephalothorax and abdomen. Antennae are absent, but chelicerae are present. Arachnids have four pairs of walking legs attached to the cephalothorax. The vast majority of arachnids are completely terrestrial. The sting of the scorpion, *Androctonus australis*, from the Sahara Desert, can be fatal to a human.

The class Pycnogonida consists of a minor group of marine

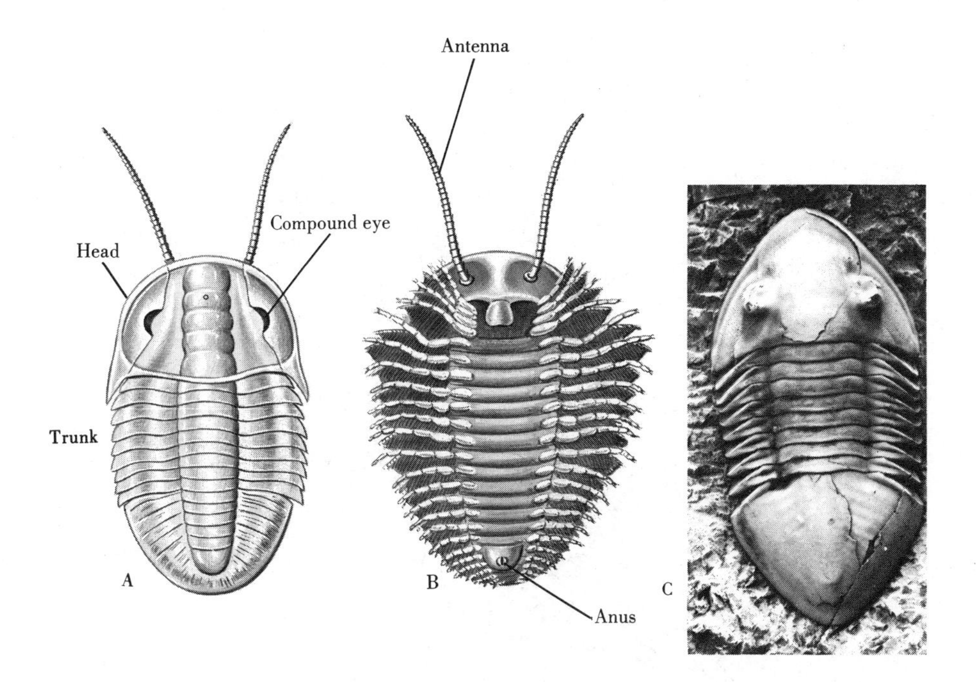

Fig. 9-1 *Trilobites. (A) Dorsal and (B) ventral views of a generalized trilobite; (C) photograph of a trilobite,* Isotelus gigas. *[(A) and (B) After Snodgrass,* A Textbook of Arthropod Anatomy. *Copyright © 1952 by Cornell University. Used by permission of Cornell University Press. (C) Courtesy of H. B. Whittington, Sedgwick Museum, Cambridge, England.*]

arthropods, commonly known as sea spiders (Fig. 1-11F). The body is generally about 4 mm long and consists of a cephalothorax (the head is fused to the first thoracic segment) and a greatly reduced abdomen. Pycnogonids possess a pair of chelicerae, but lack antennae. The cephalothorax, unlike those of horseshoe crabs and arachnids (the other arthropods that have chelicerae), is distinctly jointed posteriorly. Pycnogonids have four to six pairs of walking legs. *Nymphonella tapetis* is a Japanese species of pycnogonid whose larval stages are spent in the mantle cavity of clams.

The class Crustacea includes such arthropods as barnacles, shrimps, crayfishes (Figs. 1-11A and 9-2), lobsters, and crabs. They possess two pairs of antennae and a pair of *mandibles* (short, strong jaws) used for biting, chewing, and crushing food. They lack chelicerae. Although there are some terrestrial species of crustaceans, most are either freshwater or marine organisms, the marine ones being in the majority. The body primitively consists of a head and a trunk, the trunk being composed of a series of similar segments. But in advanced crustaceans, the trunk has differentiated into a thorax and abdomen, with the head and thorax often having united to form a cephalothorax.

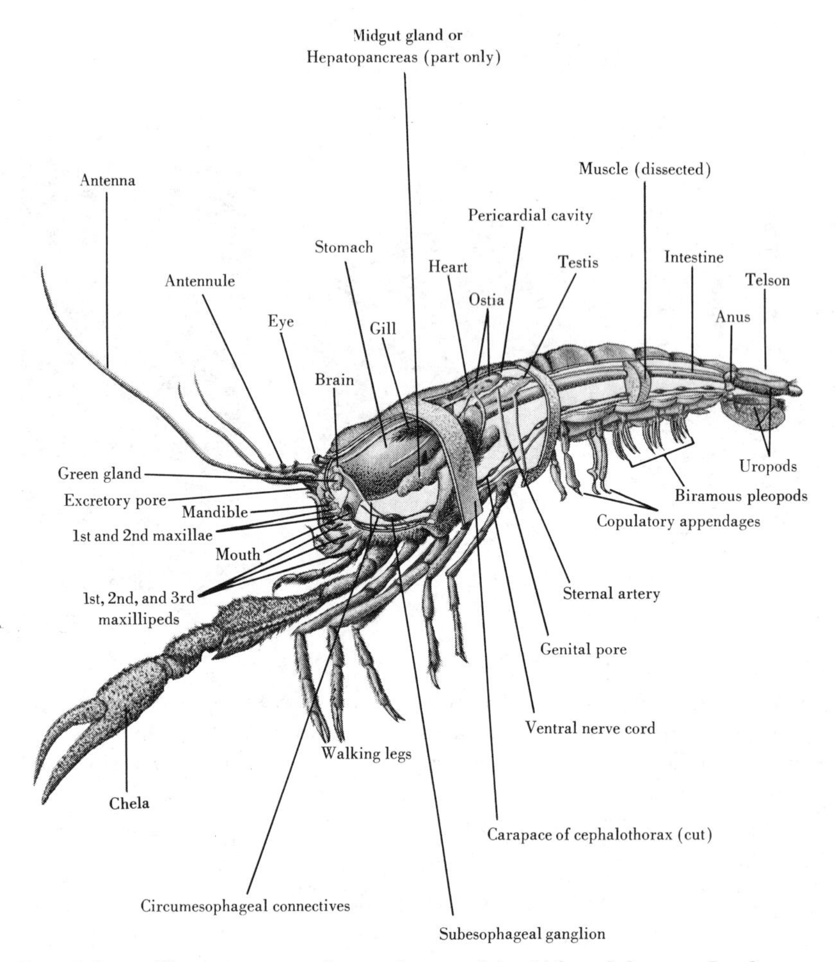

Fig. 9-2 *The structure of a male crayfish. (After Johnson, Laubengayer, DeLanney, and Cole,* Biology, *3rd ed. Holt, Rinehart, and Winston, 1966.)*

Barnacles, such as *Balanus tintinnabulum,* foul the hulls of ships, creating a fuel-wasting drag.

The class Diplopoda consists of the millipedes (Fig. 1-11E). A member of this class has a distinct head, followed by a segmented trunk having 11 to more than 100 segments. In a typical millipede the first trunk segment has no appendages, the next three segments have one pair of legs each, and each following segment, except the very last one, which is legless, possesses two pairs of legs. Each segment with four legs is derived from the fusion of two originally distinct segments and is called a *diplosegment.* The first four trunk segments are sometimes referred to as the thorax, and the rest of the trunk as the abdomen. The head has one pair of mandibles and one pair of antennae. Millipedes have no chelicerae; they occupy the same sorts of habitats as centipedes. *Fontaria georgiana* is a commonly found millipede in the southeastern United States.

The class Pauropoda (Fig. 1-11G) is a minor one, consisting of small (0.5–2.0 mm) terrestrial arthropods. The head, which bears a pair of mandibles and a pair of two-branched antennae, is followed by a segmented trunk. They do not have chelicerae. The trunk consists of 11 segments, nine of which bear a pair of legs each. *Pauropus huxleyi* occurs in the eastern and central United States and in Europe.

The class Chilopoda is represented by the centipedes (Fig. 1-11C). The chilopod body is dorsoventrally flattened and consists of a head and segmented trunk. The first trunk segment has a pair of poison claws, the duct of a poison gland opening near the tip of each of these claws; the last two segments have no appendages; and between the first and last two are 15 to about 180 segments, each of which bears one pair of walking legs. One pair of mandibles and one pair of antennae are present. Chelicerae are absent. Centipedes are terrestrial, but are restricted to damp habitats such as rotting logs. *Geophilus californicus* is a centipede found in southern California.

The class Symphyla (Fig. 1-11H) is a minor group of terrestrial arthropods, 2 to 10 mm long. The body consists of a head followed by a segmented trunk. They closely resemble centipedes but lack the poison claws of centipedes. Symphylans have mandibles, a pair of antennae, and a trunk with 12 leg-bearing segments, one pair per segment. They have no chelicerae. *Scutigerella immaculata* of the eastern United States attacks living plants and can be a severe greenhouse pest.

The class Insecta consists of arthropods that possess three well-defined body regions (head, thorax, and abdomen), one pair of antennae, and a pair of mandibles (Figs. 1-11B and 9-3). They lack cheli-

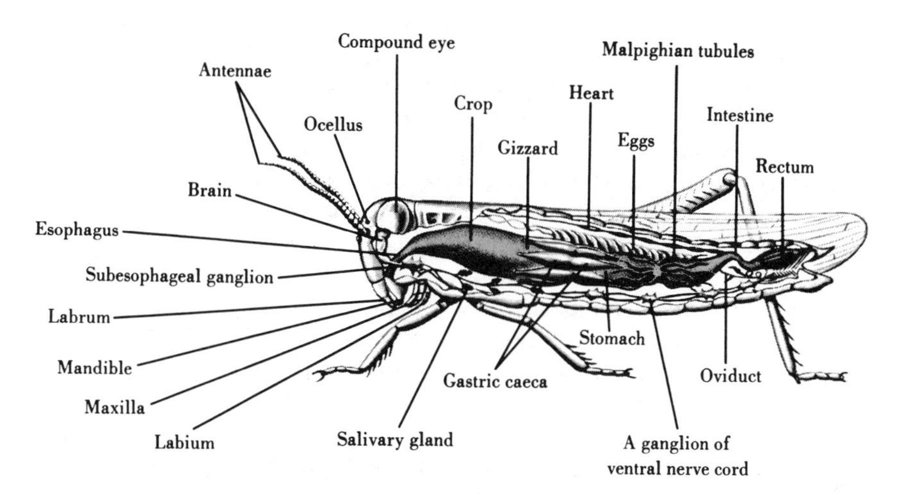

Fig. 9-3 *The structure of a female grasshopper. (After Johnson, Laubengayer, DeLanney, and Cole,* Biology, *3rd ed. Holt, Rinehart and Winston, 1966.)*

cerae. The thorax always has only three segments, each bearing only one pair of legs. This is such a constant feature that the name Hexapoda [Gr. *hex*, six ´*podos*, foot] has been used for this class. A majority of the species have two pairs of wings on the thorax. Most insects are completely terrestrial, but some develop in aquatic habitats. One of the better known insects is the fruit fly, *Drosophila melanogaster*, which has been used in genetic investigations since the early 1900s.

The arachnids, horseshoe crabs, and pycnogonids are referred to as *chelicerates* because they possess chelicerae (but not mandibles), in contrast to the other arthropods, which lack chelicerae but have mandibles and are consequently called *mandibulates*.

MORPHOLOGY AND PHYSIOLOGY

ARTHROPOD STRUCTURE

The ancestral arthropods most likely were marine organisms that had an unsegmented head, followed by a trunk that consisted of many similar segments. Each trunk segment, except for the last one, which had no appendages, is thought to have borne a pair of identical appendages, all of these trunk segments having had a similar structure. No fossil record has been found of this hypothetical ancestral type; however, a group of fossil marine arthropods known as trilobites, which became extinct at the end of the Paleozoic Era, closely approximates the hypothetical ancestral condition (Fig. 9-1). Trilobites are the most primitive arthropods known. Nevertheless, trilobites had diversified so much by the time they became extinct that paleontologists have been able to identify more than 3,900 species.

During the evolution of the arthropods, the primitive unsegmented head has incorporated one or more trunk segments, and the rest of the trunk has tended to differentiate into thoracic and abdominal portions. There has also been a tendency toward reduction in the number of trunk segments and modification of them for the performance of a variety of functions. The modifications of the trunk segments that occurred are clearly evidenced by their appendages. The trunk appendages, presumably at one time all of them having been similar, have evolved into a number of different types adapted to perform such functions as feeding, copulation, and egg brooding. In advanced crustaceans, such as the crayfish (Fig. 9-2), the abdominal appendages on all but the last segment are known as *pleopods*. The flattened appendages on the last abdominal segment are termed *uropods*.

The arthropod exoskeleton (except the calcareous plates of barnacles) is a cuticle, consisting principally of a chitin–protein complex, secreted by the underlying epidermal cells. The protein (called *sclerotin*) is a tanned, tough, horny material. In some arthropods, particularly the crustaceans, the exoskeleton is impregnated with calcium salts for extra strength. Barnacles are unique among arthropods in having a protective shell consisting of calcareous plates, also secreted by the epidermis, but the exposed parts of the body, such as the appendages, have a typical chitinous exoskeleton. At ecdysis, barnacles do not shed the plates, but, like other arthropods, they do shed the chitinous exoskeleton. Ecdysis, as defined in Chapter 2, is the shedding of the old exoskeleton. Ecdyses occur periodically to allow body growth to occur, but the plates grow more or less continuously.

Insects show tremendous diversity within the basic scheme described above for the class Insecta, as would be expected because of the huge number of species. Living insects may be divided into 28 orders; the main ones are listed in Table 9-1. Some insects, such as springtails, are wingless; others, such as the true flies, the dipterans, have only one pair of wings, but most, such as the dragonflies, have two pairs. The four orders of insects that are primarily wingless (only about 6,500 species) are known as *apterygotes*. The remaining 24 orders of insects, called *pterygotes*, are either winged or have secondarily lost their wings (fleas, for example). It seems clear that the wingless pterygotes evolved from winged stock, whereas the apterygotes arose from wingless ancestors. The second pair of wings in dipterans has been reduced to halteres, whose function will be described below; wings occur only on the last two thoracic segments. Insect wings are usually thin and membranous, but in a number of groups (the beetles, for example), the forewings have become modified into protective covers for the hindwings. The legs of insects are modified variously for running, jumping, swimming, burrowing, and holding prey. The mouth parts are likewise highly modified, depending upon the diet. The abdomen in primitive insects consists of 11 segments, but in most adult advanced insects there are only nine. However, as a result of fusion, the number may even be reduced to as few as four, as in some wasps.

The crustaceans, like the insects, show considerable diversity. The crustacean head is uniform throughout the class in possessing five pairs of appendages: two pairs of antennae, a pair of mandibles, and two pairs of maxillae. Mandibles and maxillae are feeding appendages. Crustaceans are the only arthropods which have two pairs of antennae. The second pair of antennae probably represent the appendages of a trunk segment that fused with the ancestral head. The number of trunk segments that can be counted varies from group to group among the crustaceans, and depends upon the amount of fusion that occurred in the past. In many crustaceans, the thorax or anterior trunk segments are covered by a shieldlike *carapace*, which projects posteriorly from the head and fuses dorsally with varying numbers of thoracic segments. However, in such groups as the ostracods, which will be described in more detail below, the carapace may be so large that it envelops the entire body.

Table 9-1 The Main Orders of Insects

		Order	Common Examples	
Apterygotes	Ametabolous	Thysanura	Silverfish	
		Collembola	Springtails	
Pterygotes	Hemimetabolous	Ephemeroptera	Mayflies	
		Odonata	Dragonflies	
		Orthoptera	Cockroaches, Crickets, Grasshoppers	
		Isoptera	Termites	
		Dermaptera	Earwigs	
		Mallophaga	Chewing lice	
		Anoplura	Sucking lice	
		Hemiptera	True bugs	
	Holometabolous	Neuroptera	Dobsonflies, Lacewings	
		Coleoptera	Beetles, Weevils	
		Lepidoptera	Butterflies, Moths	
		Diptera	True flies	
		Hymenoptera	Ants, Wasps, Bees	
		Siphonaptera	Fleas	

There are two types of arthropod appendages, *biramous* and *uniramous*. The biramous type is Y-shaped, consisting of a portion attached to the body *(protopodite)*, and two branches, *endopodite* (medial) and *exopodite* (lateral). Uniramous appendages are unbranched.

There are several interesting crustaceans, other than those such as lobsters and crabs, with which most nonscientists are not well acquainted. There are four in particular (Fig. 9-4). An *ostracod* (mussel shrimp), no more than a few millimeters in length, is completely enveloped in a bivalve carapace almost like the shell of a typical bivalve mollusk. The hinge is an uncalcified strip of the exoskeleton. The trunk is very reduced. Ostracods occur in both fresh water and sea water. Most of them live on or near the bottom, creeping on or through the mud and on the surface of plants. A *copepod* has distinct external segmentation with a head, thorax, and abdomen. The first, and sometimes also the second, thoracic segment is fused with the head. Characteristic of the group is a single medial eye. Most copepods are in the same size class as ostracods. Copepods are abundant in fresh water as well as in the sea; they form a major part of the diets of some whales and fishes. In fact, the copepod *Calanus finmarchicus* is probably the most abundant animal, in number of individuals, in the oceans. In addition to the free-living species, many copepods are parasitic. *Xenocoeloma brumpti*, for example, lives in the body wall of polychaetes. An *amphipod* has a laterally compressed body and averages about 1 cm in length. Some species occur in fresh water; some (beach fleas) are semiterrestrial, living along the seashore near the high-tide line; some are

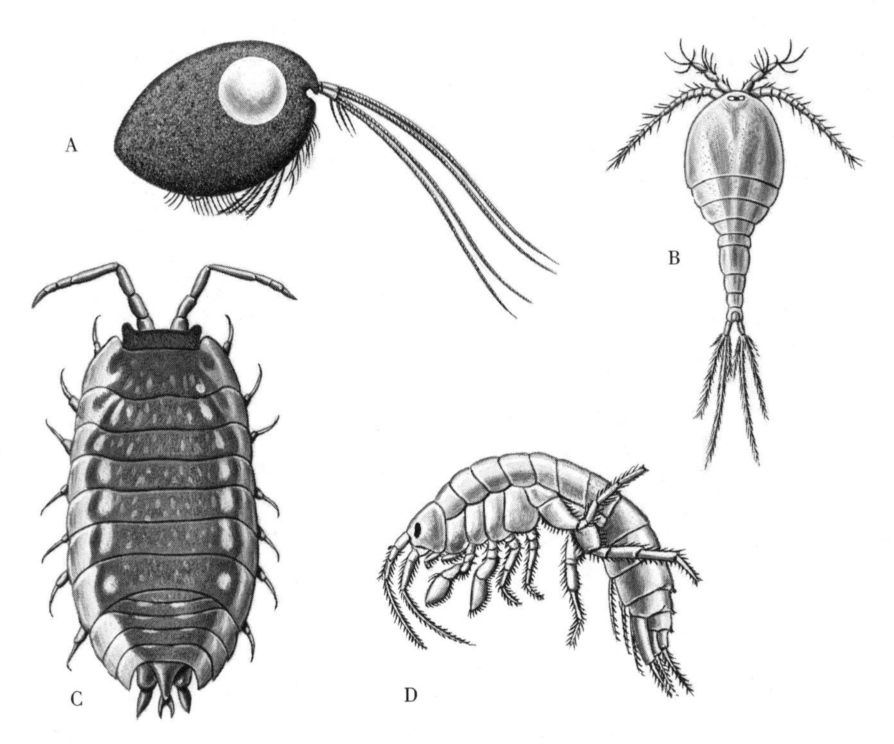

Fig. 9-4 *Selected crustaceans. (A) An ostracod; (B) a copepod; (C) an isopod; (D) an amphipod. [(D) From Pratt,* A Manual of the Common Invertebrate Animals. *Copyright 1935, McGraw-Hill Book Company. Used by permission.*]

even terrestrial, living in humid regions; but most are strictly marine. An *isopod* characteristically has a dorsoventrally flattened body with gills on the abdomen. Almost all other crustaceans, if they have gills at all, have thoracic gills. Isopods are about the same average size as amphipods. Most species are marine, but some occur in fresh water, and others are terrestrial. The terrestrial isopods (pill bugs) represent the largest group of crustaceans that successfully colonized land.

In arachnids, the evolutionary tendency has been toward progressive fusion of the cephalothorax with the abdomen, and gradual loss of segmentation in the abdomen. Scorpions show only a slight trend toward reduction of segmentation (Fig. 9-5). On the other hand, mites and ticks show the most extreme fusion; external segmentation has been lost, and there is no longer any demarcation between the cephalothorax and abdomen.

LOCOMOTION With the evolution of an exoskeleton consisting of rigid portions connected by flexible membranes, there was an ideal opportunity, through natural selection, for improvement of the general locomotor capacity of the phylum. This capability was much greater than in the animals discussed previously. The skeleton provided rigid supporting structures for muscle attachments and made the body firm enough that it could be raised off the substrate. An animal that does not have to drag its body along the substrate can move from place to place with much more facility than one

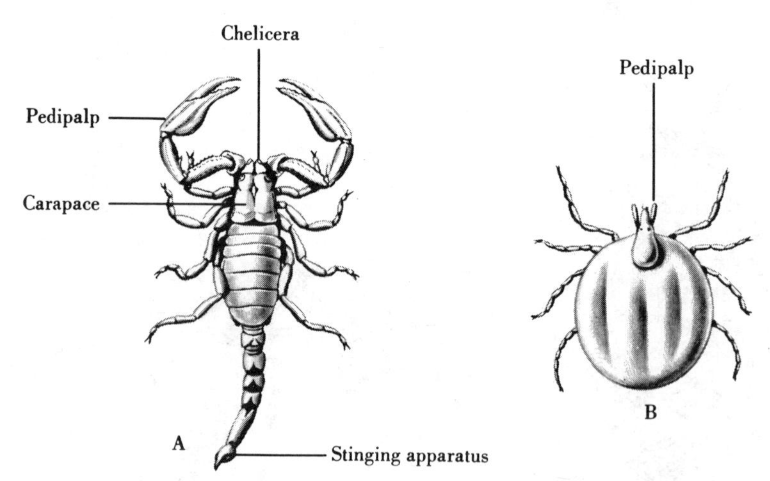

Fig. 9-5 *(A) The scorpion,* Chactas vanbenedeni, *and (B) the tick,* Ixodes ricinus. *Notice the marked reduction in external segmentation in the tick.* [*(A) After Snodgrass,* A Textbook of Arthropod Anatomy. *Copyright © 1952 by Cornell University. Used by permission of Cornell University Press. (B) After Borradaile and Potts,* The Invertebrata, *3rd ed. Cambridge University Press, 1959.*]

that does. In contrast to the parapodia of polychaetes, the legs of arthropods tend to be more slender, longer, and more ventrally located. Among the arthropods, every type of locomotion is found, including walking, swimming, running, jumping, and flying. The evolution of jointed limbs (along with the muscles that move them) has given arthropods great advantages in locomotion. Not only could the body be raised off the substrate, but also rapid movement from place to place could be accomplished by use of the limbs alone, without undulatory movements of the trunk. This is apparently a more energy-efficient mechanism than is seen in the polychaetes, for example, where fast crawling and swimming require not only the use of the parapodia but also undulatory movements of the trunk. Few arthropods stand up on their legs. In most species, the body hangs suspended from them so that the center of gravity is kept low and a high degree of stability is attained. The power of flight can alone account for much of the success insects have had. The adaptive value of wings is evidenced by the fact that there are about 100 times as many winged species of insects as wingless. Wings gave insects a big advantage over other terrestrial invertebrates by enabling these flying forms to disperse over a wider area, reach new sources of food, and avoid predators. The fact that arthropods, unlike the annelids, were no longer dependent upon the coelom for locomotion can account for the great reduction of this cavity in arthropods.

An insect wing is a flattened, roughly triangular outgrowth which develops from the laterodorsal region of either the second (mesothoracic) or third (metathoracic) segment of the thorax; most insects, as stated above, have two pairs of wings, one pair on each of these last two thoracic segments. In all insects, movement of the wings upward is accomplished by an indirect mechanism using dorsoventrally arranged thoracic muscles that originate in the ventral portion of the thorax and insert in the *tergum*, the roof of the thorax (Fig. 9-6). Contraction of these muscles pulls down the tergum, and a downward force is consequently applied to the base of the wing. This results in wing elevation, with a portion of the lateral body wall, called the pleural process, acting as a fulcrum. Downward movement of the wings is accomplished either directly by contraction of muscles (the basalar muscles) attached to the wing base as in dragonflies, indirectly as in dipterans by contraction of dorsally located longitudinal thoracic muscles, or as in beetles by both direct and indirect muscular action. When these longitudinal muscles contract, the length of the thorax decreases, causing the tergum to bulge upward and the wings consequently to pivot downward. Insects using direct muscular action alone to move their wings downward have a lower rate of wing beat than is generated by insects using at least in part indirect muscular action to lower their wings; there are about 8–30 beats per second in dragonflies versus about 200–1,046 in dipterans, as examples. The basalar muscles, which are attached directly to the wing base for lowering the wing, show a 1:1 correspon-

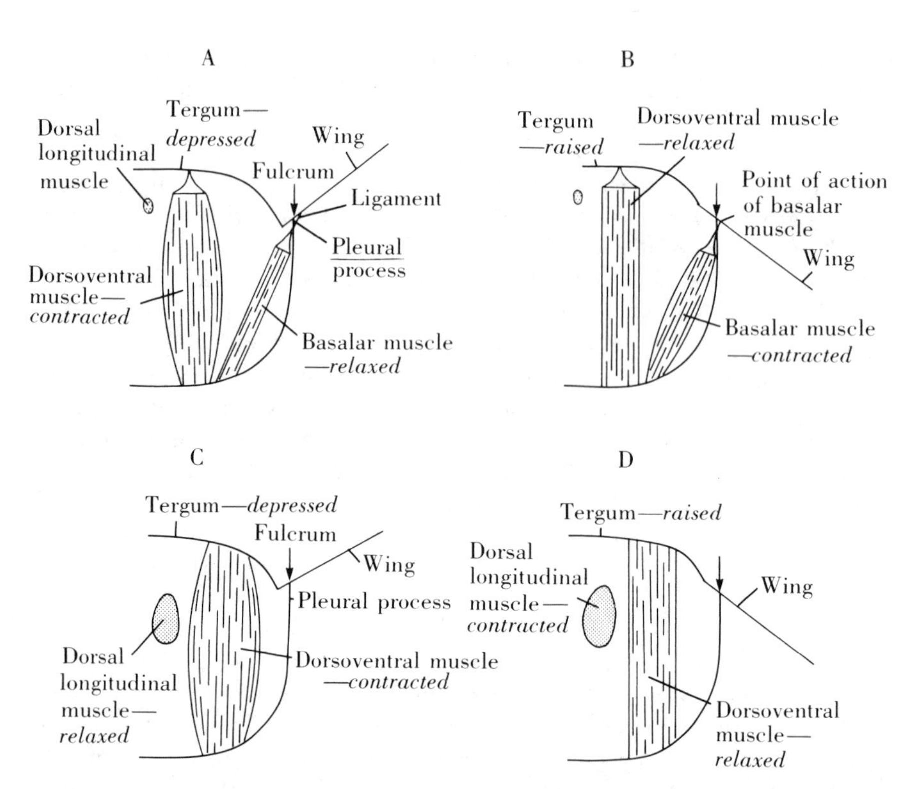

Fig. 9-6 *Diagrammatic representations of the wing movements of insects. (A and B) Cross sections through the thorax of an insect such as a dragonfly in which (A) indirect muscles produce elevation of the wings and (B) direct wing muscles cause the wings to depress. (C and D) cross sections through the thorax of an insect such as a dipteran in which both (C) elevation and (D) depression of the wings are produced by indirect muscles. (After Chapman,* The Insects: Structure and Function, *2nd ed. American Elsevier Publishing Company, 1971.)*

dence between the number of nerve impulses and the number of wing beats. That is, for every nerve impulse there is a single wing beat. Because the muscle contractions occur in synchrony with the nerve impulses, the basalar muscles are known as "synchronous muscles." In contrast, where the wings are lowered with the aid of the indirect mechanism involving the longitudinal thoracic muscles, there are usually about 12 to 20 wing beats for each nerve stimulus to these longitudinal muscles; the longitudinal muscles are contracting (and the wings are beating) at a faster rate than the muscles are being stimulated by the nerves running to them. The reason for this is that when the dorsoventral muscles that insert in the tergum and the longitudinal thoracic muscles are relaxed and suddenly stretched, they tend to contract. A muscle that exhibits a rate of contraction higher than the rate at which nerve impulses are being delivered to the muscle is called an "asynchronous muscle." These muscles are specialized to oscillate rapidly, these muscle cells differing structurally from the cells in the basalar muscles, which are directly attached to the wings. When these dorsoventral muscles that insert in the tergum contract, the longitudinal muscles become stretched, causing the longitudinal muscles to contract, stretching these dorsoventral muscles, causing them to contract, and so on. Thus each muscle is mechanically stimulated over and over again. Only enough nerve impulses are needed to maintain the rhythm of contractions. When the nerve impulses cease, the muscular contractions die out. Also, in insects with a high wing-beat frequency (those making use of longitudinal thoracic muscles to aid in lowering their wings) there is a "click mechanism," so named because the wings click or snap from one position to another. This mechanism aids in producing rapid wing movements. The wing-hinge system in those insects with high wing-beat frequencies is such that there are only two stable positions, fully up and fully down. When, through the action of the longitudinal muscles, the wings begin to lower, because of the wing-hinge arrangement when the wings are lowered beyond the midpoint, the wings snap rapidly into the fully down position. Similarly, the hinge mechanism aids in rapidly snapping the wings upward. However, flight involves more than just up and down movements of the wings. They must also be moved backwards and forwards and be rotated on the long axis by appropriate muscles to provide the lift and forward thrust essential for flight. The wings of bees, for example, trace a figure eight pattern relative to the trunk of the body. The highest wing-beat frequency of any insect is apparently that of a biting midge, a dipteran, reported to have reached 1,046 beats per second. The highest value for birds, 90 wing beats per second, was observed in hummingbirds. However, the fastest-flying insects, sphinx moths, have a maximum speed of 15 meters per second as compared with a spine-tailed swift, *Chaetura caudacuta*, which is apparently not only the fastest bird, but the fastest animal also, having been carefully clocked at 47 meters per second—26.8 meters per second is equivalent to 60 miles per hour.

FEEDING AND DIGESTION

The success of arthropods is due in large measure to their having become adapted to a remarkably diversified diet and to their having developed a variety of mouth parts and digestive organs that enable them to make use of so many different foods. It would seem that any substance capable of being a source of metabolic energy is used as food by some arthropod.

Crustaceans are filter feeders, scavengers, herbivores, or carnivores. Filter-feeding crustaceans have appendages on the cephalothorax that are heavily endowed with setae. These setae filter food particles out of the water. The water current that all filter feeders must create to bring the food particles to them is produced in crustaceans either by the beating of these filtering appendages or, more commonly, by the beating of other appendages.

The digestive system typical of crustaceans is an uncoiled tract consisting of a ventral mouth, esophagus, stomach, intestine, and anus that opens in the last abdominal segment. Digestion takes place in the stomach. In larger crustaceans, the stomach has a *gastric mill,* which consists of teeth and ridges used to grind the food particles to a fine size. Associated with the digestive tract is a midgut gland (hepatopancreas) that opens into the stomach through a pair of ducts, one on each side. The midgut gland is the source of the enzymes used in digestion and is the major site of absorption of nutrients, absorption being completed in the intestine.

Insect mouth parts fall into two major categories: those for chewing and those for sucking. In insects, such as a grasshopper, with the chewing type, the *labrum* (upper lip) and the *labium* (lower lip) hold the food, which is chewed by the mandibles and maxillae (Fig. 9-7). A tongue *(hypopharynx)* lies on the floor of the mouth. The salivary glands open at the base of the hypopharynx.

Sucking mouth parts evolved from the chewing type. Essentially, all of the chewing mouth parts became elongated and slender in

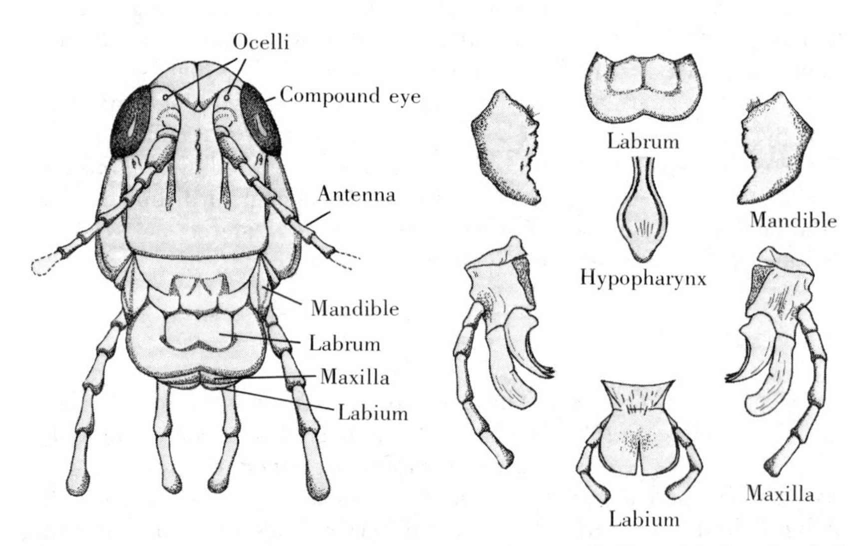

Fig. 9-7 *Front view of the head of a grasshopper (left) and isolated mouthparts (right). The mouthparts depicted are typical of the chewing type. (Modified from* Elements of Zoology, *4th ed., by Storer, Usinger, Nybakken, and Stebbins. Copyright 1977, McGraw-Hill Book Company. Used by permission.)*

order to accomplish their new functions. In a female mosquito, for example, the mandibles, maxillae, labrum, and hypopharynx became delicate, elongated stylets used to pierce the skin. In addition, the tubular hypopharynx is used to conduct saliva to the site of penetration. The saliva of blood-sucking insects contains an anticoagulant, as in the mosquito *Anopheles* that was mentioned in Chapter 2 when the life cycle of *Plasmodium* was discussed. The anticoagulant not only keeps the blood flowing out of the wound, but also prevents the blood meal from clogging the food channel. The labium became a flexible sheath that encloses the mouth parts when an insect with sucking mouth parts is not feeding. The labium does no piercing and folds back as the stylets enter the tissue pierced. The food channel is between the hypopharynx and grooved labrum, and the esophagus became a sucking organ. Muscles run from the esophagus to the rigid walls of the head. When they contract, the esophagus enlarges and sucks up the fluid through the food channel; when these muscles relax, the esophagus shrinks because of its inherent elasticity, and the fluid is forced posteriorly.

The insect digestive tract is constructed along the same basic plan as in crustaceans, except that a crop and a gizzard are interpolated between the esophagus and stomach, and there is no gastric mill nor midgut gland. Digestion is carried on in the stomach, enzymes being released from cells in its wall; absorption begins there but is completed more posteriorly. The stomach has a number of pouches, *gastric caeca.* Although not much is known about the function of these pouches, at least in some insects they appear to be an additional source of digestive enzymes. In an effort to conserve water, insects living under dry conditions often absorb so much water from the rectal contents that their feces are dry and powdery.

Millipedes are primarily herbivores, feeding on both living and decaying plant material. Centipedes are predacious, feeding primarily on small arthropods; the pair of poison claws is used to kill or stupefy prey.

In most arachnids, the chelicerae only grasp and crush prey, but in spiders they also serve as fangs for injection of poison. Each member of the first pair of postoral appendages in arachnids is called a pedipalp; these are also used by most arachnids to capture food. Among spiders, however, the pedipalps have evolved into copulatory appendages in the male, and short leglike appendages in the female.

Most arachnids are carnivores. They have evolved an interesting means of overcoming the handicap of a small mouth that, because of the exoskeleton, cannot be stretched to take in a large food mass; they partially digest their food outside the body before ingesting it. After the prey has been killed, digestive enzymes are deposited on the food, producing a partially digested broth. The broth is then ingested and passes through the esophagus to the stomach, where digestion is completed and the nutrients are absorbed. The stomach empties into a short intestine.

CIRCULATION The circulatory system in an arthropod is open. There is typically a single-chambered heart from which emerge one or more open-ended arteries. The blood flows from these arteries into the lacunae and sinuses of the hemocoel, coming into direct contact with the organs; it then slowly returns to the heart (located dorsally in relation to the digestive tract) and enters it through a number of *ostia* (openings in the wall of the heart). Arthropods have no vessels to return blood to the heart. The heart lies in a large sinus called the *pericardial cavity.* When the heart contracts, the ostia close and the blood is forced into the arteries. Relaxation of the heart allows the ostia to open, and the heart fills again with blood from the pericardial cavity. In some arthropods, the heart is an enlongated tubular structure suggestive of the dorsal blood vessel of annelids, from which it may have evolved. In others, it is reduced to a shorter, more or less spherical structure. Some of the smaller crustaceans do not have hearts; their body movements during locomotion and peristaltic contractions of the intestine are sufficient to provide a propulsive force to the blood. With respect to crustaceans, most species, including the larger ones, use hemocyanin as their oxygen-transport pigment, but the smaller-size crustaceans use hemoglobin or no oxygen-transport pigment. The horseshoe crab and a few arachnids also have hemocyanin, but apparently no oxygen-transport pigment is present in the blood of pycnogonids, centipedes, millipedes, pauropods, symphylans, and most arachnids. Also, except for the larvae of one genus of insect *(Chironomus)* which have hemoglobin in their blood, insects do not have transport pigments for oxygen. Hemoglobin of arthropods is found only in the plasma; in no animal does hemocyanin occur in the blood cells.

RESPIRATION Larger arthropods have well-developed respiratory systems, without which these highly mobile organisms would be unable to obtain sufficient oxygen to satisfy the metabolic requirements of their tissues. Many of the animals discussed in previous chapters are able to obtain a good share, if not all, of their oxygen by means of diffusion through the body surface. However, because of the relatively impermeable nature of the arthropod exoskeleton, such a mechanism of obtaining oxygen is denied to the larger arthropods. Only those arthropods that are very small or slight can obtain sufficient oxygen through the general body surface.

Three basic types of respiratory structures are found in arthropods: gills, tracheae, and lungs. The horseshoe crab and most crustaceans possess gills; insects, millipedes, centipedes, symphylans, and some arachnids have tracheae alone; some arachnids have lungs alone; and some arachnids have both lungs and tracheae. Pauropods, pycnogonids, and many species of the smaller crustaceans lack special respiratory organs.

In advanced crustaceans, the gills lie in the *branchial chambers* (spaces between the body wall and lateral portions of the carapace). Although gills in a branchial chamber are better protected than they would be if fully exposed, some mechanism is necessary to create a water current that would bring in a supply of oxygen adequate for all the animal's needs. The respiratory current is produced by the gill bailer, a paddlelike projection of the second maxilla. The second maxilla appears to have evolved initially as strictly a feeding appendage, tearing off bits of food and pushing them toward the mouth. But as a result of evolution of the gill bailer, the second maxilla appears to have become adapted secondarily for this additional respiratory role.

The five pairs of gills of the horseshoe crab, which lie on the underside of the abdomen, are called *book gills* because their structure suggests the pages of a book. Each gill consists of about 150 thin, leaflike folds of the integument, which are moved back and forth through the water, promoting gas exchange.

Lungs of arachnids are *book lungs*, which bear a striking similarity to book gills and probably evolved from them. One to four pairs of book lungs lie in a pocket on the underside of the abdomen. The opening into the pocket is called a *spiracle*. Enlargement of the pocket by muscular action draws air in; relaxation of the muscle cells results in expiration.

The tracheal system consists of internal tubes that run inward from openings on the body surface (Fig. 9-8). These openings are called spiracles also. The larger tubes, originating at the spiracles, are called *tracheae*. Tracheae divide into finer branches called *tracheoles* that terminate blindly, extending to all parts of the body. Tracheae and tracheoles have a chitinous lining that prevents their collapse. Tracheoles are in intimate contact with the tissues and even indent the cell membranes of many muscle cells, becoming functionally, although not anatomically, intracellular. Oxygen is thus carried directly to the cells. The blood is not involved in oxygen transport. In smaller insects, diffusion suffices to provide an adequate oxygen concentration within the

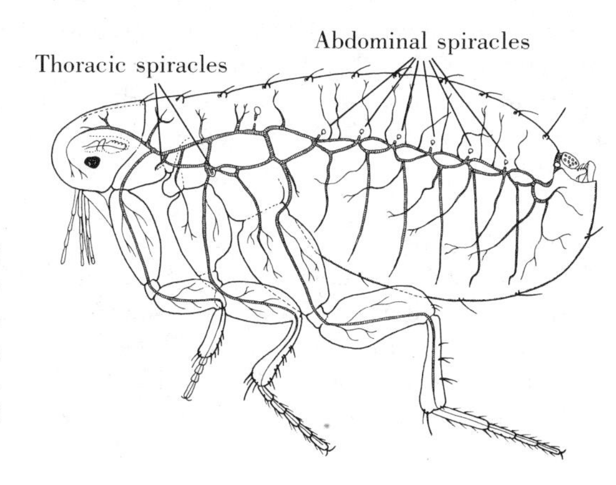

Fig. 9-8 *The tracheal system on the left side of the flea* Xenopsalla *sp. (Modified from Wigglesworth,* The Principles of Insect Physiology, *7th ed. Chapman and Hall Ltd., 1972.)*

tracheae. But larger, more active insects require an active ventilating mechanism as well, the air in the tracheae being renewed by a bellows-like action of the body wall, primarily of the abdomen.

The tracheal system is very efficient for a small organism. A system that supplies oxygen directly to the cells eliminates the necessity for the circulatory system to transport oxygen from the respiratory surface to the respiring tissues. This can account for the absence of oxygen transport pigments in virtually all animals that respire by means of tracheae. The contrast between book lungs and tracheae shows strikingly how different groups of terrestrial animals have solved the problem of obtaining oxygen.

SENSE ORGANS As would be expected, sensory mechanisms are well developed among arthropods. Most arthropods have eyes. In addition, receptors for touch, taste, and smell seem to be present in all the classes; some crustaceans and the dipterans have equilibrium receptors, and a few insect families also have phonoreceptors for sound detection.

Two basic types of eyes are found in arthropods: ocelli and compound eyes. Ocelli are relatively simple organs that can not only distinguish light from darkness but also discriminate intensity. They consist essentially of a cup-shaped layer of light-sensitive cells covered by a lens. The total contribution of the ocelli to the photoreceptive capability of arthropods is not well understood. Ocelli may also be capable of form perception and color vision (wavelength discrimination), as is the compound eye. A compound eye consists of a number of functional units or *ommatidia.* The number of ommatidia in a compound eye varies from as few as six in some ants to about 28,000 in dragonflies. Each ommatidium is elongated and has a transparent outer surface or cornea. Below this is the crystalline cone, which serves as a light guide to direct the entering light to the underlying photosensitive retinular cells. The compound eye allows a wide field of view, an arc of more than 200° in some species. Ocelli and compound eyes have no mechanism for accommodation. The compound eye is an optic system that is specialized for movement detection rather than image formation, but is nevertheless capable of forming images. Although the compound eye is readily able to detect movement of an object as its image moves from one ommatidium to the next, it is inferior to the eye of man in absolute sensitivity, intensity discrimination, and visual acuity (resolving power). Nevertheless, the compound eye, in addition to being an excellent movement detector, exhibits two other well-developed capabilities. One, the optic system of the compound eye is such that objects very close to the cornea are in focus. Being able to see objects clearly only a few millimeters away, something man cannot do, is an obvious advantage to a small animal such as a bee. Two, the compound eyes of fast-flying insects have much faster temporal resolution than do the eyes of man. That is, man ceases to see a light flickering when it flashes about 50 times per second (flicker fusion frequency), whereas the rate is up to 300 per second in some insects such as the hornet and blowfly. This high level of flicker fusion is an advantage to a fast-flying insect because it is able to recognize individual patterns in a rapidly changing field. Each ommatidium possesses screening pigments that regulate the amount of light striking the rhabdomeres of the retinular cells. The rhabdomeres are considered to be the portions of the retinular cells where the visual pigment is located. A visual pigment is a light-sensitive molecule that absorbs light, triggering the visual excitation process. In bright light the screening pigments prevent too much light from reaching the rhabdomeres; overstimulation would result in decreased photosensitivity. In dim light these pigments uncover the rhabdomeres.

The nauplius larva of crustaceans, which will be described in detail later in this chapter, has several ocelli grouped to form a single medial eye (appropriately named the *nauplius eye*). This persists in several species of adult crustaceans, along with the paired compound eyes typical of adults. The nauplius eye is, however, the only eye in adult copepods. In higher crustaceans the compound eye is on a movable stalk. Arachnids and pycnogonids only have ocelli, whereas the horseshoe crab has a pair of lateral compound eyes, and two median ocelli. Millipedes have ocelli or no eyes. There are no eyes in pauropods and symphylans. Centipedes have ocelli or compound eyes, and adult insects typically have a pair of compound eyes and three ocelli.

Taste, touch, and smell are dependent upon a variety of sensory hairs and sensory cells. In many insects, the sense of smell is confined to the antennae, but in others, olfactory receptors are also present on the mouth parts.

Insects have a highly developed ability to communicate by chemical signals. Such chemicals that are secreted by an individual in order to exert specific influences on other members of the same species are known as *pheromones*.

Sex attractants are one kind of pheromone. The sex attractant of the female silkworm moth is called bombykol; it is effective in extremely low concentrations. The amount of this sex attractant in one female silkworm moth, about one hundred millionth of a gram, could theoretically attract more than one billion males. The trail substance of ants is another insect pheromone. It is used to guide members of the colony to food. Ants, when returning to the nest after finding food, secrete this trail pheromone to mark the route so that other ants leaving the nest can find the food. Alarm substances are yet another type of pheromone. They are used to warn other members of the colony of danger. The ant *Formica polyctena* provides an example of the use of an alarm pheromone. When attacked, it sprays a mixture of formic acid and a pheromone (undecane) at the enemy, the formic acid repelling the attacker and the undecane serving to warn the other ants of the colony.

Many insects can detect sound waves. The most common mechanism for phonoreception consists simply of delicate hairs that tug on a neuron when they are displaced by sound waves, generating a nerve impulse. A more complex insect phonoreceptor is the *tympanic organ,* which consists of a thin, cuticular membrane *(tympanum)* on the body surface and a *chordotonal organ* containing sensory neurons behind it. The tympanum vibrates when struck by sound waves, and the chordotonal organ responds to these vibrations by sending impulses to the central nervous system. Tympanic organs occur on the thorax, abdomen, or forelegs, depending upon the insect. The more highly developed phonoreceptors are found among those insects, such as crickets, that are capable of producing sounds themselves.

In higher crustaceans, a *statocyst,* or equilibrium receptor that provides information concerning the animal's position with respect to gravity, is located in the basal segment of each of the antennules (the first pair of antennae). It is open to the exterior and contains a *statolith,* or concretion, that is either secreted completely by the animal or is composed of sand grains held together by a secretion from the wall of the statocyst. The statocyst is lined with sensory *hair cells.* These cells have nonmotile projections on the free surface. In spite of the fact that these "hairs" are nonmotile, one of them in each cell has the same structure as a cilium, which was described in Chapter 2. As the animal changes its position with respect to gravity, the statolith falls on and bends different "hairs," and the appropriate impulses are sent from these cells to the central nervous system, providing information concerning the animal's new position. A shrimp that uses sand grains in the formation of its statoliths was experimentally induced to put iron filings instead of sand grains into its statocysts when it replaced its statoliths, which are lost during each ecdysis. A magnet subsequently held above the shrimp caused it to invert itself, as though the magnet were the pull of gravity.

The hind (metathoracic) wings of dipterans have been reduced to a pair of gyroscopic sense organs of equilibrium, the *halteres.* Each haltere is a dumbbell-shaped organ which oscillates during flight at the same frequency as the wings beat. If both halteres are removed, some dipterans are still able to fly but cannot maintain their normal flight posture, whereas others are not able to balance well enough to fly at all; removal of only one haltere has little effect.

EXCRETION The excretory organs of arthropods are coxal glands, antennal glands (sometimes called "green glands"), maxillary glands, and Malpighian tubules. The first three have a similar basic structure, their names having been derived from the locations of the excretory pores. Some arachnids have only coxal glands, some have only Malpighian tubules, and some species have both kinds. Horseshoe crabs have coxal glands alone. In general, lower crustaceans have maxillary glands, higher ones antennal glands. Insects, centipedes, and millipedes have Malpighian tubules alone. Only two antennal or maxillary glands ever develop in an animal, but the number of Malpighian tubules and coxal glands may be greater, 2 to 250 Malpighian tubules, and 2 to 8 coxal glands. Coxal, antennal, and maxillary glands consist of a blind sac, to which is connected a tubule that leads to the excretory pore. The walls of part of the sac are sometimes greatly folded, thereby increasing the functional surface area. A portion of the tubule is enlarged to form a bladder for urine storage. A coxal gland opens on the *coxa,* the short segment of a walking leg by which the leg is attached to the body. Antennal glands open at the bases of the second pair of antennae, whereas maxillary glands open at or near the bases of the second pair of maxillae. As noted earlier, the arthropod coelom is greatly reduced. It is relatively smaller than that in any mollusk. The coelom is represented in arthropods only by the cavities of the coxal, antennal, and maxillary glands, and of the gonads and the gonadal ducts. These excretory organs work to eliminate wastes from the blood in the hemocoel. Coxal, antennal, and maxillary glands may have been derived from annelid nephridia, but because of the great reduction of the arthropod coelom, they have become highly modified. Presumably, the annelid types of nephridia are not adequate for animals with a greatly reduced coelom; on the other hand, Malpighian tubules were an innovation of terrestrial arthropods.

Malpighian tubules are hollow threadlike structures, which develop as outpocketings of the digestive tract in the region where the stomach and intestine meet. The free end in the hemocoel is closed. A filtrate of the blood consisting of amino acids, salts, water, and nitrogenous wastes passes into the lumen of the tubule and then into the intestine, ultimately reaching the rectum. The rectal wall reabsorbs the salts and amino acids from the filtrate, thereby conserving them, and also water, if the animal lives in a relatively dry habitat. The wastes remain in the rectum to be eliminated. Uric acid is the main nitrogenous waste of terrestrial insects.

The blood of freshwater crustaceans (such as the crayfishes) is hyperosmotic to the environment. Their excretory organs produce a urine that is hyposmotic to the blood. This helps to conserve salt and to eliminate the excess water that enters by osmosis. However, the urine is never as dilute as fresh water; consequently, some salt is always being lost. The loss is offset by active absorption of salt from the environment by the gills and by whatever salt is in their food.

NERVOUS SYSTEM, LEARNING, AND INSECT SOCIETIES

In order for the different parts of large, complex animals to function in an orderly relationship to one another, highly developed nervous and endocrine systems are required. The arthropod nervous system is built on the annelid plan, consisting primitively of an anterior dorsal brain, a pair of segmentally ganglionated solid ventral nerve cords with ladderlike transverse connectives between the paired ganglia in each segment, and a pair of connectives joining the brain to the nerve cords. As in the annelids, the trend has been toward elimination of the primitive condition of paired nerve cords through medial fusion. For example, crayfishes, crustaceans in the advanced order Decapoda, have a single ventral nerve cord, in contrast to the ladderlike double ganglionic chain of a tadpole shrimp, a crustacean belonging to the primitive order Notostraca. But arthropods have gone even further than annelids in that the segmental ganglia of the ventral nerve cords also tend to become consolidated longitudinally. This longitudinal consolidation tendency reaches its peak among the crabs and arachnids; in most species of crabs and arachnids, all of the ventral ganglia have fused into a single mass.

The brain of mandibulates consists of three major regions, protocerebrum, deutocerebrum, and tritocerebrum, from anterior to posterior. In contrast, the brain of chelicerates consists only of a protocerebrum and tritocerebrum. The protocerebrum contains the vision centers, association centers for integrating photoreception and movement, and centers that initiate whatever complex behavior they are capable of performing. The deutocerebrum receives impulses from the antennae alone (only the first pair in crustaceans, other mandibulates having only one pair) and contains association centers for these antennae. Chelicerates lack antennae, and correspondingly their brains have no deutocerebrum. The tritocerebrum supplies neurons to the chelicerae of chelicerates and the digestive tract of all arthropods, and it receives nerves from the second pair of crustacean antennae.

Many experiments have been performed that prove arthropods have limited capability to learn through conditioning (training). Honeybees, for example, can be conditioned to search for food at one particular time of day, demonstrating that they are capable both of being trained and of marking off 24-hour periods with precision.

Studies of the instinctive behavior patterns of insects and of their societies provide an equally fascinating body of information. A society is a group of individuals of the same species that cooperate with each other; there is a division of labor. In contrast to most human societies, an insect society consists of a single family. Insect societies have three common traits: individual members cooperate in the care of the young; there is a reproductive division of labor, with nonbreeders working on behalf of the fecund members; and there is an overlap of at least two generations capable of contributing to the colony, with offspring assisting parents. The most highly organized insect societies are found among termites, ants, wasps, and bees. Each member of the society instinctively performs its specific duties; that is, the various chores do not have to be learned.

Honeybees provide a good example of a highly advanced insect society. A honeybee colony comprises three *castes* or different types of individuals. There are, in the colony, one *queen*, usually tens or hundreds of *drones*, and usually 20,000 to 80,000 *workers*. Each caste is morphologically distinct from the other two, another instance of polymorphism, first seen in the cnidarians. The queen lays the eggs, some of which are fertilized and others not. Drones are males, serving only to fertilize the queen. Workers are sterile females, which perform a number of functions, such as rearing the young, building and guarding the hive, and providing food for the entire community. During its lifetime, a worker performs each of these tasks in a definite sequence. The younger workers do the household chores, and the older ones the foraging. The foragers visit flowers to collect nectar (a sweet liquid that is secreted by the plants, which the honeybees convert into honey) and pollen, which is rich in protein. Drones develop from unfertilized eggs, but queens and workers develop from fertilized eggs. Throughout her larval life, about six days, a potential queen is fed large quantities of a special food, *royal jelly*, that is produced by salivary glands of young workers. Larvae that will become drones and workers are fed royal jelly only during the first two days of larval life; for the rest of the larval development, about four more days, they receive only honey and pollen. This difference in diet accounts for the large size and reproductive capacity of the queen. What is in royal jelly that causes a queen, instead of a worker, to develop, is not known.

Queens live three to five years. In contrast, workers have an adult life of only five to six weeks, and drones usually live about two months, seldom as long as four months. However, the drone that mates with the queen dies in the mating process, during which his copulatory organs are torn away. The one queen, by secreting enough of an inhibitory pheromone called *queen substance*, prevents the workers from rearing a second queen. But if the queen dies, or leaves to establish a new colony, or if her fertility decreases, the quantity of queen substance in the hive quickly becomes inadequate to inhibit queen-rearing behavior, and a one- or two-day-old female larva continues to be fed royal jelly throughout her entire larval life to produce a new queen; thus the former queen is replaced.

Honeybee colonies multiply by a process called *swarming*. When a colony becomes overcrowded, the old queen, about 20,000 workers, and sometimes a few drones also, leave the old colony as a dense swarm to establish a new colony. In the old colony, a new queen is then reared from one of the female larvae, replacing her departed mother as queen. When the new queen finally hatches, she undertakes a mating flight, mating with one or more drones, and then returns to the parental hive to be queen. Curiously, honeybee colonies multiply with the parent leaving to establish a new community, whereas among most other animals it is the offspring that leave the parental community to establish a new one.

Locating food is the responsibility of those workers that are functioning as scouts. After a scout has found a rich food source, she returns to the hive and communicates her discovery; shortly thereafter foragers will leave the hive and proceed to the feeding site. They are sent, not led. K. von Frisch discovered that the returning scout informs the others of her discovery of food by performing a dance inside the hive. In the case of the subspecies of honeybee he used in his original studies, if the food is within about 50 meters of the hive the scout performs a *round dance*, in which she circles first to one side, then to the other (Fig. 9-9). The round dance does not reveal the direction the bees should fly to reach the food. It simply informs the bees that food has been discovered in the neighborhood of the hive, and they should go out to look for it. However, if the feeding site is more than about 50 meters away, the round dance becomes a *waggle dance* (Fig. 9-9B). In performing the waggle dance, the scout moves in half circles, first to one side then to the other, with a "straight run" (waggle run) intervening. During this run her body, especially the abdomen, vibrates from side to side. The length of time that the scout consumes in going through each waggle run informs the other workers of the distance to the food. Longer distances are expressed by longer waggle runs; about 0.5 second for 200 meters; about 4.0 seconds for 4,500 meters. Furthermore, if the scout is dancing on a vertical honeycomb, the angle between the path to the food and the position of the sun is shown by the angular

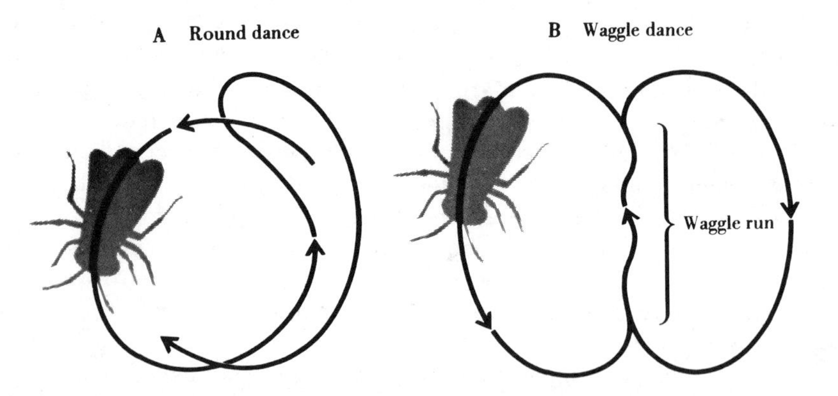

Fig. 9-9 *Diagrammatic representation of the dances of the bee. (A) Round dance; (B) waggle dance. (From Van der Kloot,* Behavior, *Holt, Rinehart, and Winston, 1968. After von Frisch,* Bees: Their Vision, Chemical Sense, and Language. *Copyright © 1950 by Cornell University. Used by permission of Cornell University Press.)*

deflection of the waggle run from the vertical. If the food source lies toward the sun, the straight run is upward. It is downward when away from the sun. If the food is 40´ to the left of the sun's position, the waggle run is 40´ to the left of the vertical, and so forth. However, if the dance occurs on a horizontal surface, then the scout uses the waggle run to point directly toward the food source. Dances are usually performed on a vertical honeycomb, but incoming scouts often begin to dance on the horizontal alighting board of the hive if there are potential forager recruits there. The dances will be performed only if the foraging will be profitable for the colony. Factors that contribute to the profitability are the sugar content of the nectar, the nearness of the food, and the rate at which the flowers replenish their nectar. For example, if the sugar content of the nectar is too low, no dances are performed, presumably because the profit would not justify the energy expended by the foragers. Furthermore, when an initially plentiful food source becomes depleted because the flowers are not able to produce nectar as fast as the bees that were recruited initially are consuming it, the dancing in the hive will cease and no additional foragers will be recruited. Investigators have found that the energy expended on the dance is proportional to the amount of food available. When a newly located food source appears to the scouts to be a profitable one that merits sending out newly recruited foragers, the degree of profitability of collecting from that food source is revealed by the vivacity with which the waggle dances are performed and the length of time they are performed. Although "vivacity" is difficult to quantify, the bees appear to perform the waggle dance more vigorously and for a longer period of time when the profitability is high; when the profitability is low, the waggle dance is sluggish and brief. For example, when the nectar is low in sugar content, but still sweet enough to evoke dancing, the waggle dances are neither vigorous nor prolonged, with the result that fewer workers arrive at the feeding site than would be seen collecting nectar there if the nectar were saturated with sugar. The more ardent the waggle dance, the more persuasive it is, and the greater is the number of bees that will go to the feeding site. Sound is also an essential part of the waggle dance message. During the straight run of the waggle dance, the signaling bee normally buzzes at a frequency of about 250 Hz by vibrating her wings. However, if she should dance without buzzing, other workers will not go to the feeding site the worker has discovered.

COLOR CHANGES Many of the higher crustaceans are capable of color changes that are among the most spectacular in the animal kingdom. The chromatophores of crustaceans are highly branched single cells with a permanently fixed outline, and the pigment migrates into and out of these branches. The common pigments are red, yellow, white, and brown-black. This type of chromatophore contrasts with that found in the cephalopods, where, as described earlier, the chromatophores are multicellular, and the shape of the pigment-containing cell depends upon the tension applied to it. Other than crustaceans, the only arthropods that show color changes by use of chromatophores are a few species of insects.

Fiddler crabs have a solar-day rhythm of color change, which persists even when they are maintained experimentally under constant conditions of illumination, temperature, and background. The crabs darken by day and pale at night. Furthermore, F. A. Brown, Jr., and his students found that a tidal rhythm of color change is superimposed upon the 24-hour solar-day rhythm. The tidal rhythm, which has a period of 12.4 hours, modulates the 24-hour cycle to produce repeating, semilunar monthly (14.8 day) fluctuations in the day-to-day patterns of color change. Fiddler crabs dig burrows for shelter along the shore. As the tide rises, the crabs enter their burrows and sit out the inundation, the burrows becoming covered by the rising water. As the tide then ebbs along the shore, and their burrows become uncovered, the crabs emerge to feed. The phase relationship between the solar-day rhythm and the tidal rhythm depends upon the time of uncovering of the burrows as the water recedes.

ENDOCRINE SYSTEM Arthropods have glands that meet the traditional criteria of an endocrine gland: a circumscribed structure that releases into the blood a hormone that exerts a coordinating effect elsewhere in the body. However, in addition to traditional endocrine glands, arthropods have other organs that release hormones into the blood, but these organs, called *neurohemal* organs, do not fit the traditional pattern. Neurohemal organs store and release hormones (often called *neurohormones*) that are produced by neurons. Neurons that produce hormones are called neurosecretory cells and are morphologically similar to conventional neurons. Both neurosecretory cells and conventional neurons have an axon, Nissl granules, and neurofibrillae, but whereas conventional neurons innervate an organ, such as a muscle, or make synaptic connections with other neurons, neurosecretory cells manufacture neurohormones. The neurosecretory cells of vertebrates, like conventional neurons, also have dendrites; but dendrites may be absent from the neurosecretory cells of invertebrates. Also, like conventional neurons, neurosecretory cells are able to receive signals from other neurons, and to generate and conduct impulses. These impulses generated by neurosecretory cells may trigger the release of their neurohormones. Neurohormones are characteristically synthesized in the cell body of a neuron, migrate down the axon, and are released from the axonal terminal. A neurohemal organ consists of a cluster of axonal terminals from several neurosecretory cells which serves as a storage and release center for the neurohormones and has a close spatial relationship to the circulatory system. However, neurosecretory cells, but not neurohemal organs, occur even where there is no circulatory system, as seen in cnidarians with their nerve-net type of nervous system. The hormones of the hydras and annelids, referred to earlier, are actually neurohormones. Presumably, the neurohormones in a cnidarian, because it lacks a circulatory system, are transported by diffusion. Higher crustaceans have three pairs of neurohemal organs: the *sinus glands*, *postcommissural organs*, and *pericardial organs*. The sinus glands lie in the eyestalks of most stalk-eyed crustaceans (Fig. 9-10), but in some stalk-eyed as well as in eyestalkless species, they are found in the head in proximity to the brain; the *postcommissural organs* lie in the vicinity of the esophagus; and the *pericardial organs* are found in the

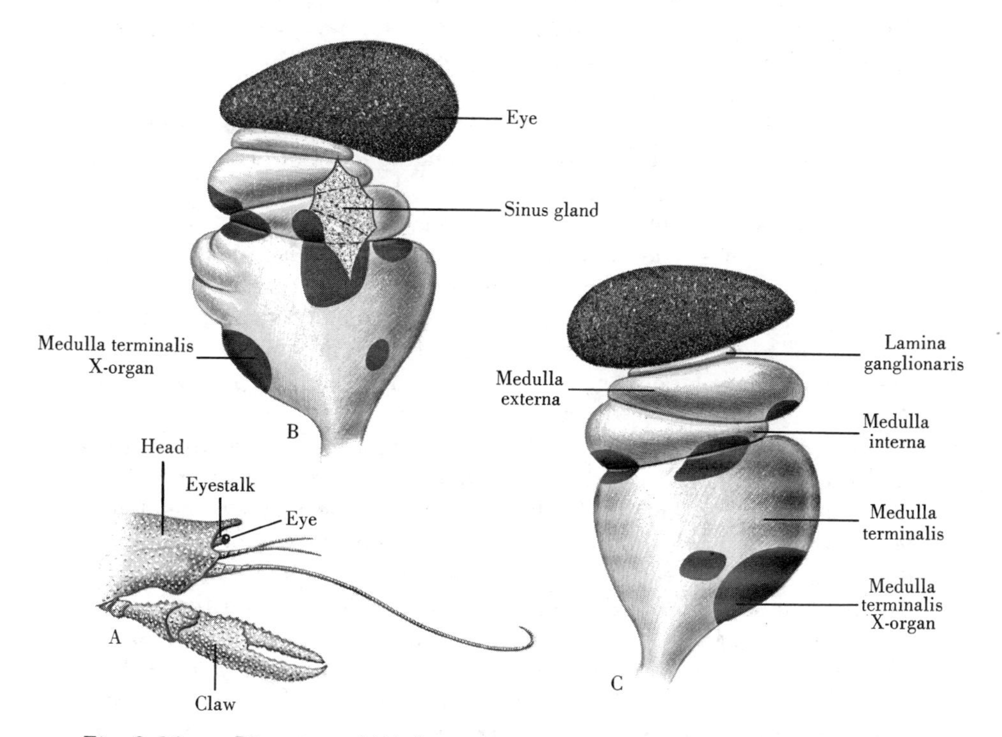

Fig. 9-10 *Diagrams of (A) the anterior end of a crayfish seen from the right side; and (B) dorsal and (C) ventral views of the structures in an eyestalk of the crayfish,* Faxonella clypeata, *showing the location of the sinus gland (stippled area) and the distribution of the cell bodies of groups of neurosecretory cell (areas in black). The lamina ganglionaris, medulla externa, medulla interna, and medulla terminalis are the ganglia that are present in the eyestalks of most stalk-eyed crustaceans. (After Fingerman and Oguro,* Transactions of the American Microscopical Society, *vol. 86, 1967.)*

pericardial cavity. Extracts of the pericardial organs always increase the amplitude of the heartbeat in an experimental situation, and in some crustaceans the frequency of beating increases also. The axonal terminals that constitute a sinus gland originate principally from the group of neurosecretory cell bodies known as the medulla terminalis X-organ. The insect neurohemal organs, about which there is the most information available, are the paired *corpora cardiaca*, which lie in the head posterior to the brain. The neurosecretory axonal terminals in the corpora cardiaca belong to the cells whose cell bodies lie in the *pars intercerebralis* of the protocerebrum. Classically, the nervous system has been considered the coordinating system designed for rapid communication, as when a nerve stimulates a muscle to contract, and the endocrine system has been looked on as being distinct from it and designed for control of long-term processes. Now, however, it appears that the nervous systems of most, if not all, animals are capable of producing hormones and the nervous and endocrine systems are no longer considered distinct entities; the terms "neuroendocrine system" and "neurohormone" are being used more and more frequently.

The color changes of crustaceans that are initiated by visual stimuli (as when an animal changes the color of its integument to match its background) involve hormones; chromatophores of crustaceans are not directly innervated. For several species, there is evidence that migration of their chromatophoric pigments is regulated by pigment-dispersing and pigment-concentrating hormones. The pigment-dispersing hormone makes the pigment more conspicuous, thereby imparting its color to the animal, whereas the pigment-concentrating hormone causes the pigment to "ball up" and consequently contribute little or nothing to the color of the animal. The color change hormones of crustaceans are released from their sinus glands and postcommissural organs.

In crustaceans, as in other arthropods, growth occurs by a series of molts. Change in form and increase in size can occur only when the restricting exoskeleton is shed and before the new exoskeleton is hardened. The term "molting" includes not only the actual shedding of the old exoskeleton (ecdysis) but also the preparations for ecdysis, such as initiation of the formation of the new exoskeleton under the old one prior to ecdysis, and tissue growth and completion of the new exoskeleton after ecdysis. The interval of relative quiescence, during which the physiological processes normally associated with the molting process are substantially absent, is called the *intermolt period.* The molting cycle in crustaceans is regulated by the interaction of two oppositely acting hormones. One, the *molt-inhibiting hormone,* is released from the sinus glands. The second, from ecdysial (molting) glands located in the cephalothorax, is the molt-promoting hormone, *ecdysone.* In fact, ecdysone appears to be the hormone that induces molting activity in all arthropods.

Growth and differentiation in postembryonic insects are regulated by hormones in a very interesting manner. The insect brain produces a neurohormone, the *brain hormone*, which is stored in the corpora cardiaca. Its function is to activate a pair of ecdysial glands in the prothorax, known appropriately as the *prothoracic glands*, which are the source of ecdysone in insects. A third hormone, *juvenile hormone*, is also involved. It is produced by the corpora allata, located near the corpora cardiaca. During the juvenile stages, juvenile hormone suppresses pupal and imaginal differentiation. A *pupa* is a nonfeeding, quiescent stage toward the end of the developmental life of some insects. When a pupa molts it transforms into an *imago* or adult insect. Juvenile hormone ensures that the insect will not develop into an imago prematurely. It allows growth through the actions of the brain hormone and ecdysone, but not maturation. However, juvenile hormone is not released by the last instar, and because of the absence of this hormone, an imago is formed instead of another instar. *Instar* is the name given to the growth stages (that is, before an insect becomes an adult) between molts of all insects.

The giant banded chromosomes in the salivary glands of some larval insects develop localized swellings or puffs following injection of ecdysone. These puffs presumably result from activation by ecdysone of the genes involved in growth and differentiation and may reflect changes in the amount and variety of specific messenger ribonucleic acids. These messengers would then carry into the cytoplasm information that is necessary for the proper alignment of amino acids in the synthesis of specific enzymes that would evoke within these target cells the biochemical changes involved in the molting process.

The sex of an individual is genetically determined, but in a male crustacean the expression of his genetic sex as evidenced by his development into a normal male, both anatomically and functionally, requires a hormone from the androgenic glands. These glands develop to a functional state only in males and are typically attached to the distal end of each sperm duct. The primordia of the androgenic glands appear in young crustaceans of both sexes, but in females they do not develop into functional organs. After removal of the androgenic glands from very young individuals which would otherwise have developed into normal males, their gonads, instead of becoming sperm-producing testes, become egg-producing ovaries. The androgenic glands are responsible for normal differentiation not only of the testes but also of the sperm ducts and male copulatory appendages. Crustacean testes themselves do not seem to be the source of any hormone. Immature genetic females that receive implants of androgenic glands become masculinized; they develop the copulatory appendages characteristic of males and sperm ducts, and their ovaries become transformed into sperm-producing testes. There is no female equivalent of the androgenic gland. But in contrast to the testes of crustaceans which do not appear to produce any hormone, the ovaries of crustaceans secrete a hormone that promotes the development of female structures, such as a brood pouch, that are used for incubating developing eggs.

REPRODUCTION Arthropods, with few exceptions, are dioecious; barnacles are the only large group of monoecious arthropods. In all terrestrial arthropods, fertilization of the eggs occurs within the body of the female, but in aquatic species, fertilization may be external. The sexual dimorphism of some arthropods is striking. For example, male fiddler crabs (Fig. 9-11) have one claw (chela) that is extremely enlarged, in contrast to the two small, equal claws of the female.

Arthropods typically have centrolecithal eggs with superficial cleavage. There are two types of postembryonic development in arthropods, *ametabolous* and *metabolous*. If the animal, after hatching from the egg, closely resembles a diminutive adult and attains the adult form by a gradual process in which at every molt the external changes (other than increase in size) are extremely slight, then the animal is referred to as being ametabolous. Most arthropods, however, are metabolous; that is, there is a distinct change of body form during growth and development. Such a change of body form is called a *metamorphosis*.

Crustaceans characteristically have early in their development a free-swimming nauplius larva (referred to earlier), which has an unsegmented body, the single medial eye mentioned earlier, and three pairs of appendages, namely two pairs of antennae and one pair of mandibles (Fig. 9-12). As development progresses, the young crustaceans go through a series of molts, leaving the nauplius larva stage behind them, while gradually acquiring a segmented trunk and the full complement of

Fig. 9-11 *Male (top) and female (bottom) specimens of the fiddler crab,* Uca pugilator. *Note the enlarged chela of the male. (Courtesy of Ward's Natural Science Establishment, Inc.)*

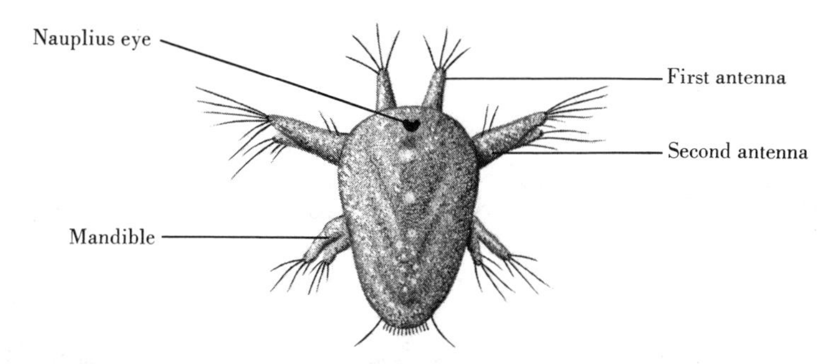

Fig. 9-12 *A nauplius larva of* Cyclops fuscus. *(After Green,* A Biology of Crustacea. *Used by permission of Quadrangle Books, Inc., 1961.)*

appendages, eventually taking on the adult form. Even the true crabs, which as adults have a greatly reduced abdomen bent under the cephalothorax, go through a developmental stage in which the abdomen is relatively large and not turned under.

Apterygote insects, such as the silverfish, are ametabolous. In contrast, the pterygote insects are all metabolous. Many authors, as herein, distinguish two types of metabolous development among these pterygotes: hemimetabolous and holometabolous (Table 9-1, Fig. 9-13).

Hemimetabolous development is shown by 15 orders of insects, including orthopterans and hemipterans. In hemimetabolous development, the newly hatched insects resemble the adults in many ways, important differences being the small size, underdeveloped wings, immature genitalia, and different body proportions of the young. Hemimetabolous development is characterized by a progressive transition to the adult, during which the instars come to resemble their parents more and more with each molt. The instars of hemimetabolous insects are often specifically called nymphs. The final nymphal molt, leading to the production of the adult, is associated with the most dramatic changes and is consequently referred to as the metamorphic molt. The eggs of holometabolous insects (nine orders), such as lepidopterans, dipterans, and hymenopterans, hatch into wormlike larvae, often called maggots or caterpillars, that have very little resemblance to the adult. At each molt, the larvae show very little developmental progress other than an increase in size. The molt of the last larval stage is associated with the formation of the pupa. The pupal stage may be prolonged because of an extended period of quiescence *(diapause)*. The major difference, then, between hemimetabolous and holometabolous development is in the timing of imaginal differentiation. In hemimetabolous species, it tends to be spread out over the entire developmental period, but in holometabolous species it is largely confined to the transformation of the last larval stage to the pupa and of the pupa to the imago.

Some biologists recognize three types of metabolous development in insects instead of the two adopted here. Even with the three

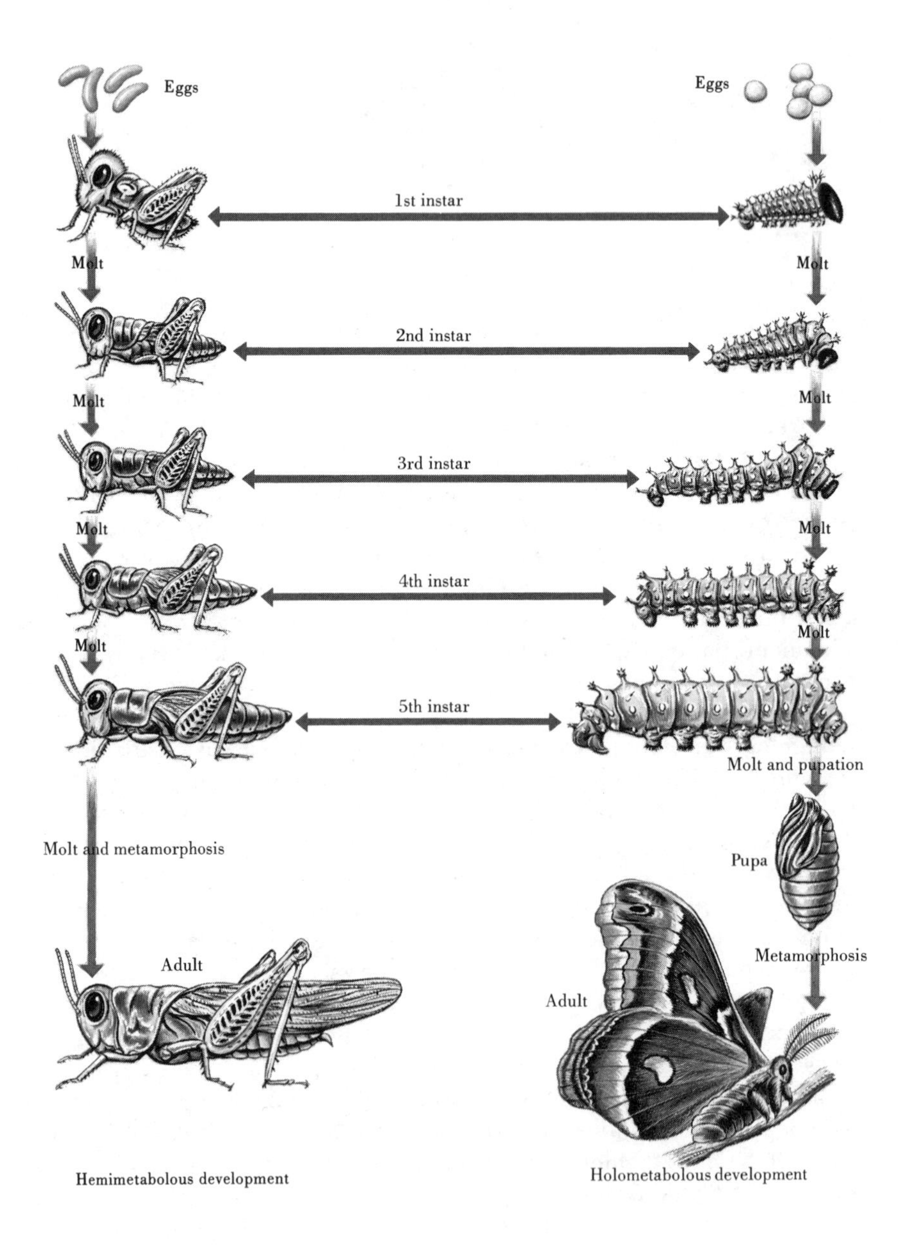

Fig. 9-13 *The hemimetabolous development of a grasshopper and the holometabolous development of a moth. (After Turner,* General Endocrinology, *3rd ed. W. B. Saunders Company, 1960.)*

types, the holometabolous type remains the same as described herein; but the hemimetabolous pattern is divided into two types, which are named *hemimetabolous* and *paurometabolous.* Both the hemimetabolous and the paurometabolous insects undergo gradual metamorphosis; according to these authors, hemimetabolous insects are those, such as the dragonflies, that have an aquatic nymphal stage, whereas the paurometabolous insects are those, such as the grasshoppers, that have terrestrial nymphs. The aquatic nymphs are often more specifically called *naiads.* Naiads show adaptations to aquatic life. For example, presumably because wings would interfere with swimming, the wings of naiads are generally slower growing than those of terrestrial nymphs.

There are about ten times as many species of holometabolous insects as of other types, which suggests there is a selective advantage in having this type of development. The advantage seems to arise from the fact that, as is untrue of other insects, the developmental stages of holometabolous insects are so strikingly different from the adults that the larvae and adults almost never compete with each other for habitats and food. The larvae of holometabolous insects are highly adapted for growing, whereas the adults are specialized for reproduction and dispersal.

In spite of the great diversity in form among insects, they appear to fall into four major evolutionary grades. The most primitive are the wingless insects, the apterygotes, which are ametabolous. The pterygotes or winged insects, all of which are metabolous, form the three more advanced groups. The second major type consists of those insects, such as dragonflies, that cannot fold their wings against their bodies. This inability is a disadvantage in that the insect requires a greater surface area for landing than does a flying insect that can fold its wings against its body. The third level of organization is occupied by those insects, such as grasshoppers, that can fold their wings against their bodies, but have hemimetabolous development. Those insects having holometabolous development, the fourth group, are the most advanced; all of them can also fold their wings against their bodies.

PHYLOGENY The current consensus among biologists is that arthropods evolved either directly from a primitive polychaete stock or from a segmented ancestral form that gave rise to both the polychaetes and arthropods. There is considerable evidence for a close relationship between annelids and arthropods. As examples, both phyla have a primary segmentation, but it has been secondarily reduced in a number of arthropods. In polychaetes and arthropods, the primitive arrangement of appendages is one pair on every segment. The nervous systems in both phyla have the same basic plan:

an anterior dorsal brain, primitively two ventral nerve cords with paired ganglia in each segment, and connectives joining the brain to these solid ventral nerve cords. Finally, as mentioned earlier, the body of the ancestral arthropod was probably composed of a head and segmented trunk, which is typical of polychaetes.

The phylum Onychophora, which is discussed further in the next section of this chapter, is sometimes referred to as the "missing link" between the annelids and arthropods because onychophorans have characteristics of both. However, although the onychophorans do provide support for the theory that annelids and arthropods are phylogenetically related, the use of the term "missing link" may not be valid for the reason that onychophorans were probably not a direct link between them.

Although there is general agreement on the origin of the arthropods, there is a considerable difference of opinion among the experts in the field of arthropod phylogeny as to whether arthropods are a monophyletic or polyphyletic group. Also, the phylogenetic interrelationships of the arthropod classes are not agreed upon. An older theory, based primarily on the comparative anatomy of adults, holds that the arthropods are monophyletic, that a single, primitive arthropod stock was the common ancestor of all living arthropods. On the other hand, a more recent theory, based to a large extent on evidence from comparative functional morphology and developmental biology, declares that the arthropods are polyphyletic, three different lines of ancestral arthropods having evolved independently of each other from polychaete stock. Furthermore, in 1973, the main proponent of this polyphyletic theory, S. M. Manton, stated that the differences among living arthropods are so great that they do not constitute a single phylum. Instead, she suggested that they be divided into three phyla, each representing one of the suggested ancestral lines. The three proposed phyla are the Crustacea, the Chelicerata (which would consist of the arachnids, horseshoe crabs, and pycnogonids), and the Uniramia. The phylum Uniramia would include not only the insects, centipedes, millipedes, pauropods, and symphylans, but also the onychophorans. Centipedes, millipedes, pauropods, and symphylans are clearly closely related to each other. The bodies of all are composed of a head and a segmented trunk; many of the segments are leg-bearing. Previously, however, other biologists had concluded that onychophorans represent a separate phylum that arose as an offshoot either of a line that led from the annelids to the arthropods or of the line that led to the arthropods from some ancestral stock that gave rise to both the annelids and arthropods. Recent support for this older, traditional view that onychophorans are not arthropods has come from electron microscopic study of tissues from several invertebrates. The Golgi complex in the cells of all the traditional arthropods, such as the horseshoe crab, the honeybee, and a crayfish, that were tested showed a positive staining reaction with bismuth. In contrast, bismuth-staining of the Golgi complex was not detected in cnidarians, flatworms, a roundworm, annelids, mollusks, or an onychophoran. The name "Uniramia" refers to the fact that the appendages of onychophorans, insects, centipedes, millipedes, pauropods, and symphylans are primitively uniramous. In contrast, the appendages of crustaceans, trilobites, and chelicerates are or probably were originally biramous, and where the uniramous condition occurs in these animals, it appears to have been a secondary modification.

The relationship of the trilobites to the rest of the arthropods is also not agreed upon. It has been suggested that they were (a) ancestral only to the chelicerates, (b) ancestral only to the chelicerates and crustaceans, (c) ancestral to none of the living arthropods, being merely a dead-end group, and (d) ancestral to all living arthropods.

ONYCHOPHORA

The phylum Onychophora [Gr. *onychus*, claw + *phorus*, bearing], as mentioned above, is a significant group because its members exhibit a combination of annelid and arthropod characteristics (Fig. 9-14). The fossil record reveals that the early onychophorans were marine forms, their descendents later moving inland to occupy the humid environments, as in tropical rain forests, where onychophorans occur today. About 70 living species have been described. Onychophorans have the appearance of slugs with legs. As a matter of fact, when first discovered, they were thought to be mollusks. They are bilaterally symmetrical, segmented protostomes with a size range of about 1.5 to 15.0 cm. The only obvious external sign of segmentation is the serial repetition of 14 to 43 pairs of short, unjointed, conical legs. The number of pairs depends upon the species, and also in some species the females have more pairs than do the males. The legs, like those of arthropods, are positioned more ventrally than are the parapodia of annelids. Each leg terminates with a pair of claws, and the number of legs depends upon both the sex and species. Unlike the body of an arthropod, the body of an onychophoran is covered with a thin, flexible cuticle, which, as in annelids, is not divided into separate plates. The head has a pair of arthropodlike mandibles, one pair of antennae, and a small eye at the base of each antenna.

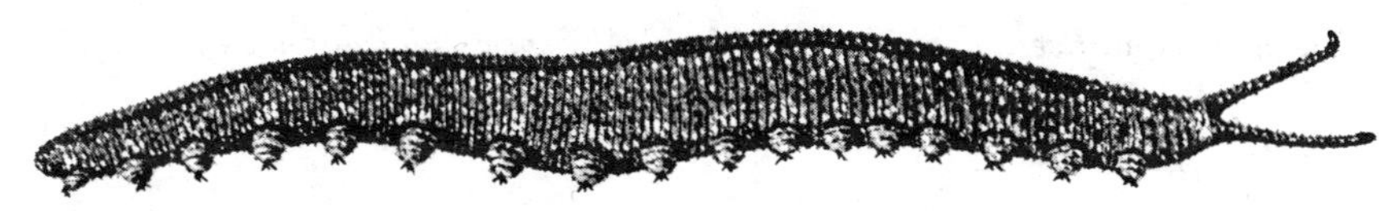

Fig. 9-14 *An onychophoran,* Peripatus *sp. (After a photograph, courtesy of Carolina Biological Supply Company.)*

Internally, the segmented body plan is obvious from the segmentally arranged ganglionic swellings on the paired ventral nerve cords and the segmentally arranged annelidlike metanephridia. The schizocoel of onychophorans is greatly reduced, as in arthropods. The metanephridia open internally into remnants of this coelom. Oxygen is obtained by means of arthropodlike tracheae. The circulatory system is arthropodlike, being an open one with a hemocoel and a heart with ostia. The eggs of onychophorans are centrolecithal and undergo superficial cleavage. No annelid has such an egg (or cleavage pattern) but, as stated earlier, this type of egg and cleavage pattern are typical of arthropods.

Onychophorans have more arthropod than annelid characteristics. If onychophorans comprise a phylum of animals different from arthropods and if onychophorans arose as an offshoot of an ancestral line that led from the annelids to the arthropods, the fact that onychophorans have more arthropod than annelid characteristics indicates that onychophorans arose along this ancestral line closer to the arthropod end of the line than to the annelid end. The recent suggestion, mentioned above, to include onychophorans in a new phylum, the Uniramia, along with such traditional arthropods as the insects has two faults. It downgrades seemingly important differences between onychophorans and the rest of the animals proposed for inclusion in the Uniramia, such as the onychophorans' lack of a jointed exoskeleton, and it emphasizes other aspects of the biology of onychophorans, such as their embryology and the primitive unbranched (uniramous) condition of their appendages. It will be interesting to learn whether this proposed new classification scheme will be accepted by many arthropod experts after they have had sufficient opportunity to scrutinize it.

FURTHER READING

Anderson, D. T., *Embryology and Phylogeny in Annelids and Arthropods.* New York: Pergamon Press, 1973.

Cisne, J. L., "Trilobites and the Origin of Arthropods," *Science*, vol. 186 (1974), p. 13.

Esch, H., "The Evolution of Bee Language," *Scientific American*, vol. 216:4 (1967), p. 96.

Fingerman, M., *The Control of Chromatophores.* New York: Pergamon-Macmillan, 1963.

Frisch, K. von, *Bees, Their Vision, Chemical Senses, and Language* (rev. ed.). Ithaca: Cornell University Press, 1971.

Gupta, A. P., ed., *Arthropod Phylogeny.* New York: Van Nostrand Reinhold, 1979.

Horridge, G. A., "The Compound Eye of Insects," *Scientific American*, 237:1 (1977), p. 108.

Locke, M., and P. Huie, "Bismuth Staining of Golgi Complex Is a Characteristic Arthropod Feature Lacking in Peripatus," *Nature*, vol. 270 (1977), p. 341.

Manton, S. M., "Arthropod Phylogeny—A Modern Synthesis," *Journal of Zoology*, vol. 171 (1973), p. 111.

Manton, S. M., *The Arthropoda: Habits, Functional Morphology, and Evolution*. New York: Oxford University Press, 1977.

Savory, T. H., "Daddy Longlegs," *Scientific American*, vol. 207:4 (1962), p. 119.

Snodgrass, R. E., "Evolution of the Annelida, Onychophora, and Arthropoda," *Smithsonian Miscellaneous Collections*, vol. 97:6 (1938), p. 1.

Tiegs, O. W., and S. M. Manton, "The Evolution of the Arthropoda," *Biological Reviews*, vol. 33 (1958), p. 225.

Warner, G. F., *The Biology of Crabs*. London: Paul Elek (Scientific Books) Ltd., 1977.

Waterman, T. H., ed., *The Physiology of Crustacea, Metabolism and Growth*, vol. I. New York: Academic Press, 1960.

Waterman, T. H., ed., *The Physiology of Crustacea, Sense Organs, Integration, and Behavior*, vol. II. New York: Academic Press, 1961.

Wigglesworth, V. B., *The Principles of Insect Physiology*, 7th ed. London: Chapman and Hall, 1972.

Williams, C. M., "The Metamorphosis of Insects," *Scientific American*, vol. 182:4 (1950), p. 24.

Wilson, E. O., *The Insect Societies*. Cambridge: Belknap Press of Harvard University Press, 1971.

chapter 10

Echinodermata, Bryozoa, Brachiopoda, Chaetognatha, and Hemichordata

ECHINODERMATA

The phylum Echinodermata [Gr. *echino,* spiny + *derma,* skin] includes the sea stars (starfishes), brittle stars, sea urchins, sand dollars, sea cucumbers, and sea lilies. As stated in Chapter 1, the larvae of echinoderms are bilaterally symmetrical. This is a primary characteristic of the phylum. But most adult echinoderms, after metamorphosis, exhibit a secondary pentamerous radial symmetry; that is, the body can ordinarily be divided into five parts arranged around a central axis that runs from the oral to the aboral surface. No other animals begin life as bilaterally symmetrical organisms and metamorphose into radially symmetrical ones. However, tertiarily, some echinoderms have gone on to become bilaterally symmetrical again. The most distinguishing feature of echinoderms is their unique hydraulic system, called the *water vascular system*, which consists of a series of tubes and body-wall appendages. These appendages are known as *podia*. The entire water vascular system is derived from the enterocoel and is fluid-filled. Typically, the body surface has projecting calcareous spines or tubercles, usually covered by a ciliated epidermis, which are part of the skeleton and which are the basis for the name of the phylum.

All members of this phylum are unsegmented and have no excretory organs. Echinoderms are largely bottom-dwelling. None of them lives in fresh water. They are found along shorelines and out to great depths in the oceans (6,000 meters). The largest echinoderms, some sea cucumbers, are about 2 meters long. In contrast, some sea stars are only about 1 centimeter wide, including the length of their arms.

CLASSIFICATION The phylum is commonly divided into five classes:

The class Crinoidea consists of the sessile, flowerlike sea lilies (Fig. 1-12A), which are attached by a stalk to the substrate, and the stalkless, free-moving feather stars. Crinoid have a pentamerous body that is drawn out into arms. The primitive number of arms is five, but most crinoids have ten or more. Along each side of every arm is a row of short, jointed appendages which give the arms a feathery appearance. *Antedon bifida* is a common feather star of the west coast of Europe.

The class Holothuroidea consists of the sea cucumbers (Fig. 1-12B). They have no arms and are cucumber-shaped as their common name implies. The body is elongated along its oral–aboral axis. The mouth is surrounded by 10 to 30 tentacles. Sea cucumbers, particularly *Thelenota ananas*, are used as human food in the Indo-Pacific region of the world.

The class Echinoidea consists of the radial or regular echinoids, known as sea urchins (Figs. 1-12D and 10-1), and the irregular or

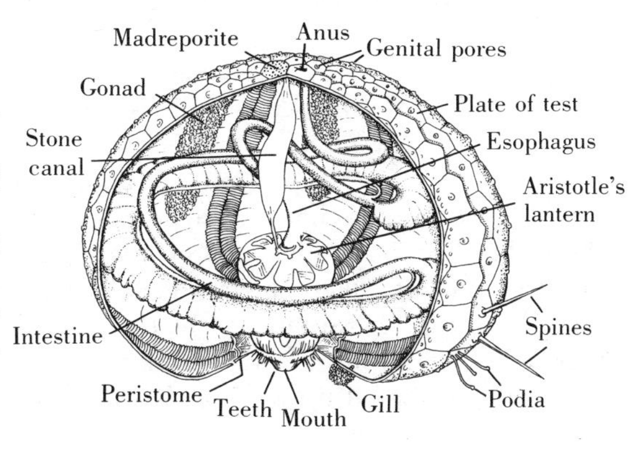

Fig. 10-1 *The structure of a sea urchin. Most of the spines, podia, and pedicellariae have been omitted. (Modified from* Elements of Zoology, *4th ed. by Storer, Usinger, Nybakken, and Stebbins. Copyright 1977, McGraw-Hill Book Company. Used with permission of McGraw-Hill Book Company.)*

bilaterally symmetrical echinoids, the heart urchins and sand dollars. The body wall contains a continuous skeletal structure, the *test*, consisting of closely fitted, immovable calcareous plates. The spines are movable. Arms are absent. Developmental biologists have used the eggs and sperm of the sea urchin *Arbacia punctulata*, found from Cape Cod to Yucatan, in many experiments.

The class Asteroidea is represented by the sea stars (Fig. 1-12C). They are flattened along the oral–aboral axis. Most of them have five arms that radiate from a central disk, but some have more, up to 50. There is no distinct line of demarcation between the central disk and the arms. The sea star, *Asterias forbesi*, commonly found on the eastern coast of the United States from Cape Cod southward, feeds primarily on bivalve mollusks and consequently is a constant problem to oyster fishermen.

The class Ophiuroidea consists of the brittle stars, also called serpent stars (Fig. 1-12E). Ophiuroids typically have five arms (a very few species have six or seven). The arms are sharply demarcated from a central disk and are unbranched in most species of ophiuroids. The brittle star, *Ophioderma brevispina*, is found along the coast of the United States from Massachusetts to Florida.

Some biologists combine the Asteroidea and Ophiuroidea into a single class called the Stelleroidea because all the asteroids and ophiuroids are star-shaped, free-moving, and have arms that jut outward from a central disk—characteristics which exclude the rest of the echinoderms. Herein, however, the older scheme of treating the Asteroidea and Ophiuroidea as separate classes has been retained, as is still done by many other biologists, because the numerous differences between the two groups seem to justify continuing to separate them. Those biologists who lump them into a single class treat the Asteroidea and Ophiuroidea as subclasses of the Stelleroidea.

MORPHOLOGY AND PHYSIOLOGY

ECHINODERM STRUCTURE

The main part of the body of crinoids is a pentamerous structure called the *crown* or *corona*. It consists of a central mass plus the arms. In the more primitive crinoids (the sea lilies) the aboral side of the crown is attached to a stalk that often bears flexible, jointed appendages (*cirri*). Sea lilies are the only living, attached echinoderms, being fixed to the substrate by the stalk. The more advanced crinoids (the feather stars) are stalkless and free-moving. Feather stars usually have cirri, which are attached directly to the aboral surface of the crown and are used to grasp the substrate. Unlike in all other living echinoderms, the oral surface of crinoids is directed upward, which is also the orientation seen in the oldest fossil echinoderms. Holothuroids are elongated and lie sideways. In the rest of the echinoderms, the oral surface is applied to the substrate. Some primitive crinoids have five arms, but in most of the rest of the crinoids, the arms branch immediately after emerging from the central portion of the crown to give a total of 10, and in the remaining few there is repeated branching, sometimes giving a total of 200. The arrangement of the skeletal elements in the arms gives them a jointed appearance. As stated earlier, along each side of every arm is a row of short, jointed appendages; these are called *pinnules*. The mouth and anus are both on the oral surface of the crown with the mouth usually at the center of the crown and the anus usually near the base of one of the arms. From the mouth, five ciliated furrows, the *ambulacral grooves*, extend over the oral surface of the central mass to the arms and then continue along the entire length of the oral surface of the arms and also of most of the pinnules. Those pinnules which lack an ambulacral groove are located at the base of each arm and are called the *oral pinnules*. They are usually longer than the rest of the pinnules. Along the margins of the grooves are podia used by crinoids for feeding. The oral pinnules not only lack an ambulacral groove, but podia also.

A holothuroid has an elongated body with the mouth usually at one end and the anus usually at the other. However, the mouth and anus are sometimes displaced slightly away from their terminal positions. The tentacles surrounding the mouth are modified podia. The podia on the body proper are locomotory. Because of the elongated shape of holothuroids they are forced to lie on one side instead of on the oral or aboral surface, which has led to the differentiation among holothuroids of "dorsal" and "ventral" surfaces, as evidenced by changes that occurred in the distribution and structure of the podia on the body surface. In the primitive species, the podia are restricted to five bands. However, in more advanced species, the podia have spread out of the bands and are scattered over the entire body surface. In the most advanced species, the scattered podia on the upper or dorsal surface have become reduced or even absent, well-developed podia occurring only on the lower or ventral surface. When the podia on the upper (or dorsal) surface became different from those on the lower (or ventral) surface, the holothuroids became bilaterally symmetrical again, an example of tertiary bilateral symmetry. The larvae have a primary bilateral symmetry; primitive holothuroids have a secondary radial symmetry, and the most advanced holothuroids have taken on a tertiary bilateral symmetry.

Echinoids have rounded bodies, which are more or less spherical, as in sea urchins, or greatly flattened along the oral–aboral axis, as in sand dollars. The spines of urchins are elongated, whereas in sand dollars they are short but more numerous. In sea urchins the mouth, in the center of the oral surface, is surrounded by a membranous structure, the *peristome*, that is thickened at its inner edge, forming a lip.

The anus of the sea urchin is in the *periproct,* a membranous area at the aboral pole. In the heart urchins and sand dollars, the mouth has migrated away from the center of the oral surface, and the periproct, including the anus, has moved in the opposite direction on the aboral surface, producing bilaterally symmetrical organisms. Scattered over the general body surface of all echinoids are *pedicellariae,* minute appendages that have a head composed of two to five movable jaws. Asteroids are the only other echinoderms that have pedicellariae (Fig. 10-2). These are used for defense, capturing small prey, and cleaning the body surface. When used for cleaning the body surface, pedicellariae crush debris into small particles, which are then swept away by the ciliated epidermis covering their body.

The asteroid mouth is in the center of the oral surface, and an ambulacral groove containing two or four rows of podia extends from the mouth to the tip of each arm. An anus is lacking in some asteroids. When it is present it is inconspicuous, and lies on the aboral surface at, or close to, the center of the disk. Short spines, pedicellariae, and *papulae* (tiny fingerlike projections of the coelom that are used for oxygen uptake) are found on the asteroid body surface. Papulae occur only in asteroids.

Ophiuroids resemble asteroids; the most obvious difference between them is that the arms of ophiuroids are more clearly set off from the flattened central disk than are the arms of asteroids. Ophiuroids have received the common names "brittle star" because of the tendency of their arms to break off when handled, and "serpent star" because their arms are also sinuous. Relative to the diameter of the central disk, the arms of ophiuroids are much longer than those of asteroids. Ophiuroid arms are protected on all sides by four longitudinal rows of skeletal calcareous plates, which give the arms a jointed appearance. There are no ambulacral grooves; the podia simply project outward along the arms between neighboring calcareous plates. A mouth is in the center of the oral surface, and no member of this class has an anal opening.

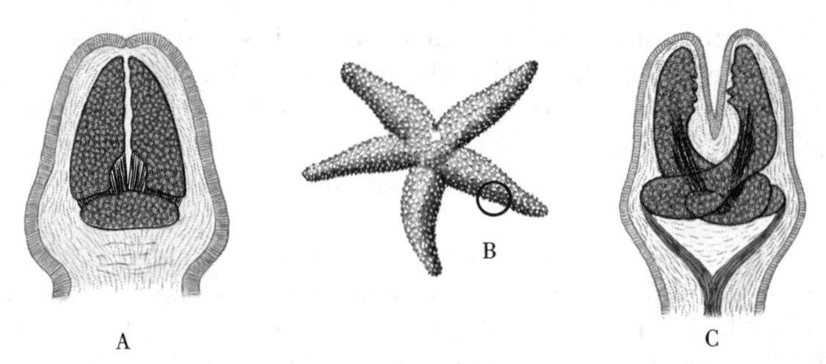

Fig. 10-2 *(A) Straight and (C) crossed types of pedicellariae found in (B) the sea star,* Asterias. *(After Hyman,* The Invertebrates, *Vol. IV. Copyright © 1955, McGraw-Hill Book Company. Used by permission.)*

The echinoderm has an internal skeleton, termed an *endoskeleton*. The portion of the skeleton embedded in the dermis of the body wall consists either of small separated pieces (*ossicles*), or of closely fitted plates forming a firm test. Only an echinoid has a test. Ossicles vary in size and shape from class to class, being smallest in the holothuroids, where they are microscopic. The projecting spines, which, as mentioned earlier, are also part of the skeleton and covered with epidermis, are either protuberances that are extensions of dermal ossicles or separate entities perched on underlying ossicles.

As was noted above, the water vascular system is unique to echinoderms (Fig. 10-3). It appears to have evolved originally as a device for capturing food but is now used additionally for other functions. Crinoids have retained not only the primitive orientation of echinoderms (the upward-directed oral surface) but also (in using their podia to obtain food) what seems to have been the original function of the water vascular system. In crinoids, this system is constructed in the following manner. The mouth is encircled by a canal (the *ring canal*), from which five *radial canals* originate, each extending and branching, if there are more than five arms, into the arms just below the ambulacral grooves. Branching radial canals ultimately enter each pinnule, including those pinnules that lack an ambulacral groove. Finally, the radial canals are connected to the hollow podia by *lateral canals*. The oral pinnules lack not only ambulacral grooves and podia, but also lateral canals. The ring canal of crinoids also gives rise to about 150 short tubes (*stone canals*), the free ends of which open directly into the coelom. The water vascular system in each of the other classes follows the general plan outlined for the crinoids, differences being principally in the number and site of termination of the stone canals. In all echinoderms, except crinoids, each stone canal is connected to a *madreporite*. A madreporite is a porous plate of the exoskeleton through which fluid could conceivably enter the water vascular system, and when present, it is always connected to a stone canal. Furthermore, only one stone canal

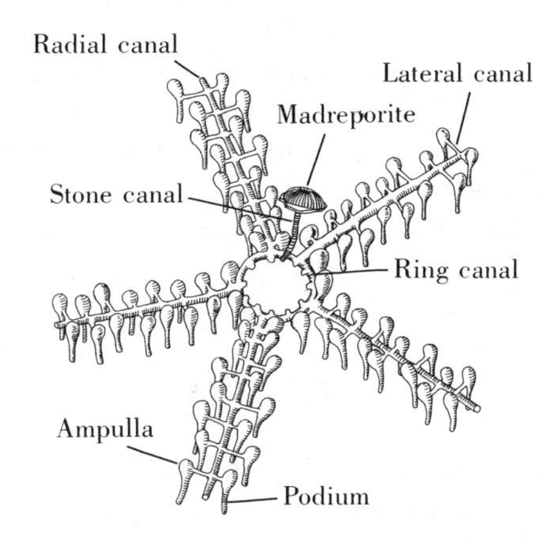

Fig. 10-3 *The water vascular system of an asteroid. (After Coe,* Echinoderms of Connecticut, *Connecticut Geological and Natural History Survey, 1912. Used by permission.)*

terminates in each madreporite. In some of the holothuroids and the ophiuroids, more than one madreporite is present, but in the other echinoderm species, there is only one. Holothuroids have up to about 30 madreporites, ophiuroids one or five, but one is the usual number in both classes.

The stone canals connect to the body wall, always by means of a madreporite in all echinoderms except the crinoids and most holothuroids. In most holothuroids, the madreporite and stone canal have lost connection with the body wall and simply hang free in the coelom; coelomic fluid presumably enters the water vascular systems of those holothuroids in which there is a madreporite suspended in the coelom, and in crinoids. In those few holothuroids with a madreporite in the body wall, it lies on the animal's upper surface. Whenever a holothuroid has more than one madreporite, they are always in the coelom, no holothuroid having more than one madreporite in its body wall. In ophiuroids, each madreporite is on the oral surface of the central disk, near the mouth. In echinoids, one of the skeletal plates around the periproct serves as the madreporite. The asteroid madreporite, a circular, grooved plate, is readily seen on the aboral surface of the central disk, located between the bases of two arms. When the madreporite is on the body surface, it would seem that the fluid in the water vascular system is water that has been pumped in from the outside through this porous plate, by ciliary activity. However, recent studies of the fluid in asteroid podia revealed that it was nearly identical to the external sea water, except that the podial fluid had about 60% more potassium and was consequently slightly hyperosmotic to the external sea water. No evidence could be obtained showing that any of the fluid in the podia had entered the system *via* the madreporite. Perhaps any water movement that might occur through the madreporite serves only to equalize pressure differentials between the system and the environment. The epidermis of the podia seemed to be, in large measure, responsible for the pumping in of the potassium. The hyperosmotic state in the podia would itself then create a driving force for entry of water into the system through the podia rather than in through the madreporite. Because these studies were done with asteroids, the possibility still exists that in some other echinoderms, sea water does have free access into the water vascular system when the madreporite is located in the body wall. Likewise, coelomic fluid may have free entry into the water vascular systems of crinoids and of those holothuroids in which the madreporite hangs in the coelom.

LOCOMOTION No echinoderm has the ability to move about rapidly. Sea lilies, being attached to the substrate, cannot move from place to place but can bend their stem and arms. A feather star, maintaining its oral surface upward, can crawl on the tips of the arms or swim by raising and lowering its arms in the water. Holothuroids and asteroids crawl on the substrate through use of their podia, whereas echinoids use their podia and spines to move about. Some holothuroids, in addition to crawling on the substrate, also burrow. When burrowing, they not only make use of circular and longitudinal muscles in the body wall, much as earthworms do, but also use their podia as anchoring devices, like setae of earthworms. Alternate contractions of these two groups of muscles enable the holothuroid to dig slowly into the substrate. Podia used in locomotion are sometimes called *tube feet*. Holothuroids will also use the tentacles surrounding the mouth as an aid in burrowing. These tentacles, which are almost always branched, are modified podia and are used to help push aside the substrate as the animal burrows through it. The ophiuroids often use rowing or undulatory movements of their arms to propel themselves over the substrate, and a few ophiuroids are reportedly even able to swim through use of their arms. Ophiuroids move about with little or no aid from their podia. They usually do not move from place to place by use of their podia alone. Their podia are small compared with those of asteroids.

Food gathering, as noted earlier, was probably the original function of the water vascular system, locomotion being a secondarily acquired function for it. The fluid-filled water vascular system functions as a hydraulic system to force water into the podia and extend them. Creation of a hydraulic pressure in the water vascular system for extension of the podia in crinoids and ophiuroids is accomplished by contraction of circular muscles in the canals of the water vascular system; crinoids do not use their podia for locomotion and ophiuroids no more than slightly, if at all. On the other hand, those echinoderms that are highly dependent upon their podia for locomotion (the asteroids, holothuroids, and echinoids) have *ampullae* (Fig. 10-4), which are a more effective device for providing the increased hydraulic pressure and vol-

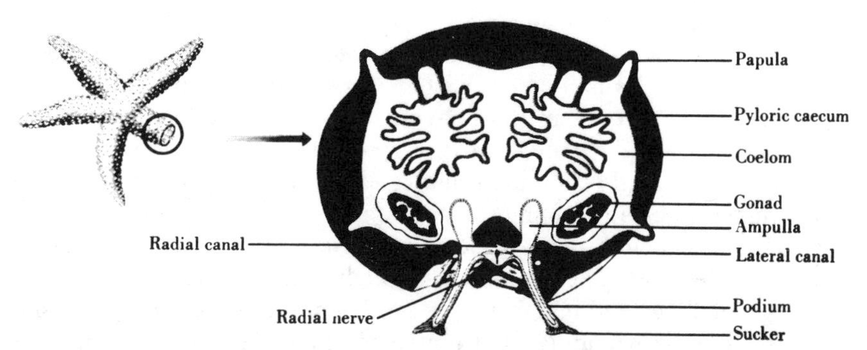

Fig. 10-4 *Cross section through the arm of a sea star. (Modified from Cuénot, from Barnes,* Invertebrate Zoology, *W. B. Saunders Company, 1963.)*

ume required if podia are to be truly useful in locomotion. An ampulla is a small muscular bulb that bulges into the coelom at the end of a lateral canal. A valve in each lateral canal prevents backflow of fluid into the larger canals when the corresponding ampulla contracts. There is one ampulla for each podium, each lateral canal terminating in one podium and one ampulla. When an ampulla contracts and the valve in the corresponding lateral canal closes, fluid is forced into the corresponding podium, elongating it. If a podium contacts the substrate, the center of its terminal disk or sucker is withdrawn, resulting in suction and adhesion; in addition, the tips of the podia secrete an adhesive substance. Podia of crinoids and ophiuroids lack suckers, which are present only where the podia are important in locomotion. Asteroid and echinoid podia push, not pull, the animal forward. During locomotion each podium swings forward in a stepping motion, adheres, and then draws itself backward (the longitudinal muscle on only one side of the podium contracting), thereby applying leverage to the organism, moving it forward. In contrast, holothuroid podia pull the animal forward. Their podia become attached, all the longitudinal muscles to the podia shorten, pulling the holothuroid ahead. This difference between the action of the podia in holothuroids versus the asteroids and echinoids is probably due to the fact that holothuroids are bulkier, heavier-bodied animals.

FEEDING AND DIGESTION Echinoderms have a variety of feeding mechanisms. A crinoid with its oral surface directed upward, which represents the primitive orientation, feeds on particles suspended in the water. Any small particle, living or not, that contacts a crinoid podium is trapped in mucus secreted by the podium and is thrown by a whiplike action of the podium into an ambulacral groove. Cilia in the groove then move the material along the arms and finally into the mouth. This type of feeding, in which suspended particles serve as the food, but (unlike filter feeding) where there is no current-producing mechanism, is called *suspension feeding*.

The digestive tract of a crinoid leads from the mouth into a short esophagus, which directly joins an intestine that opens at the anus, digestion and absorption presumably occurring in the intestine. The other classes show modifications of this basic plan; for example, the stomach in asteroids and ophiuroids is well developed, and an intestine and anus are absent in ophiuroids.

Most holothuroids make use of the tentacles surrounding the mouth for feeding. These tentacles, which secrete mucus, are utilized to trap small food items, living or not. Some species are suspension feeders, simply extending their tentacles into the sea water, collecting suspended food particles that come in contact with and become trapped on the tentacles; in other species, tentacles are swept along the surface of the substrate, collecting food that adheres to them. The tentacles are then periodically drawn one at a time into the mouth, and any adhering food is removed from them. Still other species use their tentacles for feeding in quite a different way: they use them to shovel into their mouths some of the substrate, out of which they digest the nutrients. Burrowing holothuroids do not use their tentacles in feeding. They simply swallow the substrate as they advance through it, digesting whatever nutrients may be present in it. A holothuroid has a small stomach and a long, coiled intestine. The intestine is the main site of digestion and absorption.

Sea urchins and sand dollars have a complex chewing apparatus (Aristotle's lantern) that consists essentially of five large teeth and the muscles necessary to move them. Heart urchins have no Aristotle's lantern. The teeth protrude through the mouth a bit. When feeding, sea urchins are capable of causing the lantern to protrude even further through the mouth than when at rest, but sand dollars cannot. Sea urchins feed by moving over the food, holding it with spines and podia, and chewing it by means of the lantern. They are omnivorous, consuming any organic matter they can obtain, living or dead. Presumably because the lantern of a sand dollar cannot be protruded further through the mouth for chewing on larger food masses, it feeds on small food particles carried to the mouth by the action of cilia and podia. Heart urchins feed on food particles delivered to the mouth through the cooperative action of podia and spines. An echinoid has a long intestine but no stomach; digestion and absorption occur in the intestine.

Many asteroids are carnivorous, feeding on both living and dead snails, bivalve mollusks, crustaceans, and even small fishes. The asteroid stomach is well-developed, consisting of an oral cardiac portion and a smaller aboral pyloric portion. A horizontal constriction separates the two portions. Each arm contains a pair of hollow digestive glands, called *pyloric caeca,* which are attached to the pyloric stomach. The caeca not only secrete the digestive enzymes, but also seem to be the main region of absorption of digested food, and, in addition, they probably carry on intracellular digestion of small particles; but digestion is primarily extracellular. Ciliary tracts in the ducts of the pyloric caeca provide for movement of materials into and out of the caeca. The stomach is also involved in absorption of nutrients but apparently not to the extent that the pyloric caeca appear to be. In those sea stars with short, relatively inflexible arms, the prey is swallowed whole; larger-sized undigested material is eliminated through the mouth. In contrast, those sea stars with long flexible arms are able to evert the cardiac stomach through the mouth and engulf the prey. The cardiac stomach may then be retracted, immediately bringing in the prey, or digestion may proceed with the cardiac stomach still outside. If digestion has been going on outside the body, when the digestible parts of the prey have been reduced to a thick broth, the broth is taken into the body; stomach muscles then contract, retracting the cardiac stomach, and undigestible matter, such as clam shells, is left behind.

Ophiuroids have a simple digestive system—lacking, as stated above, an intestine and anus. Unlike asteroids, they have no pyloric caeca. Some ophiuroids feed on bottom detritus and small dead or living animals. Larger material is pushed into the mouth by the long arms, smaller particles by the podia. Other ophiuroids are suspension feeders. They wave one or more arms through the water, and suspended food particles become trapped in mucus on the arms. The trapped food particles are then moved by cilia and podia to the mouth. The ophiuroid digestive tract is very simple. A short esophagus leads to a saclike stomach where digestion and absorption occur.

CIRCULATION The fluid-filled coelom provides the main means of internal transport, and movement of the coelomic fluid is produced by the ciliated lining of the coelom. However, an open circulatory system, called the *hemal system,* is also present; but its role in internal transport is secondary to that of the coelom. The hemal system consists of a series of interconnecting channels running through the connective tissue of the body and the axial complex. The axial complex consists of a spongy portion, called the axial gland, and a small pouch in the wall of the coelom, called the dorsal sac. The dorsal sac is contractile, and part of the axial gland is lodged in it. Hemal channels enter and leave the axial gland. Contraction of the dorsal sac exerts pressure on the portion of the axial gland it encloses, causing the hemal fluid (blood) to circulate. Consequently, the axial complex is sometimes called the echinoderm heart. The hemal channels are essentially elongated blood sinuses and lacunae, inasmuch as they lack distinct walls of their own. It has been suggested that the main function of the hemal system is probably to distribute nutrients absorbed by the digestive organs, because the hemal channels that course through the organs of the digestive system are especially large and numerous. The coelomic and hemal fluids of holothuroids have cells that contain hemoglobin, but no other echinoderm has an oxygen transport pigment.

RESPIRATION Echinoderms do not have highly developed respiratory mechanisms, in keeping with the inability of those which can move about to do so rapidly. (Sea lilies, as stated, are attached to the substrate.) The characteristic, relatively inactive mode of existence of echinoderms makes so small a metabolic demand that their poorly developed respiratory system suffices. Crinoids simply make use of the thin parts of the body surface, especially the podia, to obtain oxygen. They have no special respiratory structures.

The rest of the echinoderms not only make some use of the thin parts of the body surface for oxygen uptake, but, in addition, have more specialized respiratory mechanisms. Most holothuroids possess a pair of *respiratory trees* for the uptake of oxygen. A respiratory tree is a highly branched tube that lies in the coelom and that originates in an expanded terminal portion of the intestine called the *cloaca*. The respiratory tree is filled with oxygen-laden water by a pumping action of the cloaca. Oxygen-depleted water is then forced out by contraction of the tubules of the respiratory tree. Holothuroids that do not have respiratory trees obtain oxygen through the general body surface. Sea urchins have gills, which are thin-walled outpouchings of the body wall, located in the region of the mouth; heart urchins and sand dollars have leaflike podia on their aboral surface for taking up oxygen. Asteroids obtain oxygen by diffusion through the surfaces of the papulae and the podia. The nonpodial respiratory adaptations, gills and papulae, are thin-walled evaginations of the body surface that contain coelomic fluid. These surface amplifications bring the coelomic fluid closer to the water outside the animal, facilitating respiratory exchange. In ophiuroids, to each side of each arm base on the oral surface of the central disk there is a slit that opens into a bursa (saclike invagination) in which respiratory exchange occurs. Water is brought into the bursae through the action of cilia and in many cases by a pumping action of the central disk.

SENSE ORGANS All echinoderms appear to have photosensitive nerve endings scattered throughout the epidermis. Most echinoderms show a negative response (avoidance reaction) to light, although a few react positively. In addition to these photosensitive nerve endings, asteroids have a pigmented photosensitive spot at the tip of each arm consisting of an aggregate of ocelli, and a few holothuroids have a cluster of pigmented photoreceptor cells at the base of each tentacle. Some sea urchins show an interesting shadow reflex: when an object casts a shadow over them, their spines become erect, presumably a defense mechanism. Some sea urchins and holothuroids have statocysts that enable these animals to orient themselves with respect to gravity.

EXCRETION Echinoderms, as stated earlier, have no excretory organs. Soluble wastes are eliminated by diffusion through the body surface. Ammonia is the main nitrogenous waste product of protein catabolism. The fluids in the coelom and hemal system of these animals are always isosmotic with the external environment because echinoderms have no means of regulating the osmotic concentration of the fluids that bathe their internal organs. Consequently, an echinoderm that invaded an area of overly diluted sea water would not survive the internal dilution that would occur. Because of their inability to osmoregulate, echinoderms are restricted to a marine environment or to one where, because of the influx of fresh water, the salinity of the medium is only somewhat less than that of full-strength sea water.

NERVOUS SYSTEM The echinoderm nervous system consists essentially of a circumoral nerve ring that gives off a radial nerve for each ambulacral area or arm, and the radial nerves in turn connect with a peripheral nerve net in the podia, epidermis, and if ampullae and pedicellariae are present, in them also. Echinoderms have nothing comparable to a brain. The nervous system seems to be in transition from a nerve net, where the control is peripheral, to a system where the control is central. The nerves regulating locomotion illustrate very well the dual control (peripheral and central) of the activities of echinoderms. For example, cutting a radial nerve of a sea star at the base of an arm destroys the coordination of that arm relative to the other arms; the podia in the arm with the severed nerve continue to step in coordination with each other but not in harmony with the podia of the other arms.

REPRODUCTION Echinoderms are, with few exceptions, dioecious. Sea urchin eggs have been used in many basic studies of development; the experiments of H. Driesch (1892), for example, demonstrated indeterminate cleavage for the first time. Echinoderms have a number of different larval types, but the *bipinnaria larva* of asteroids (Fig. 10-5) is of special significance because it bears a very strong resemblance to the *tornaria larva* of hemichordates, a phylum whose characteristics will be described later in this chapter. The resemblance is, in fact, so strong that for several years after the tornaria larva was discovered, it was considered to be the larva of an echinoderm. This similarity of larval types is regarded as evidence for a close phylogenetic relationship between these two phyla.

With most echinoderms, the eggs and sperm are shed directly into the sea, where fertilization occurs. However, an interesting exception, coelomic incubation, was alluded to in the discussion of the coelom in Chapter 1. The eggs of the sea cucumber, *Thyone rubra*, pass

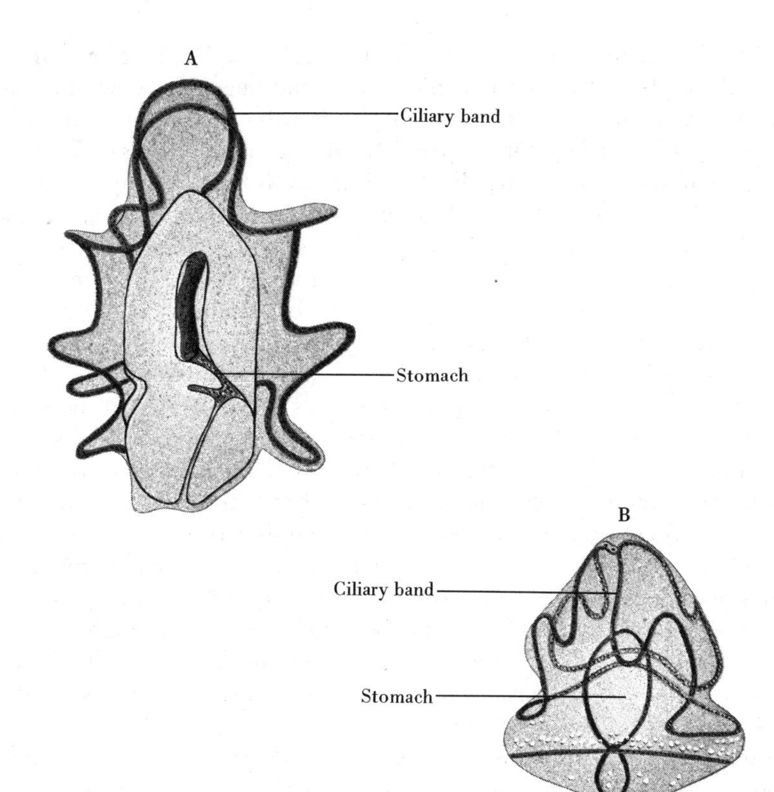

Fig. 10-5 *(A) The bipinnaria larva of the asteroid,* Astropecten auranciacus *and (B) the tornaria larva of the hemichordate,* Balanoglossus clavigerus. [*(A) After Hyman,* The Invertebrates, *Vol. IV. Copyright © 1955, McGraw-Hill Book Company. Used by permission. (B) After Hyman,* The Invertebrates, *Vol. V. Copyright © 1959, McGraw-Hill Book Company. Used by permission.*]

from the ovary into the coelom, where they are fertilized. How the sperm reach the eggs is not known. Some asteroids and ophiuroids can reproduce asexually also; the central disk divides in two and each half subsequently regenerates the missing parts.

The radial nerves of asteroids and of at least one species of echinoid contain a neurohormone called the *gonad-stimulating substance*. It induces release of a second substance, 1-methyladenine, from within the gonad. The shedding of eggs and sperm is induced by 1-methyladenine.

PHYLOGENY It seems clear that echinoderms arose from a bilaterally symmetrical, unattached ancestor, because of the bilateralism of their larvae. But how can the secondary radial symmetry in the adults be explained? It is known from studies of fossils that the early echinoderms had become attached to the substrate by means of a stalk originating on the aboral surface as in present-day sea lilies (the most primitive living echinoderms). Furthermore, the oldest stalked fossils had bilaterally symmetrical bodies, whereas the more recent stalked fossils had bodies that were either imperfectly radially symmetrical, which suggests that they were in the process of evolving from bilaterally to radially symmetrical animals, or bodies that had already become truly radially symmetrical. These attached bilaterally symmetrical fossils are the most primitive echinoderms known. Bilaterally symmetrical animals are not well adapted for a sessile existence. They have their sense organs concentrated at the head end, whereas the advantageous arrangement of the sensory cells in a sessile individual is an even (radial) distribution around the body. With a radial arrangement of its sensory cells, a sessile animal is adequately prepared to receive, equally well, stimuli coming from all directions. Consequently, it would appear that radial symmetry offered an adaptive advantage during the evolution of modern echinoderms. The advantage of radial symmetry to sessile animals was first discussed in Chapter 4. After the radial condition appeared, most echinoderms lost their attachment to the substrate but retained their secondarily acquired radial symmetry; the oral surface then became applied to the substrate, as in sea stars and sea urchins. When these echinoderms became free-moving again, some of them, some sea cucumbers for example, took on a bilaterally symmetrical condition again. Although the radial symmetry that appeared in this phylum can be explained easily as an adaptive mechanism, the reason why pentamerism became established among echinoderms is not so evident. However, because, according to the fossil record, echinoderms adopted the pentamerous condition very early in their history, there must be some advantage to retaining it. One hypothesis of the origin of pentamerism suggests that it arose in conjunction with a skeleton, such as the test of sea urchins. According to this hypothesis a skeleton which consisted of five plates would have allowed for radial symmetry in an echinoderm's test yet would have kept the sutures between the skeletal plates to a minimum number, the sutures being planes of weakness. With five equal-sized skeletal plates forming the circumference of the test, no two sutures would have been opposite each other, thereby producing a stronger skeleton than there would have been with an even number of plates, where the sutures would have been opposite each other. Furthermore, the skeleton is stronger when the number of skeletal plates is small. Five is the smallest number of plates around the circumference that could have produced a genuinely radially symmetrical animal. A more recent hypothesis of the origin of pentamerism emphasizes the five arms of an echinoderm, such as a typical sea star, rather than skeletal structure. It suggests that an ancestral echinoderm lived in flowing water and that pentamerism was essential for survival in that habitat. One arm would need to be extended directly upstream to detect the approach of danger. A pair of arms, one on each side of the upstream arm, would be required for capturing food carried downstream by the current. If only one arm were directed downstream for detection of danger from that direction, it would interfere with the current carrying away waste products. Consequently, two arms are required downstream with the anus between them, giving a total of five arms. The functions of these arms would be carried out best by having the arms at equal angles to their neighbors.

The tentacles that surround the mouth of holothuroids are suggestive of a *lophophore* (a tentaculated food-catching organ) found in three phyla of protostomes, the Phoronida, Bryozoa, and Brachiopoda. The animals of these three phyla are collectively known as the lophophorates. They have traits that reveal their affinities both to the protostomes and to the deuterostomes. It will be recalled from Chapter 1, for example, that all deuterostomes are enterocoelomates, but some brachiopods are the only enterocoelous protostomes, while the rest of the brachiopods are schizocoelomates. However, neither echinoderms nor any other deuterostomes appear to have evolved directly from the lophophorates. More likely the earliest deuterostomes evolved from the same ancestral schizocoelous, protostome stock that gave rise to the lophophorates. This ancestral stock probably had a lophophorelike organ for feeding that evolved not only into the lophophore as we know it today but also into the echinoderm water vascular system. It will be recalled that the original function of the water vascular system is thought to have been feeding. But not all biologists agree that there ever was a splitting-off of deuterostomes from protostomes into the two main lines shown in Fig. 1-3. J. Hadži, for one, who has been referred to earlier in this volume, devised a phylogenetic scheme that is radically different from the views of most biologists. According to his scheme, the phyla evolved, in the main, one after the other along a single straight line, and deuterostomes evolved from protostomes through an annelid that lost its segmentation, the lower deuterostomes being unsegmented. In contrast to J. Hadži's concept, the generally accepted viewpoint is that neither annelids, mollusks, nor arthropods were ancestral to the echinoderms.

BRYOZOA

The phylum Bryozoa [Gr. *bryon*, moss + *zoon*, animal], also known as the Ectoprocta [Gr. *ekto*, outside + *proktos*, anus] consists of animals, commonly called moss animals, that are bilaterally symmetrical, unsegmented, colonial, schizocoelous protostomes (Fig. 10-6). About 4,000 species exist; all but 50 of them are marine. The entoprocts (Table 1-2) were at one time included in this phylum, but the discovery that the entoprocts are pseudocoelomates resulted in their being placed in a new phylum. The moss

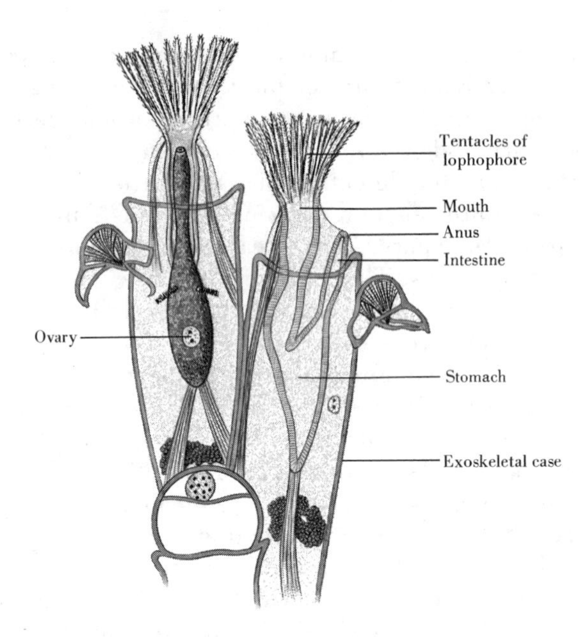

Fig. 10-6 *A bryozoan,* Bugula avicularia. *(After Parker and Haswell,* A Text-Book of Zoology, *6th ed., Vol. I. Macmillan, 1949.)*

animals are permanently fastened in exoskeletal cases of gelatinous, chitinous, or calcareous material. The individual moss animals are usually less than 0.5 mm long. This phylum is one of the three phyla of lophophorates. A *lophophore* is a circular or horseshoe-shaped fold of the body wall that bears ciliated tentacles and embraces the mouth. The tentacles of the lophophore are hollow; their cavity is an extension of the coelom.

BRACHIOPODA The phylum Brachiopoda [Gr. *brachion,* arm + *podos,* foot] is composed of animals that bear an external resemblance to bivalve mollusks because both possess a bivalved shell (Fig. 10-7). However, whereas the valves in mollusks are

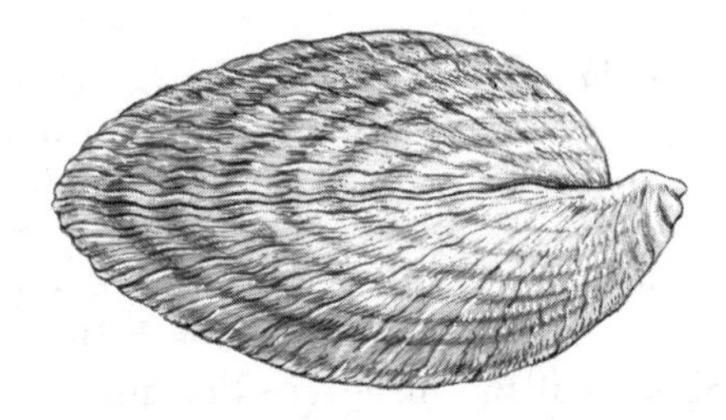

Fig. 10-7 *A brachiopod,* Magellania australis.

left and right, in brachiopods they are dorsal and ventral. There are 260 known living species, all of which are marine. Brachiopods are bilaterally symmetrical lophophorates; the shell may be up to 8 cm long. The ventral valve is larger than the dorsal. These animals show no trace of segmentation. Their common name is "lamp shells."

Brachiopods are an interesting group of protostomes because, as mentioned earlier, although some members of this phylum are schizocoelomates, other members of this phylum are the only protostomes that are enterocoelomates; all other enterocoelous organisms are deuterostomes. Because both the ectoprocts and brachiopods have a lophophore, they would appear to be closely related to each other.

CHAETOGNATHA The phylum Chaetognatha [Gr. *chaeton*, bristle + *gnathos*, jaw] consists of transparent, bilaterally symmetrical, unsegmented, enterocoelous marine animals which are usually about 3 cm long (Fig. 10-8). They are com-

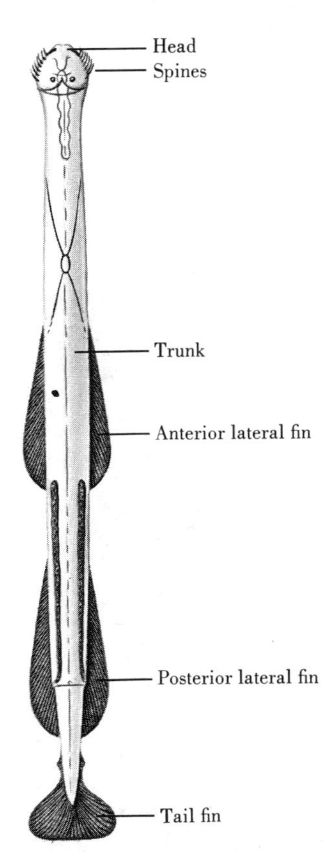

Fig. 10-8 *A chaetognath,* Sagitta elegans. *(After Ritter-Zahony.)*

monly called "arrow worms." Chaetognaths are highly predacious carnivores. Only about 50 living species comprise this phylum. The body is torpedo-shaped, with one or two pairs of lateral horizontal fins, a rounded head armed with spines for capturing prey, and a horizontal tail fin. A thin cuticle covers the body.

Chaetognaths are an evolutionarily significant group because they are the most primitive deuterostomes. This phylum was apparently an early offshoot from the deuterostome line that also gave rise to the echinoderms, hemichordates, and chordates. Chaetognaths are not closely related to the rest of the deuterostomes. For example, there is no larval stage among chaetognaths resembling those of echinoderms and hemichordates. The newly hatched chaetognath closely resembles the adult; there is no metamorphosis.

HEMICHORDATA The phylum Hemichordata [Gr. *hemi*, half + *L. chorda*, cord] consists of two classes; the Pterobranchia (pterobranchs) (Fig. 10-9) and the Enteropneusta (acorn worms) (Fig. 10-10), the pterobranchs being the less advanced

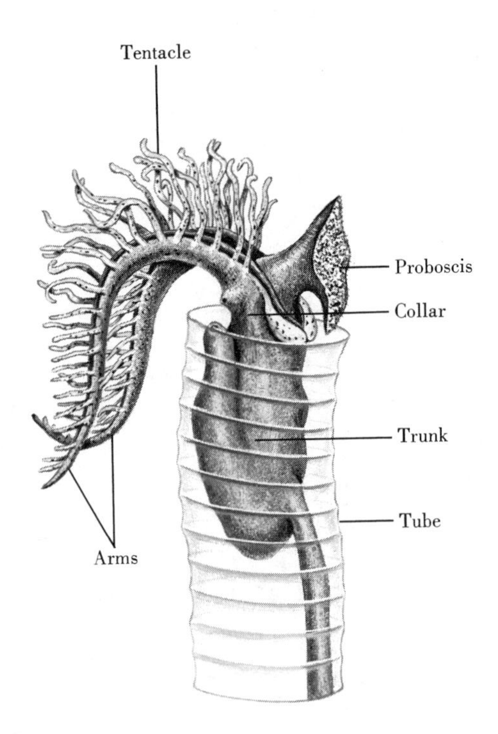

Fig. 10-9 *A pterobranch,* Rhabdopleura *sp. (After Dawydoff, after Delage and Hérouard, in* Traité de Zoologie, *Vol. XI, P.-P. Grassé, ed. Masson et Cie., 1948.)*

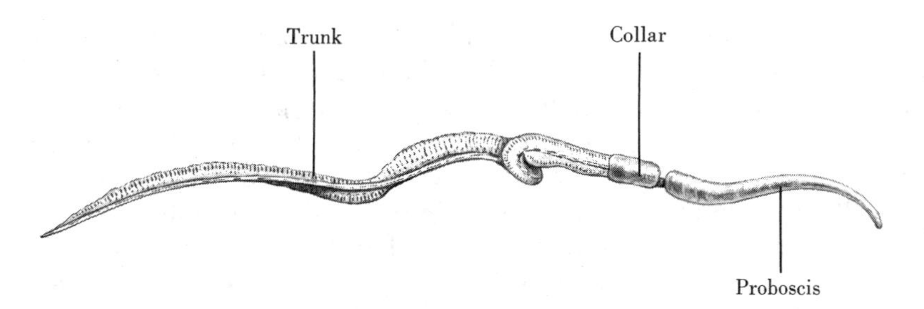

Fig. 10-10 *An acorn worm,* Saccoglossus kowalevski. [*After Hyman,* The Invertebrates, *Vol. V. Copyright © 1959, McGraw-Hill Book Company. Used by permission.*]

class. The majority of the 80 species in this phylum of enterocoelomates are acorn worms, so called because the shape of the proboscis suggests an acorn. Hemichordates are unsegmented, bilaterally symmetrical, wormlike, marine deuterostomes. The high probability that the hemichordates are phylogenetically closely related to echinoderms because of the similarity of their larval forms has been mentioned earlier in this chapter. The body of hemichordates is divisible into three regions: proboscis, collar, and trunk. Almost all pterobranchs live in tubes that they secrete, forming colonies in which the individuals are connected together or aggregates in which the members are independent of each other. In contrast, acorn worms are always solitary and never live in tubes, either burrowing in the substrate or living under stones and shells. The most prominent feature of pterobranchs is the tentaculated arms which are attached to the collar. These arms and tentacles are hollow, containing extensions of the coelom, and were probably phylogenetically derived from a lophophorelike organ, just as was suggested earlier in this chapter for the water vascular system. However, acorn worms lack these tentaculated arms. Presumably, such tentaculated arms are a primitive hemichordate feature that the acorn worms lost. Pterobranchs feed by using their tentaculated arms to capture suspended particles, the food becoming trapped in mucus secreted by these tentaculated arms themselves. Some species of acorn worms feed by ingesting the substrate through which they burrow, digesting the organic matter, whereas other acorn worms are suspension feeders; food coming in contact with the proboscis becomes trapped in mucus thereon and is carried to the mouth by ciliary action. The pharynx of hemichordates has paired slits that open externally on the surface of the anterior portion of the trunk and internally inside the pharynx. These slits, called pharyngeal slits or gill slits (although in this phylum there are no gills associated with them), are like those found in the phylum Chordata, the next phylum to be discussed. In hemichordates, a water current, created solely by ciliary action, enters through the mouth and leaves through these slits. The main role of this water current in hemichordates now appears to be to provide a fresh supply of oxygen and to carry away carbon dioxide. Probably, the original function of pharyngeal slits in hemichordates and chordates was to provide an exit for a water current that brought in suspended particulate food, and their role in respiration apparently evolved only secondarily. Chordate pharyngeal slits will be discussed in greater detail in the following chapter.

The nervous system of hemichordates consists essentially of a nerve ring at the base of the proboscis, another nerve ring at the junction of the collar and the trunk, a dorsal longitudinal nerve cord, and a ventral longitudinal nerve cord. In some acorn worms, the portion of the dorsal nerve cord in the collar is hollow. Chordates, as will be discussed in more detail in the next chapter, are the only other animals which have pharyngeal slits and a cavity in at least one portion of the nervous system during some part of their life cycle. In fact, the hemichordates and chordates are the only animals which have the characteristics of having pharyngeal slits or a nervous system that has a hollow portion.

An extinct class of marine animals, the graptolites, is known from the fossil record only by tubelike remains. Opinion on what phylum graptolites belong to is divided. Some investigators have decided that these fossilized remains are perisarcs from colonies of hydrozoan cnidarians, but others have concluded that they are tubes of pterobranchs.

FURTHER READING

Anderson, J. M.,"Aspects of Digestive Physiology among Echinoderms," *Proceedings of the XVI International Congress of Zoology,* vol. 3 (1963), p. 124.

Barrington, E. J. W., *The Biology of Hemichordata and Protochordata.* San Francisco: Freeman, 1965.

Binyon, J., *Physiology of Echinoderms.* New York: Pergamon Press, 1972.

Boolootian, R. A., ed., *Physiology of Echinodermata.* New York: Interscience, 1966.

Decker, C. E., and I. B. Gold, "Bithecae, Gonothecae and Nematothecae on Graptoloidea," *Journal of Paleontology,* vol. 31 (1957), p. 1154.

Driesch, H., "The Potency of the First Two Cleavage Cells in Echinoderm Development. Experimental Production of Partial and Double Formations." Reprinted in B. H. Willier and J. M. Oppenheimer, eds., *Foundations of Experimental Embryology.* Englewood Cliffs, N.J.: Prentice-Hall, 1964.

Gutmann, W. F., K. Vogel, and H. Zorn, "Brachiopods: Biomechanical Interdependences Governing their Origin and Phylogeny," *Science,* vol. 199 (1979), p. 890.

Hyman, L. H., *The Invertebrates, Echinodermata,* Vol. IV. New York: McGraw-Hill, 1955.

Hyman, L. H., *The Invertebrates, Smaller Coelomate Groups,* Vol. V. New York: McGraw-Hill, 1959.

Kanatani, H., and H. Shirai, "On the Maturation-Inducing Substance Produced in Starfish Gonad by Neural Substance," *General and Comparative Endocrinology, Supplement 3* (1972), p. 571.

Kozlowski, R., "Les Affinités des Graptolithes," *Biological Reviews*, vol. 22 (1947), p. 93.

Millott, N., "Animal Photosensitivity, with Special Reference to Eyeless Forms," *Endeavour*, vol. 16 (61), (1957), p. 19.

Millott, N., ed., *Echinoderm Biology. Symposia of the Zoological Society of London*, No. 20. New York: Academic Press, 1967.

Nichols, D., *Echinoderms*, rev. ed. London: Hutchinson and Co., 1966.

Prusch, R. D., and F. Whoriskey, "Maintenance of Fluid Volume in the Starfish Water Vascular System," *Nature*, vol. 262 (1976), p. 577.

Stearns, L. W., *Sea Urchin Development: Cellular and Molecular Aspects*. Stroudsburg, Pennsylvania: Dowden, Hutchinson and Ross, 1974.

Stephenson, D. G., "Pentamerism and the Ancestral Echinoderm," *Nature*, vol. 250 (1974), p. 82.

Wilkie, I. C., "Arm Autotomy in Brittlestars (Echinodermata: Ophiuroidea)," *Journal of Zoology*, vol. 186 (1978), p. 311.

chapter 11

Chordata—Lower or Invertebrate Chordates

The phylum Chordata [L. *chorda,* cord], to which man belongs, is the most highly evolved phylum in the animal kingdom. Early embryos of all chordates share three cardinal characteristics: a notochord, a dorsal hollow nerve cord, and pharyngeal slits, which, as mentioned in Chapter 10, are also called gill slits. All three of these characteristics may, however, disappear or be modified in the adult. The *notochord* is a flexible rod that runs longitudinally, dorsal to the intestine, and functions to support and prevent telescoping of the body. A notochord is found only in chordates and is, in fact, the feature from which the phylum name was derived. The *dorsal hollow nerve cord* develops dorsal to the notochord and is markedly different from the solid ventral nerve cords of annelids and arthropods. *Pharyngeal slits*, or *gill slits*, are particularly interesting structures. As stated in the previous chapter, they probably first arose as part of a feeding mechanism—small food particles were collected from a current of water that entered the pharynx via the mouth and then left the pharynx through the slits in its walls; gills for respiration became associated with these slits only secondarily. The gills of fishes develop from the walls bordering the slits, and this explains the unique association of these respiratory organs and the digestive tract in vertebrates. These slits form even during the development of strictly air-breathing chordates, but are never functional because they close before these animals have completed their development. The average adult size of the chordates is greater than that of any other phylum. Among the members of this phylum is the blue whale, the largest animal that has ever existed, larger even than any of the dinosaurs. It attains a length of about 34 meters and a weight of about 115,000 kilograms.

The living members of the phylum Chordata are customarily separated into three subphyla, the Urochordata, the Cephalochordata, and the Vertebrata. The first two comprise the lower chordates. They are exclusively marine animals and are the subject of this chapter; vertebrates provide the subject matter of the next two chapters. The lower chordates are invertebrates; they never develop vertebrae. The urochordates are commonly called tunicates because the body is covered by a saclike case, or tunic. Neither larval nor adult tunicates exhibit segmentation. Adult tunicates bear little resemblance to the rest of the chordates. The cephalochordates are anatomically a much more uniform group of organisms than are the urochordates. There are about 2,000 living species of urochordates, but only about 30 extant species of cephalochordates. Cephalochordates, although invertebrate, have a fishlike shape which suggests their affinity with the vertebrate fishes. Also, cephalochordates are segmented animals, like the vertebrates. Cephalochordates have two common names. One is lancelet, but their most frequently used common name is amphioxus. Amphioxus will be used herein.

CLASSIFICATION The subphylum Urochordata consists of three classes.

The class Ascidiacea (Figs. 1-13A, 11-1A) consists of those solitary and colonial tunicates that are commonly called sea squirts. This common name is derived from the fact that when an ascidian is removed from water the body contracts (there are muscles in the body wall), forcing water to squirt out of its siphons. Ascidians are sessile as adults, one end being attached, usually to rocks, shells, or wharf pilings. The body has two external openings at the unattached end, the *buccal siphon* and the *atrial siphon* for entrance and exit respectively

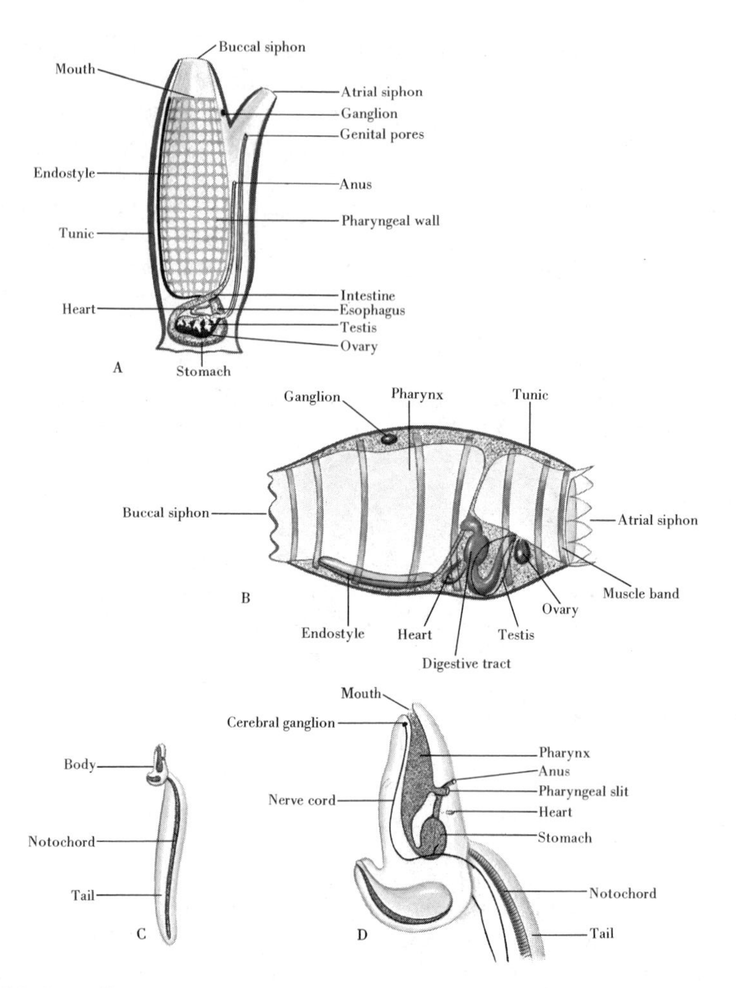

Fig. 11-1 *Representative tunicates. (A) The ascidian* Ciona intestinalis; *(B) the thaliacean,* Doliolum mulleri; *(C) the larvacean,* Oikopleura albicans, *removed from its gelatinous tunic; (D) enlargement of a portion of C.* [*(A) After Hegner and Engemann,* Invertebrate Zoology, *2nd ed. The Macmillan Co., (B) After Brien, after Neumann, in* Traite de Zoologie, *Vol. XI, P.-P. Grasse, ed. Masson et Cie., 1948; (C,D) After Borradaile and Potts,* The Invertebrata, *3rd ed. Cambridge University Press, 1959.*]

of a water current. Adult ascidians have neither a hollow nerve cord nor notochord; these structures are lost when their larvae undergo metamorphosis. *Amaroucium stellatum* is a colonial ascidian, common from Cape Cod to North Carolina, that fisherman call "sea pork" because its tunic has the color and feel of salted pork.

The class Thaliacea (Fig. 11-1B) consists of free-swimming, solitary, and colonial tunicates, commonly called salps, whose buccal and atrial siphons are always at opposite ends of the body. Adults retain neither the notochord nor hollow nerve cord of their larvae. The thaliacean, *Salpa democratica*, is often found in great numbers off the coast of New England.

The class Larvacea (Fig. 11-1C,D) consists of free-swimming solitary tunicates which as adults retain some structures (such as a tail and notochord) which the larvae of all tunicates have but which adult ascidians and thaliaceans lose when their larvae metamorphose. Another name for this class is Appendicularia. Adult larvaceans have the appearance of a large tunicate tadpole larva (called a tadpole, presumably because it resembles an amphibian tadpole) that has been bent, depending upon the species, either 90° or into a "U." Larvaceans have the smallest average size of the three classes of tunicates, being only a few millimeters long. The larvacean, *Oikopleura flabellum*, occurs in both the Atlantic and Pacific Oceans.

The subphylum Cephalochordata consists of only one class.

The class Leptocardii consists of amphioxus alone (Fig. 1-13B). Amphioxus usually lies buried in mud at the bottom of the sea, with only its anterior end protruding, but it is capable of swimming. It retains the three basic chordate characteristics throughout life. The body, laterally compressed and streamlined, is usually about 5 cm long. Its segmentation shows up best in the muscles along the sides of the body. Like the tunicates, it is jawless and lacks paired appendages. *Branchiostoma caribaeum* is a species of amphioxus that lives in the West Indies.

MORPHOLOGY AND PHYSIOLOGY

LOWER CHORDATE STRUCTURE

Adult ascidians and thaliaceans show little resemblance to amphioxus or the vertebrates, but tunicate larvae and adult larvaceans, as noted in the classification section, resemble amphibian tadpoles having an ovoid trunk and a long tail, amphibians being vertebrates. Ascidians constitute the majority of the tunicate species. Solitary ascidians range in shape from spherical animals to cylindrical ones; one end is attached to the substrate and the siphons occur at the free end. Solitary thaliaceans are somewhat cask-shaped. There are also highly organized colonies of ascidians and thaliaceans in which many individuals share a common tunic. The tunic is secreted by the epidermis of the underlying body wall. Each member of such a colony has its own buccal siphon opening to the outside, but the atrial siphon of each individual opens into a common atrial chamber which has a single opening to the outside. In general, the solitary forms are larger than the individuals that make up a colony. Surprisingly, in ascidians and thaliaceans, a large component of the tunic is almost always a type of cellulose named *tunicine*. Cellulose is more commonly associated with plants than with animals. Proteins are other constituents of the tunic of ascidians and thaliaceans. Furthermore, proteins are apparently the only constituent of the larvacean tunic. Larvaceans have a tunic that is gelatinous, not containing cellulose. An interesting feature of the larvaceans is that they can discard the tunic ("house"), whenever the passages through it get clogged by large food particles, and quickly secrete a new one.

Amphioxus has a fishlike appearance (Fig. 11-2). Its body is bluntly pointed at both ends with no distinct head, and, as stated above, is flattened side to side. The mouth is surrounded by the oral hood that bears fingerlike projections, called *buccal cirri*, that prevent the entry of oversized particles that might clog the digestive tract. The ventral surface is flattened and bounded laterally by a pair of longitudinal folds of the body wall, the *metapleural folds*.

Amphioxus is semitransparent; its integument is not pigmented. Consequently, the segmental character of the muscles along the sides of the body can be easily seen.

The larvae of all tunicates have notochords, providing them with axial skeletons; but larvaceans are the only tunicates which retain a notochord as adults. Adult ascidians and thaliaceans do not have a comparable solid skeleton. Instead, their blood-filled hemocoel functions in part as a hydrostatic skeleton. The notochord of amphioxus, like that of larvaceans, persists throughout life, as mentioned earlier.

LOCOMOTION Ascidians spend their adult lives attached to the substrate, whereas the adults in the other classes of tunicates are capable of swimming, but only very slowly. The tunicate tadpole larva, before metamorphosis, leads a free-swimming existence for a brief period of time, using its tail to provide the force for movement. This free-swimming larva not only makes possible the dispersal of ascidians (the adults, as stated earlier, are sessile), but also has the burden of locating a suitable substrate to which it can ad-

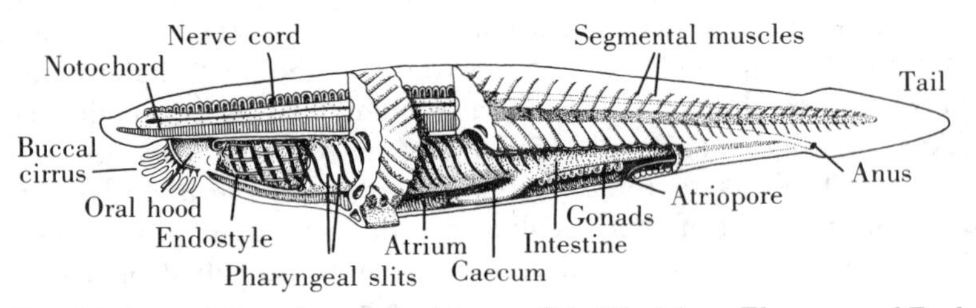

Fig. 11-2 *A lancelet or amphioxus. (Modified from* Elements of Zoology, *4th ed., by Storer, Usinger, Nybakken, and Stebbins. Copyright © 1977, McGraw-Hill Book Company. Used by permission.)*

here and metamorphose into a successful adult. Adult thaliaceans and larvaceans can be propelled forward in reaction to the stream of water that exits posteriorly from the tunic. The water current for locomotion in larvaceans is created by the beating of the muscular tail, whereas in thaliaceans, this current is produced by the contraction of circular bands of muscle in the body wall.

Amphioxus, as stated above, is capable of swimming, although it usually lies in the substrate with only its anterior end protruding. This segmented animal swims by rapid lateral bending movements of its body. Segmentation in chordates appears to have originated as an adaptation for locomotion in water. Segmentation allows for more precise control of the undulatory movements of the trunk muscles than would be possible if these muscles were not segmentally arranged.

FEEDING AND DIGESTION

The tunicates and amphioxus are filter feeders. They produce a water current that carries suspended food particles into the pharynx. As the water passes out of the pharynx through the slits in its walls, the food particles become entangled in a mucus sheet. In ascidians, thaliaceans, and amphioxus the water passes from the pharynx into an *atrium*, a chamber between the body wall and the pharynx. In larvaceans, however, the water passes directly from the pharynx to the exterior; a larvacean has no atrium. The atrium in ascidians and thaliaceans opens to the exterior through the atrial siphon. However, in amphioxus, the opening from the atrium is simply a pore, called the *atriopore*. It is located posteroventrally, anterior to the anus. A pore not only opens flush with the body wall of an animal, in contrast to a siphon, which is a tubular extension outward from the body, but a pore is also relatively smaller in diameter in proportion to the total size of the animal than is a siphon. The mucus sheet in which the food particles become trapped is secreted by the cells in a longitudinal ciliated groove, the *endostyle*, in the floor of the pharynx of both tunicates and amphioxus. The food-laden mucus is then propelled into the esophagus by ciliary action. In most ascidians, the water enters the pharynx by the action of pharyngeal cilia alone, but in some species rhythmic contractions of the body wall assist. As in locomotion, the feeding current in thaliaceans is created by muscular pulsations of the body wall, while in larvaceans, it is created by the beating of the tail. In amphioxus, the water current is produced by the cooperative effort of cilia on the oral hood and in the pharynx. The postpharyngeal portion of a tunicate's digestive tract consists of an esophagus, stomach, intestine, and anus. Enzyme secretion, digestion, and some absorption occur in the stomach, but the intestine is the main site of absorption.

In amphioxus, the pharynx joins a short esophagus, which leads into a straight intestine that ends at the ventral anus; there is no stomach. Near the beginning of the intestine in amphioxus is a fingerlike outpocketing, the caecum, that emerges from the intestine and extends forward along the right side of the digestive tract. The caecum is not only a source of digestive enzymes, but some of its cells are capable of engulfing and digesting small particles of food. Enzyme-secreting cells are also present in the wall of the anterior portion of the amphioxus intestine. Extracellular digestion goes on in the anterior portion of the intestine. Amphioxus thus combines intracellular and extracellular digestion. Absorption of the products of extracellular digestion occurs in the posterior portion of the intestine.

CIRCULATION The circulatory systems of tunicates and amphioxus are open, whereas in the vertebrates they are closed. The tunicate heart is a short, more or less U-shaped tube that is open at both ends and is found close to the intestine. The circulatory pathways in tunicates lack true walls; they are simply sinus channels in mesenchyme. A unique feature of the tunicate circulatory system is the periodic reversal of the direction of blood flow through the heart. Blood flows in one direction, and there is a brief period of rest followed by a reversal in the direction of the contractile wave. There is no oxygen transport pigment in tunicates.

The circulatory system of amphioxus shows the basic pattern of blood flow that was adopted by the vertebrates, anterior flow ventrally and posterior flow dorsally. The blood flows forward in the ventral aorta beneath the pharynx. Vessels called *aortic arches* originate from the ventral aorta and carry the blood up through the *pharyngeal bars* (the columns of tissue between the pharyngeal slits). The blood from the aortic arches is then collected by a pair of dorsal aortae, which unite posterior to the pharynx to form a single dorsal aorta. This single dorsal aorta delivers the blood to the system of lacunae that supplies the tissues with blood. There are no true capillaries. From the lacunae in the tissues the blood is collected in vessels that lead back to the ventral aorta. The blood does not contain an oxygen transport pigment. Furthermore, there is no heart. Instead, several of the blood vessels are pulsatile. This pattern is the reverse of that seen earlier in the annelids, where the blood flowed anteriorly to the dorsal blood vessel and posteriorly in the ventral blood vessel.

RESPIRATION The large volume of water passing through the pharynx of tunicates and amphioxus provides not only food but also oxygen. Most of the oxygen that tunicates and amphioxus need appears to be picked up by the blood as it flows through the blood-filled spaces inside the walls of the pharynx. Some oxygen also seems to be obtained by diffusion through body surfaces other than that of the pharynx.

SENSE ORGANS An eye occurs in tunicate tadpoles and adult larvaceans. But ascidian and thaliacean adults lack such localized sense organs. They do, however, have scattered individual sensory cells in their siphons that are sensitive to chemical stimuli, light, and touch. These individual sensory cells apparently provide sensory input for the reflexes which not only help to clear the siphons of particles that might clog the filtering apparatus of the pharynx but also regulate the rate of flow of water through the animal. A rapid contraction of the body wall can cause water to be shut out of both siphons with sufficient force to unclog them, should the need arise. Also, bands of muscle in the walls of the siphons can constrict the siphons, reducing the rate of water flow, should the water become contaminated. Illumination of the siphons or touching them will also cause constriction of the siphons.

Amphioxus has numerous photoreceptors embedded in the walls of its hollow nerve cord. Each photoreceptor consists of only two cells, a photosensitive cell and a cell containing black pigment. The position of the pigment cell relative to the photosensitive cell varies depending upon where, along the body, these cells are located. Presumably, this variation in the position of the pigment cell provides amphioxus a capacity for directional orientation.

EXCRETION Tunicates lack discrete excretory organs. Instead, their excretory system consists simply of amoebocytes that accumulate wastes while floating freely in the blood and then become fixed on the surfaces of several organs, particularly the stomach, intestine, and gonads. The deposition of these amoebocytes continues throughout the life of the individual, constantly increasing in bulk. Much of the waste material deposited in these cells is uric acid. However, the major nitrogenous waste product of tunicates is ammonia, which diffuses out across the body surfaces, mainly into the seawater passing through the pharynx.

The excretory system of amphioxus, unexpectedly, consists of protonephridia with solenocytes, as in some polychaetes. About 200 protonephridia lie above the pharynx in close proximity to blood vessels, from which they presumably receive most of the waste products that they excrete. The nephridiopores open into the atrium. As is characteristic of protonephridia (see Chapter 5), those of amphioxus do not open internally. Amphioxus is the only deuterostome which has protonephridia. The coelom of amphioxus is much reduced in size relative to that of vertebrates, presumably because of crowding by the large-sized atrium. Curiously, as will be discussed further in this chapter, a tunicate has no coelom, either during development or as an adult.

NERVOUS SYSTEM The nervous system reaches its highest level of development among the chordates, culminating in humans, who have the capacity of articulate speech for communication. The mental development of humans has enabled them to dominate all other animals and to survive in every climate.

Adult ascidians and thaliaceans do not retain the larval hollow nerve cord. In both of these classes it has become reduced to a single, solid ganglion. The situation is more complex in larvaceans. From their larval hollow nerve cord are formed a ganglion, called the cerebral ganglion, and a nerve cord that emerges from it and extends into the tail. Authorities are divided on the question of whether the nervous system of adult larvaceans retained the hollow character of the larval nervous system. It is doubtful that there is a cavity anywhere in the larvacean nervous system. From these ganglia (and from the larvacean nerve cord also) arise nerves that innervate various parts of the body.

Amphioxus retains a hollow nervous system throughout life. The central canal widens only slightly at the anterior end but, in contrast to vertebrates, in amphioxus the wall there is not thickened. For this reason the expanded anterior end of the nerve cord of amphioxus is most frequently referred to as the *cerebral vesicle* and not simply as the "brain," thus implying that it has not evolved into a true brain that oversees the activity of the more posterior portions of the nervous system. Vesicles are, by definition, thin-walled, hollow structures. At least seven pairs of nerves that carry messages to and from the various structures in the anterior end of the body emerge from the cerebral vesicle. Posterior to the cerebral vesicle, each side of the nerve cord gives off two sets of nerves. One set emerges dorsally and alternates with a set that emerges ventrally; the dorsal nerves are *sensory*, carrying incoming messages, whereas the ventral nerves are *motor* ones, carrying messages outward from the central nervous system. This alternating dorsal sensory-ventral motor pattern is not exhibited by the nerves that originate from the cerebral vesicle.

REPRODUCTION Tunicates, with few exceptions, are monoecious; the rest of the chordates are dioecious. A tunicate usually has only one ovary and one testis. Colony formation among the ascidians and thaliaceans occurs by asexual budding. Larvaceans reproduce only sexually. Their inability to reproduce by budding presumably accounts for the absence of colonial larvaceans. The tadpole larva of tunicates was described earlier in this chapter. During the metamorphosis of this larva is when, for example, the notochord is lost.

Amphioxus has about 25 pairs of segmentally arranged gonads, the sexes being separate. Fertilization in amphioxus, as among tunicates, is external.

PHYLOGENY The phylogenetic relationships among the phyla that constitute the deuterostomes hold much interest for man because he is a deuterostome. Hemichordates, deuterostomes also, show close relationships to both the echinoderms and the chordates. The similarity of the echinoderm bipinnaria larva to the hemichordate tornaria larva was cited in Chapter 10 as evidence for a close phylogenetic relationship between these two phyla. With respect to evidence for a close relationship between hemichordates and chordates, as stated in the previous chapter, hemichordates have pharyngeal slits similar to those found in chordates, and in some hemichordates one nerve cord is sometimes hollow, reminiscent of the tubular nervous system of chordates. Hemichordates and chordates are the only animals that have, as stated earlier, pharyngeal slits or a hollow section in their nervous systems at least during some portion of their lives. In fact, at one time, the hemichordates were actually included in the phylum Chordata because a preoral diverticulum of the hemichordate oral cavity was mistaken for a notochord. But it is now clear that hemichordates do not have a notochord. If it had been a notochord, and with hemichordates already possessing pharyngeal slits and a hollow nerve cord (at least in some species), these animals would have met the basic criterion for inclusion among the chordates, possession of all three. It appears that an ancestral deuterostome stock, after giving rise to the echinoderms as a blind side-branch, continued along the main line, next giving rise to the hemichordates and finally then to the chordates. Only chordates, however, evolved a notochord. Hemichordates do not seem to have given rise directly to the chordates.

The tunicates probably diverged early from the main chordate line of evolution because neither during their development nor as adults do they show a sign of segmentation or, as stated earlier, have a coelom, both of which are characteristics of all other chordates. The coelom (an enterocoel in chordates as in all deuterostomes) was already present in the deuterostome line that gave rise to the chordates, but segmentation was not. The segmentation of amphioxus and the vertebrates does not appear to have been inherited from any other phylum because they are the only segmented deuterostomes. Segmentation apparently evolved anew among the chordates only after the tunicates departed from the line leading to amphioxus and the vertebrates. Tunicates presumably inherited a coelom from an ancestral form and then lost it.

In the following way chordates likely originated from an ancestral, coelomate, larval stock that resembled the free-swimming tadpole larva of present-day tunicates. The adults of this ancestral stock were sessile bottom dwellers feeding by means of a lophophore. The mobile larval stage became independently successful, advancing along the road to becoming a chordate. As a result of the success of the larvae, the sessile adult stage was abandoned (dropped from the life cycle). One of the early changes that occurred as this ancestral larval stock began the long journey to becoming the chordates as we know them today had to have been a shift in its feeding mechanism, from the tentaculate method of lophophorates to the filter feeding mechanism, associated with pharyngeal slits, that we find today in the lower chordates. Also, to be successful, especially with the abandonment of the reproductive adult stage, this larval stock had to have acquired the ability to reproduce. The seemingly anomalous attainment of sexual maturity by a larva, as occurs even today in some salamanders, is called *paedogenesis*. The hollow nerve cord and notochord could have evolved in this tadpolelike larval stock as aids to swimming, the notochord providing support for the tail while the longitudinal nerve cord enabled the larva to control better the swimming movements of its tail. This ancestral, originally larval coelomate stock would then have split in two, one portion evolving into the tunicates. In becoming the tunicates, this free-swimming, originally larval stock would have had to lose its coelom and adopt the sessile mode of life of adult ascidians and the relatively inactive modes of life of the thaliaceans and larvaceans. As a consequence, there was no selection pressure on tunicates to evolve the more highly organized nervous system found in the rest of the chordates. The second portion of this originally larval, coelomate ancestral stock would have continued its active, free-swimming existence while becoming segmented, and ultimately became amphioxus and the vertebrates. In fact, it is generally agreed that the larvaceans evolved from the portion of the ancestral stock that give rise to the tunicates in much the same way as suggested here for amphioxus and the vertebrates. That is, larvaceans retained essentially a larval structure (hence their name), but on it was superimposed a reproductive system. Amphioxus is generally considered to be an offshoot of the line leading from the ancestral coelomate larval chordate stock to the vertebrates, not a direct ancestor of the vertebrates.

A group of organisms which had been considered for many years as an extinct class of primitive echinoderms, called the Stylophora, may have actually been chordates. The stylophorans lived from the Cambrian Period into the Devonian Period. In 1967, R. P. S. Jefferies reexamined many of these fossils and decided that they were not echinoderms but instead were chordates, more primitive than any other known chordates. Furthermore, not only did he take the position that these extinct animals were chordates, but also that they represent one more subphylum of chordates in addition to the Urochordata, Cephalochordata, and Vertebrata, and named it "Calcichordata." Also, he hypothesized that free-swimming members of the subphylum Calcichordata were the ancestors of the urochordates, cephalochordates, and vertebrates.

FURTHER READING

Alldredge, A., "Appendicularians," *Scientific American,* vol. 235: 1 (1976), p. 95.

Barrington, E. J. W., *The Biology of Hemichordata and Protochordata.* San Francisco: Freeman, 1965.

Barrington, E. J. W., ed., *Protochordates, Symposia of the Zoological Society of London,* Number 36. New York: Academic Press, 1975.

Eaton, T. H., "The Stem-Tail Problem and the Ancestry of Chordates," *Journal of Paleontology,* vol. 44 (1970), p. 969.

Jefferies, R. P. S., "The Subphylum Calcichordata (Jefferies, 1967) Primitive Fossil Chordates with Echinoderm Affinities," *Bulletin of the British Museum (Natural History) Geology,* vol. 16 (1968), p. 243.

chapter **12**

Chordata Continued—Vertebrates

The vertebrates represent the most advanced subphylum of chordates. Not only do vertebrates have the three salient chordate characteristics (notochord, dorsal tubular nerve cord, and pharyngeal slits), but, as vertebrates, they also have a *vertebral column* consisting of a longitudinal series of segmentally arranged cartilaginous or bony vertebrae. The vertebral column reinforces or replaces the notochord.

CLASSIFICATION

The subphylum Vertebrata consists of seven classes.

The class Agnatha, or jawless fishes, includes the lampreys (Fig. 1-13C), which are found in both fresh and salt waters, and hagfishes, which are strictly marine. An agnathan has an elongated cylindrical body; the skeleton is cartilaginous. They lack not only jaws but also scales and paired fins. However, like higher classes of fishes, agnathans do have unpaired fins, such as the tail fin. Adult lampreys have a sucker around the mouth for attachment to prey, but hagfishes lack such a sucker, their mouths being surrounded by short, sensory tentacles instead. Agnathans are sometimes commonly called cyclostomes. *Petromyzon marinus,* the sea lamprey, invaded the Great Lakes, doing much damage to the commercial fish resources by injuring or destroying the fish there, but recent control measures have been to a large degree successful and the commercial catches are increasing.

The class Elasmobranchiomorphi consists of the sharks (Fig. 1-13D), skates, and rays. The name Chondrichthyes is frequently also used for this class. These fishes also have a cartilaginous skeleton, and, in addition, they have jaws and paired fins. There are two pairs of lateral fins, pectoral and pelvic. Elasmobranchiomorphs are largely marine, but some live in fresh water. The integument has many scales of dermal origin. The vertebrate integument consists of two parts, the outer epidermis and the inner dermis. The great white shark, *Carcharodon carcharias,* found in all warm temperate and tropical oceanic waters, attains a length of 12 meters.

The class Osteichthyes consists of the fishes, such as the perch (Fig. 12-1), which have a skeleton of bone, a characteristic of the higher classes of vertebrates as well. Scales are present in most species, and, when present, are of dermal origin. A few species are scaleless. Osteichthyans have jaws and two pairs of lateral fins (pectoral and pelvic), as do the elasmobranchiomorphs. Marine and freshwater species exist. The top minnow or mosquitofish, *Gambusia affinis,* which feeds voraciously on mosquito larvae, is widely used to control the mosquitoes that transmit malaria and yellow fever.

The class Amphibia consists of the frogs (Fig. 1-13E), toads, salamanders, and the snakelike, legless caecilians. A few species of salamanders spend their entire lives as aquatic organisms, never emerging onto land; but the adults of the remaining large majority of the species of amphibians have assumed a terrestrial life—the first chordates to do so—but they are restricted to a moist habitat. Furthermore, almost all of these terrestrial amphibians enter fresh water to reproduce. There are no marine amphibians. Those few that do not shed their eggs directly into water must find a moist area in which to lay their eggs to minimize the chance of desiccation, inasmuch as the amphibian egg does not have a shell. Amphibians are tetrapods (having four legs rather than fins). The integument of amphibians is covered by a layer of mucus secreted by glands in the integument. Amphibians have no external scales. The leopard frog, *Rana pipiens,* is the most commonly occurring frog in the United States.

The class Reptilia consists of the turtles (Fig. 1-13F), snakes,

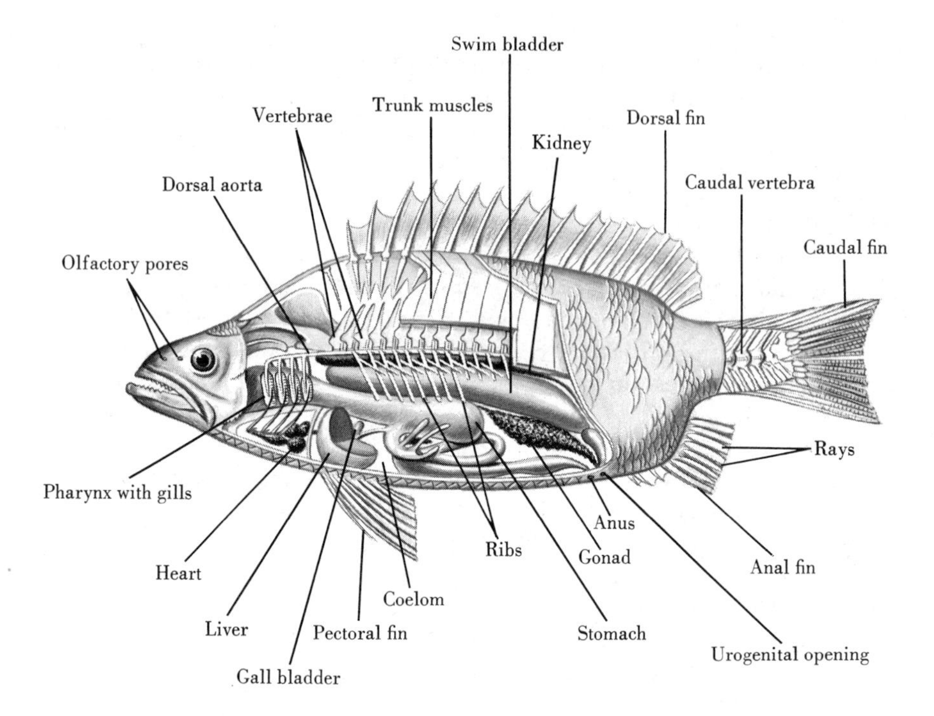

Fig. 12-1 *Diagram of the internal organs of a typical member of the class Osteichthyes. (After Dillon,* Principles of Animal Biology. *Macmillan, 1965.)*

lizards, crocodilians, and the tuatara of New Zealand. The tuatara is the most primitive reptile alive today, the only survivor of an order that has existed from ancient times, the Rhynchocephalia. A tuatara has a lizard-like appearance, but differs from other reptiles in numerous ways, including the fact that it lacks a male copulatory organ. Reptiles were the first vertebrates to become adapted for life in dry places on land. The reptile integument has horny epidermal scales that minimize water loss. Furthermore, reptiles were the first chordates to have a shelled (*cleidoic*) egg. Such an egg can be laid in dry places on land and the embryo developing inside it is protected from desiccation, thereby freeing the adults of any need to return to an aquatic environment to breed. The number of green turtles, *Chelonia mydas*, has been greatly reduced because they have been hunted so much in the past for their meat. This is a marine species that occurs in many places throughout the world including the Gulf of Mexico, Caribbean Sea, China Sea, Gulf of Aden, and the east coast of Australia. An adult may attain a weight of about 225 kg and can yield about 45 kg of edible muscle.

The class Aves consists of the only animals with feathers, the birds. Their forelimbs are modified as wings. Birds are *endotherms*, that is, they are able to maintain a relatively constant body temperature even when the environmental temperature fluctuates widely. The reproductive rate of the brown pelican, *Pelecanus occidentalis*, has diminished greatly because of the accumulation of the chlorinated hydrocarbon insecticide DDT and its breakdown products, DDD and DDE, in the fishes it eats. The consequent high levels of these compounds in the pelicans interfere with the deposition of calcium in their egg shells. As a result, the thin-shelled eggs crack easily and do not hatch. The number of brown pelicans in the wild has decreased dramatically.

The class Mammalia consists of those animals that have hair and mammary glands. The mammary glands of females supply milk for nourishing the young. Mammals include such animals as the duck-billed platypus, whale, elephant, and man. They are endotherms also. The jaguar, *Felis onca*, is the largest species of the cat family (Felidae) in the western hemisphere, occurring today from Patagonia in South America northward through Mexico, and even is found, on rare occasions, in the southwestern United States.

MORPHOLOGY AND PHYSIOLOGY

VERTEBRATE STRUCTURE

The agnathans are the most primitive vertebrates. Adult lampreys, as stated, have a sucker surrounding the mouth, but they lack jaws. In spite of the lack of jaws, the adults of most species of lampreys are predacious. However, in some species of freshwater lampreys, the adults never feed after undergoing metamorphosis to the adult stage. As adults they simply reproduce and die. The sucker of agnathans is armed with horny epidermal teeth, which aid in attachment to the higher fishes that they feed on. Teeth of higher vertebrates begin development in the dermis, and are not homologous to the epidermal teeth of lampreys. In keeping with the lack of jaws among agnathans, they also never develop dermal teeth. Hagfishes, as stated, lack this sucker. They are scavengers rather than predators, feeding on dead or dying fishes. The lack of a sucker is presumably the reason hagfishes are scavengers and not active predators. The integument of agnathans has many mucous glands, but no scales.

Elasmobranchiomorphs have two different body outlines. The sharks are typically streamlined and "fishlike," whereas the skates and rays are flattened, and their large pectoral fins are broadly joined to the head and trunk. The pectoral fins are those that correspond to the forelimbs of a tetrapod. The streamlined shape of sharks is an excellent adaptation for a pelagic (living in the open sea), predacious hunter. On the other hand, skates and rays are bottom dwellers, feeding largely on invertebrates. Their flattened body is apparently an adaptation for bottom dwelling. All elasmobranchiomorphs, as stated earlier, also have paired pelvic fins. The pectoral and pelvic fins of fishes evolved into the legs of tetrapods. Obviously, the evolution of paired fins was a major advance. The mode of origin of the paired fins is not known with certainty, but it is likely that they evolved from outwardly projecting integumentary folds similar to the metapleural folds of amphioxus (this is the *fin-fold theory*). The scales on the skin of elasmobranchiomorphs are minute, with a backward-pointing spine. They are known as denticles or placoid scales and their structure resembles that of simple dermal teeth.

Osteichthyans have a large variety of body shapes, ranging from that of the long and highly flexible eel to that of the sea horse, which swims erect and has a prehensile tail. In the vast majority of living osteichthyans, the skeleton and flesh barely extend into the fins; the fins are virtually only mere webs of skin supported by rays. Such fishes are called *ray-finned.* A ray consists of a row of narrow scales. In contrast, a small group of fishes are *lobe-finned;* at the base of their fins is a fleshy lobe containing skeletal elements and these fins are edged with skin-covered fin rays. Lobed fins are especially well developed in *Latimeria chalumnae*, which belongs to an order of bony fishes, the Coelacanthiformes (coelacanths). This order was believed to have been extinct for 75 million years until a living specimen was found in 1938 off the east coast of South Africa (Fig. 12-2). Lobed fins are biologically significant because it is from a fin of this type that the tetrapod limb most likely evolved. Furthermore, the vertebrate that first made the

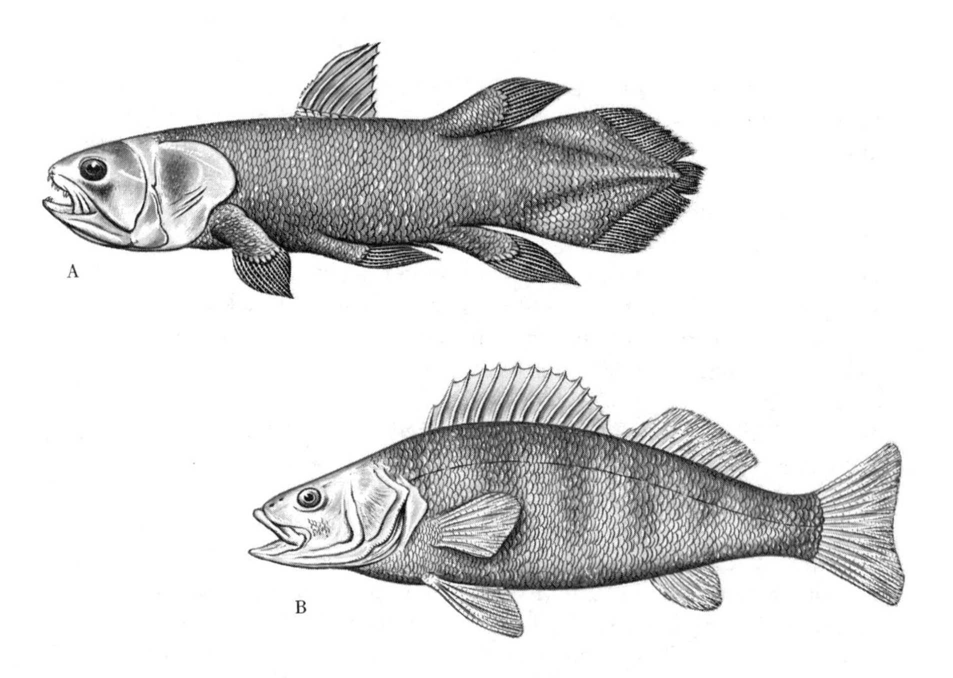

Fig. 12-2 *(A) The living coelacanth,* Latimeria chalumnae, *is lobe-finned; (B) the yellow perch,* Perca flavescens, *is ray-finned.*

transition from an aquatic existence to terrestrial life was probably a lobe-finned fish that used its fins to crawl out of the water onto land. In addition to coelacanths, the lungfishes (three genera of bony fishes) are the only other living lobe-finned fishes. Lungfishes, as their name implies, have lungs and can breathe air and so survive during a drought when their ponds dry up. They also have gills for aquatic respiration. The piscine ancestor of tetrapods probably also had lungs along with gills.

The amphibians can be divided into two groups on the basis of whether as adults they retain the larval tail. One, exemplified by frogs and toads, loses the larval tail during metamorphosis; the second group (for example, salamanders) retains the tail. When amphibians are compared with fishes, the most striking difference is the presence of four legs in place of the paired pectoral and pelvic fins. Amphibians have appendages adapted for walking, running, and hopping on land instead of fins for swimming. A majority of the frogs and toads have an eardrum, *tympanic membrane,* on each side of the head just posterior to the eye, marking the first appearance of this structure in the chordates. The rest of the amphibians do not have tympanic membranes. Larval amphibians respire by gills, but when metamorphosis occurs, lungs for air-breathing differentiate. However, some salamanders do not undergo a metamorphosis. As suggested in the previous chapter for the ancestor that gave rise to the chordates, these salamanders exhibit *paedogenesis,* becoming sexually mature and reproducing, but essentially retaining the rest of their larval characteristics throughout life (being permanently aquatic and continuing to respire with gills). Legs and lungs are clearly important adaptations for terrestrial life.

Reptiles show more diversity of body form than do amphibians. There is the turtle with its roughly oval shell and four limbs on one extreme and on the other there is the elongated, legless snake. Intermediate between them are the lizards and crocodilians, whose body shape is reminiscent of that of salamanders. Reptilian skin is characteristically dry, and has horny scales that are formed by the epidermis. The scales may lie flat, side by side, or they may overlap, like shingles. A turtle's shell consists of dermal bony plates that meet in immovable sutures, and over these plates are large epidermal scales. The absence of legs in caecilian amphibians and snakes is a degenerate, not a primitive condition.

Birds are characterized by their unique body covering of feathers, which are epidermal structures, although the legs of birds have epidermal scales similar to reptilian scales. It appears that feathers evolved from reptilian scales. Birds are toothless; the horny beak is a lightweight substitute for teeth. Weight-lightening adaptations are important to flying organisms. However, not all birds are able to fly. Penguins lost their ability to fly when their wings became modified into paddlelike flippers for swimming; in the process, penguins became specialized for aquatic life rather than a flying existence. The kiwi from New Zealand, on the other hand, is flightless because its wings are apparently *vestigial*, that is, they are small and degenerate as a result of descent with modification from ancestors in which they were at one time better developed and functional. The wings of a kiwi are so small that they are normally completely hidden by the bird's shaggy plumage. Kiwis presumably descended from flying birds, but they live where there are few terrestrial enemies. If there are no enemies on the ground, then there is no longer a selective advantage in having wings and being able to fly away from a predator on the ground. Natural selection would, therefore, no longer favor wings, which would explain their having become vestigial in the kiwi. With the evolution of birds from reptiles, the shift from reptilian four-legged locomotion to the avian stance on two legs and the appearance of flight dictated changes in the body proportions. The tail was reduced, and the posterior portions of the body were compressed. These changes brought the center of gravity (a) close to the wings, which could then more easily bear the weight of the body during flight, and (b) above the legs, which bear the weight when the bird is standing.

Mammals range in size from shrews about 5 cm long to the blue whale. Although most mammals have two pairs of limbs that are adapted for running, there are some obvious modifications. The forelimbs of whales are modified as flippers, and there is no external sign of hindlimbs; in seals both pairs of appendages are modified into flippers; and bats have forelimbs that are modified as wings for flight. Hair, found only in mammals, develops from the epidermis. Both birds and mammals are endotherms. Feathers and hair act as insulation to aid in body temperature regulation.

Comparative studies of the skeletons of many different vertebrates have revealed that a progressive replacement of the notochord by vertebrae has occurred. In agnathans, the vertebrae are at best poorly developed, and the notochord persists throughout their lives as the major antitelescoping and supporting structure. The Aves and Mammalia are the only classes of vertebrates in which the notochord is completely obliterated during the development of every member. In elasmobranchiomorphs, primitive osteichthyans, primitive amphibians, and primitive reptiles, the notochord persists in adults, surrounded by the vertebrae. The vertebrae are stronger and provide more support for the body than does a notochord, but they allow less flexibility of the trunk.

The most striking aspects of the evolution of the skeleton of the vertebrate head are (a) the appearance of jaws and (b) the increase in the size of the cranial part in proportion to the facial part in order to accommodate the larger brain that was evolving. The relative increase that has occurred in the cranial portion is clearly evident when, for example, the head skeletons of a human and frog are compared. The major advances in the evolution of the vertebrate limbs have been their conversion from fins to legs, and their gradual movement from the sides of the body to a position below it. Limbs below the body give greater support, raise the body higher off the ground, and allow for more rapid locomotion. Many of the bones in the skeletons of birds are hollow, containing continuations of the air sacs that originate in the lungs. This adaptation facilitates flight because the skeletal weight is minimized with little reduction in strength. The significance of the air sacs in respiration will be explained below.

LOCOMOTION In only two phyla, the Arthropoda and Chordata, has the ability to fly evolved. Vertebrates are, of course, also able to swim and move from place to place on land. In most fishes, the propulsive force for swimming arises from lateral movements of the body, caused by contraction of the muscles along each side of the body, that produce waves which move backwards along the trunk and tail; the fish is pushed forward in reaction to the backward thrust of the muscularly induced waves. The paired and unpaired fins (other than the tail fin) of most fishes are used only for steering, not for propulsion. In contrast, the tail fin is involved in both forward movement and steering. In skates and rays, however, the major propulsive force arises from waves passing over the enlarged pectoral fins.

A snake, being limbless, must use its trunk and cutaneous musculature to provide the necessary propulsive force over land. All snakes can, and most often do, propel themselves forward by lateral undulatory movements of the body which exert pressure against the substrate; thus they push themselves forward. Their ventral scales, in particular, are rough and prevent slippage. A few snakes are also able to move forward in other ways. Some snakes will, on occasion, move forward in a straight line: the ventral scales are advanced a bit and pressed against the substrate, and the body is pulled forward within the skin by rib action and by cutaneous muscles. In still another type of forward locomotion, some snakes make use of the tail as an anchor: after drawing the body up by means of a series of bends, the tail is placed firmly on the ground and the body is straightened, pushing the head end forward. To move forward efficiently, snakes need a rough surface for anchoring their scales.

In birds, lift in flight is provided chiefly by large pectoral muscles that make up as much as one-fifth of the bird's weight. Contraction of the muscles moves the wings downward. Smaller muscles are used to raise the wings.

FEEDING AND DIGESTION Those agnathans, lampreys, and hagfishes that do feed as adults have methods of feeding that compensate for their lack of jaws. As stated earlier, some adult lampreys do not feed, whereas others are predacious. The sucker of predacious lamprey attaches to a fish, and the tongue, which has epidermal teeth that are sharper and larger than those on the sucker, begins to rasp away the tissue until the lamprey is able to suck out the blood on which it mainly feeds. Hagfishes, in contrast, are scavengers. They burrow into the flesh of dead or dying fishes, using the rasping tongue to gouge out the soft parts. Pharynx, esophagus, and intestine are present in agnathans, but there is no stomach. The agnathan digestive tract is uncoiled. The internal surface area of the intestine is increased by the presence of a typhlosole, reminiscent of that found in the intestine of most earthworms. The agnathan typhlosole spirals along the length of the intestine. A liver and bile duct are also present in all agnathans. This duct empties bile produced by the liver into the intestine. However, hagfishes have a gallbladder in which the bile may be stored before it enters the intestine. Most vertebrates have a gallbladder but many lampreys, birds, and some mammals, such as whales, do not. The role of bile is to emulsify and facilitate the digestion of fats. Agnathans lack a discrete enzyme-secreting pancreas. Instead, in the wall of the anterior part of the intestine are clusters of cells that resemble the enzyme-secreting cells in the discrete pancreas of higher vertebrates. Unlike an adult lamprey, a larval lamprey, called an ammocoete, not only structurally resembles amphioxus but is also a filter feeder with an endostyle in the floor of the pharynx that secretes mucus in which the food particles are trapped (as is the case with the tunicates and amphioxus) until the larva metamorphoses. An ammocoete has no sucker or rasping tongue. This larva produces its feeding and respiratory curents by muscular pumping, the water entering through the mouth and leaving through the pharyngeal slits. The earliest known fishes, the now-extinct ostracoderms, first mentioned in Chapter 1, were agnathans, and (as stated in that chapter) were probably filter feeders also. Hagfishes do not appear to have an ammocoete stage. Young hagfishes apparently hatch as juveniles, no larvae, and feed in the same manner as do the adults.

With the advent of jaws and dermal teeth among the elasmobranchiomorphs and osteichthyans, the feeding behavior of the chordates underwent marked changes. The fossil record suggests that the ancient agnathans were bottom dwellers that were able to feed only on small particles of food that accumulated on the bottom, but even the ancient jawed fishes were able to bite off large chunks or even ingest an entire organism. At the same time, the effectiveness of jawed fishes in catching prey was enhanced considerably by the appearance of pectoral and pelvic fins as additional aids in locomotion. Agnathans are poor swimmers compared with those fishes that possess paired fins, expending much effort but making only slow forward progress. The vertebrate stomach first appeared among the elasmobranchiomorphs. It probably developed by enlargement of the end of the esophagus, which served as a place of storage of the large food masses that were consumed. This stomach is an adaptation that provided predators that fed periodically the opportunity to ingest a large quantity at one time. The ability of the stomach wall to secrete hydrochloric acid probably developed later, as a means of reducing spoilage of the stored food; it is a good bactericidal agent. Finally, the stomach of elasmobranchiomorphs added an enzyme, pepsin, to digest proteins in the acid solution. The distance that the food must move within the short, uncoiled intestine of elasmobranchiomorphs is considerably greater than the intestine's external length because of a membrane (the *spiral valve*) that creates a spiral passageway through the intestine. This increases the functional length over which digestion and absorption can occur. Only elasmobranchiomorphs have a spiral valve in the intestine. Elasmobranchiomorphs acquired a distinct, elongated pancreas, lying in a mesentery near the beginning of the intestine. Although elasmobranchiomorphs, as a group, are characterized as predators, two of the larger sharks, the basking shark and whale shark, have abandoned the predatory lifestyle in favor of filter feeding. From water currents entering their mouths and leaving *via* the pharyngeal slits, they strain *plankton*, the minute floating or feebly swimming organisms, by means of combs on their gills, called *gill rakers*. This water current is produced by opening the mouth and lowering the floor of the oral cavity, thus expanding its volume, causing the water to rush in through the mouth. The mouth is then closed and the floor is raised, forcing the water out the slits.

The major structural difference between the digestive tracts of elasmobranchiomorphs and osteichthyans is the type of intestine that is present. Osteichthyans possess an intestine that is an elongated, slender tube, which is highly coiled, in contrast to the short, uncoiled intestine of elasmobranchiomorphs. Tetrapods likewise have an elongated, coiled intestine.

Although fishes have many individual mucus-secreting cells in the oral epithelium, apparently mainly for lubrication of the food being swallowed, amphibians were the first vertebrates to evolve a multicellular oral gland, an event presumably associated with their new terrestrial existence. Saliva, the product of this multicellular oral gland, not only lubricates food to facilitate swallowing it, as does the mucus in the oral cavity of fishes, but the saliva also moistens the food, allowing the taste receptors in the oral cavity to function. Chemicals from food can be tasted only after they have entered into solution. Furthermore, the saliva of some frogs and of a number of birds and mammals contains amylase, the starch-hydrolyzing enzyme. With the appearance of salivary glands, all of the kinds of glands associated with the digestive tract of even the highest vertebrates were then established. The tongue of frogs and toads is a unique food-capturing organ. It is long and attached anteriorly so that it can be flicked out a considerable distance. The tip is coated with a sticky substance to which insect prey adheres. The intestine of tetrapods, unlike that of fishes, is divided into two distinct portions, a small intestine that is longer and narrower than the hind portion, the large intestine.

The reptilian digestive system shows a variety of modifications from what might be considered the typical higher vertebrate condition. As is well known, some reptiles have modified salivary glands, the secretions of which are poisonous, and a pair of teeth (fangs) on the upper jaw are modified for injecting this poison. In some cases there is a groove on one side of each fang along which the poison runs into the wound created by a bite; in others, the fangs are hollow like a hypodermic needle. In turtles, teeth have been replaced by a horny beak. In crocodilians, part of the stomach is modified into a gizzard for grinding food. A snake swallows its food whole, neither tearing nor chewing it. It has the ability to open its mouth wide. It can swallow prey even wider than the usual diameter of its own body because of several structural adaptations. The three main ones are: (1) the two halves of the lower jaw are not firmly united but are joined anteriorly by an elastic ligament; (2) the lower jaw is attached loosely to the rest of the skull; and (3) the bones of the *primary palate,* the roof of the oral cavity, are movable.

Birds, like turtles, have a beak instead of teeth. The shape of a bird's beak usually indicates its food habits. The beak is hooked for piercing and tearing prey in carnivorous birds, such as hawks; short and thick for cracking seeds, as in cardinals; or serrated, as in the mergenser, for capturing and holding fishes. This adaptation of the beak to the diet is advantageous, provided that the particular food source does not become scarce.

A classic example of adaptive modification of the avian beak to a particular kind of food is Darwin's finches, 13 species of birds found only on the Galápagos Islands, located roughly 970 km west of Ecuador. Anatomically, the species differ mainly in the sizes and shapes of their beaks. Darwin's observations of these birds convinced him of the reality of evolution and helped him formulate his theory of natural selection. These 13 species are considered to have evolved from a single species that reached the islands from the mainland of South America, perhaps blown there by a storm. The new arrivals found a variety of habitats and foods that they could exploit without competition from other, similar birds. Darwin's finches were probably the first land birds to reach the islands. Populations derived from the original colonizers ultimately became isolated from each other on the several islands that make up the Galápagos, and new species ultimately evolved. The role of geographic isolation in speciation was explained in Chapter 1. Because the species now differ in the food they seek and the habitats they occupy, there is little competition among them. The differences in their beaks reflect the differences in the kinds of food they eat. Some of the types of beaks Darwin's finches now have are adapted for seed-eating, wood-boring, and nectar-feeding. Even among the seed-eaters, there are differences in the sizes of their beaks; those with larger beaks are able to crack open and feed on larger seeds, thereby probably avoiding competition with small-beaked species that can feed only on smaller seeds. Darwin's finches provide an example of the phenomenon known as adaptive radiation, the spread of populations into different habitats with the subsequent evolution of adaptations that are specialized for the different modes of life in the new environments.

Many birds, mainly grain-eaters, have a crop, an expanded portion of the esophagus, used for food storage. A bird's stomach is divided into an anterior portion, called the *proventriculus*, that secretes the gastric juice and a posterior, more muscular portion, the gizzard, that grinds and mixes the food with the gastric juice. The gizzard helps to compensate for the lack of teeth.

The endothermy of birds and mammals imparts an additional metabolic demand which ectotherms (those animals whose body temperature corresponds to that of the environment) do not have. Energy must be expended to maintain a relatively constant body temperature in the face of changing environmental temperatures. However, the advantages of endothermy far outweigh the disadvantages because the internal organs of endotherms are not subjected to the temperature fluctuations of the environment. The rate of work output of muscles is greater at higher temperatures than at lower ones, and the usual chemical reactions that occur in the cells of animals can be finely tuned to operate most effectively within the particular narrow temperature range found in the body of an endotherm. In cold weather, as the muscles of an ectotherm become sluggish, an endotherm enjoys a distinct advantage in competing for food. A hot day can also present a problem to ectotherms. On such a day their body temperature is elevated, causing them to have a high metabolic rate, but the respiratory and circulatory systems of these animals may not be able to provide their cells with sufficient oxygen to meet the demand, which could be critical. Aquatic ectotherms especially have this problem when their habitat heats up, because warm water holds less dissolved oxygen than does cold water. The normal body temperatures of birds tend to be about 3°C higher than those of mammals. The key to endothermy is the fact that endotherms have metabolic rates that are about four times greater than those of ectotherms of the same weight at the same temperature. The higher metabolic rates of endotherms enable them generally to produce enough heat to maintain their body temperature above the ambient temperature with only the minimal rate of heat production. However, in particularly cold temperatures, endotherms do have the ability to produce additional heat as needed. For example, the muscle contractions associated with shivering produce extra heat. Ectotherms, because of their lower metabolic rate, must use external heat sources, such as sunshine, to raise their body temperature above the air temperature. They are thus more vulnerable to changes in climatic conditions than are endotherms.

Paleontologists are currently engaged in a lively dispute over the recently published hypothesis that dinosaurs, unlike their now-living reptilian relatives, were endotherms. However, many, if not most, paleontologists have concluded that this hypothesis is incorrect. One of the arguments in favor of this hypothesis that dinosaurs were endotherms is a geographical one; dinosaur fossils have been found in the rocks of northern Canada where, it is argued, only endotherms could have supplied themselves enough body heat to survive in the Arctic climate, especially during the winter, when there is little sunshine. Opponents suggest that dinosaurs were mobile enough to migrate southward for the winter. Because there is no living dinosaur and both sides have to rely on fossils and analogies with living animals, the controversy may never be resolved to everyone's satisfaction.

As is true of the avian beak, mammalian teeth show a diversity of size and shape, and also of number, that can be correlated with the animal's diet. Carnivorous mammals, especially members of the cat family, do little or no chewing, and have a reduced number of cheek teeth (the premolars and molars). On the other hand, the canine teeth, being adapted for biting and piercing, are well developed in carnivorous mammals but insignificant or absent in most herbivorous mammals. Herbivorous mammals, however, do have numerous cheek teeth with a large grinding surface for macerating their food. Man has a mixed diet. Consequently, his teeth are structurally and functionally intermediate between the extreme specializations of the carnivores and herbivores.

The incisors, the teeth at the anterior of the jaws, of rodents are large and chisel-shaped for gnawing, whereas the incisors of moles are conical for grasping insects. Ruminants, such as cows, sheep, and deer, have lost their upper incisors; the lower incisors of these herbivores close against a fleshy pad on the upper jaw to nip off vegetation, which is then ground by the cheek teeth. The incisors of carnivores are sharp and pointed for cutting and nipping.

The length of the intestine in bony fishes and tetrapods reflects both the diet and the adult size of the species. Relative to the size of the animal, herbivores have a longer intestine than do carnivores. Plant food is difficult to digest because of the cellulose cell walls that surround the protoplasm of these cells, and a long intestine is, therefore, an adaptive feature that provides for extraction and absorption of a greater amount of nutrients from a diet of vegetable matter than would be possible in a shorter intestine. Very few animals, none of the chordates, are able to digest cellulose. Even if the cell walls have been broken by chewing, they remain mixed with the protoplasmic content of these cells and tend to shield it from the enzymes secreted by the animal. This masking makes digestion and absorption of plant food less efficient than of animal food. Furthermore, to compensate for this inability to digest cellulose, herbivorous mammals harbor colonies of bacteria, and sometimes also protozoans, in a portion of the digestive tract, as in the stomach of a cow, and these microorganisms digest the cellulose plant walls for them, helping to expose the portions of the plant cells that can be digested to enzymes the animals produce. The nutrients made available to the host by the digestive activity of the microorganisms, such as energy-rich molecules derived from cellulose, can be absorbed by the host. Consequently, use of these microorganisms by herbivores makes available to the animals a higher proportion of the nutrients from their food than would otherwise be available. That is, microorganism-aided digestion is an adaptation of herbivores to make them digestively more efficient.

With respect to the relationship between the length of the intestine and the adult size of the animal, as the size of an organism increases, the volume of tissue that requires a supply of nutrients increases at a faster rate than does the surface area of the intestine. Consequently, a disproportionate elongation of the intestine is necessary to provide sufficient absorptive area. The result is, therefore, that small carnivores have the shortest intestines, and large herbivores the longest. This effect of diet on the length of the intestine is even seen in the different stages of the life cycle of frogs, where the herbivorous tadpole has a much longer intestine relative to the size of the animal than does the carnivorous adult.

CIRCULATION In all vertebrates, the circulatory system is a closed one; the arteries connect to the capillaries, which in turn connect to the veins. The vertebrate heart is ventral to the digestive tract, whereas, in contrast, in arthropods the heart is located dorsal to the intestine. Hemoglobin is the oxygen-transport pigment in the blood of all vertebrates except some Antarctic fishes that do not possess an oxygen-transport pigment. When hemoglobin is present in the blood of vertebrates, this pigment is always enclosed in cells. The hearts of vertebrates, like those of mollusks, are chambered.

Major changes occurred in the vertebrate circulatory system as the transition was made from aquatic to terrestrial life. As a result of these modifications, a system of vessels suitable for aquatic respiration in conjunction with gills became transformed into one suitable for aerial respiration in conjunction with lungs. The primitive plan of the vertebrate circulatory system can be seen in fishes. The heart of the fishes, agnathans, elasmobranchiomorphs, and osteichthyans has one auricle and one ventricle. The auricle receives deoxygenated blood from the tissues of the body; the blood then passes into the ventricle, and when the ventricle contracts, the blood flows anteriorly through the ventral aorta, which is bifurcated anteriorly. The aorta emerges from the ventricle, passing anteriorly ventral to the pharynx, giving off a series of paired vessels (the *aortic arches*). The aortic arches are located in the *visceral arches*, one aortic arch in each visceral arch (each mass of tissue just anterior to the first pharyngeal slit, just posterior to the last pharyngeal slit, and in between the other pharyngeal slits of vertebrates is called a visceral arch). The slits form when the pharyngeal wall produces evaginations that meet ectodermal invaginations, ultimately breaking through. These aortic arches proceed dorsally through the visceral arches, breaking up into capillaries in the gills, which are attached to these visceral arches (oxygen is taken up here). The aortic arches then reform dorsally and join the paired dorsal aortae, which, as occurs in amphioxus, fuse posteriorly into a single dorsal aorta that carries the blood posteriorly. Having no gills to supply, the aortic arches of reptiles, birds, and mammals develop as continuous tubular vessels. However, in amphibians, because of the gills that their larvae have, each aortic arch gives off a small branch that supplies the gills. The amphibian aortic arches themselves, unlike those of fishes, do not break up into capillaries. At metamorphosis, when the gills are lost, so are the branches from the aortic arches. These branches, of course, persist throughout the life of those salamanders that do not lose their gills.

Adult agnathans have 6 to 15 pairs of aortic arches, depending upon the species. Six pairs of aortic arches develop in the embryos of the jawed vertebrates, from elasmobranchiomorphs through mammals; this was presumably the format in the ancestral jawed vertebrate (Fig. 12-3A). The fate of these six pairs of aortic arches in each of the classes of jawed vertebrates reveals the likely evolutionary pathway that these arches followed. In elasmobranchiomorphs, the first pair disappears during development (Fig. 12-3B); in most osteichthyans (Fig. 12-3C) and all higher vertebrates, the first two pairs disappear during development. Some osteichthyans lose only the first pair.

The following major modifications of the primitive plan of the vertebrate circulatory system occurred with the acquisition of lungs by the amphibians. A pulmonary trunk to supply the lungs split off from the ventral aorta, both emerging from the one ventricle (Fig. 12-3D,E). The amphibian heart evolved a second auricle, thus having two auricles but only one ventricle. Blood from the lungs is received by the left auricle and deoxygenated blood from the rest of the body is received by the right auricle. With only a single ventricle, however, some mixing of oxygenated and deoxygenated blood occurs. Consequently, some of the deoxygenated blood, without having gone to the lungs for oxygenation, is sent to the dorsal aortae for recirculation through the body. Modifications of the aortic arches of amphibians involved loss of the portions of the dorsal aortae between the third and fourth arches in frogs, toads, and some tailed amphibians, thereby separating the cranial circulation from that of the lower part of the body. The fifth pair of aortic arches disappeared completely in frogs and toads and all reptiles, birds, and mammals, but persists in some salamanders, along with the portions of the dorsal aortae between the third and fourth arches, as shown in Figure 12-3D. As a result of the loss of the fifth pair of arches by all the tetrapods (except for a few salamanders), the two fourth aortic arches,

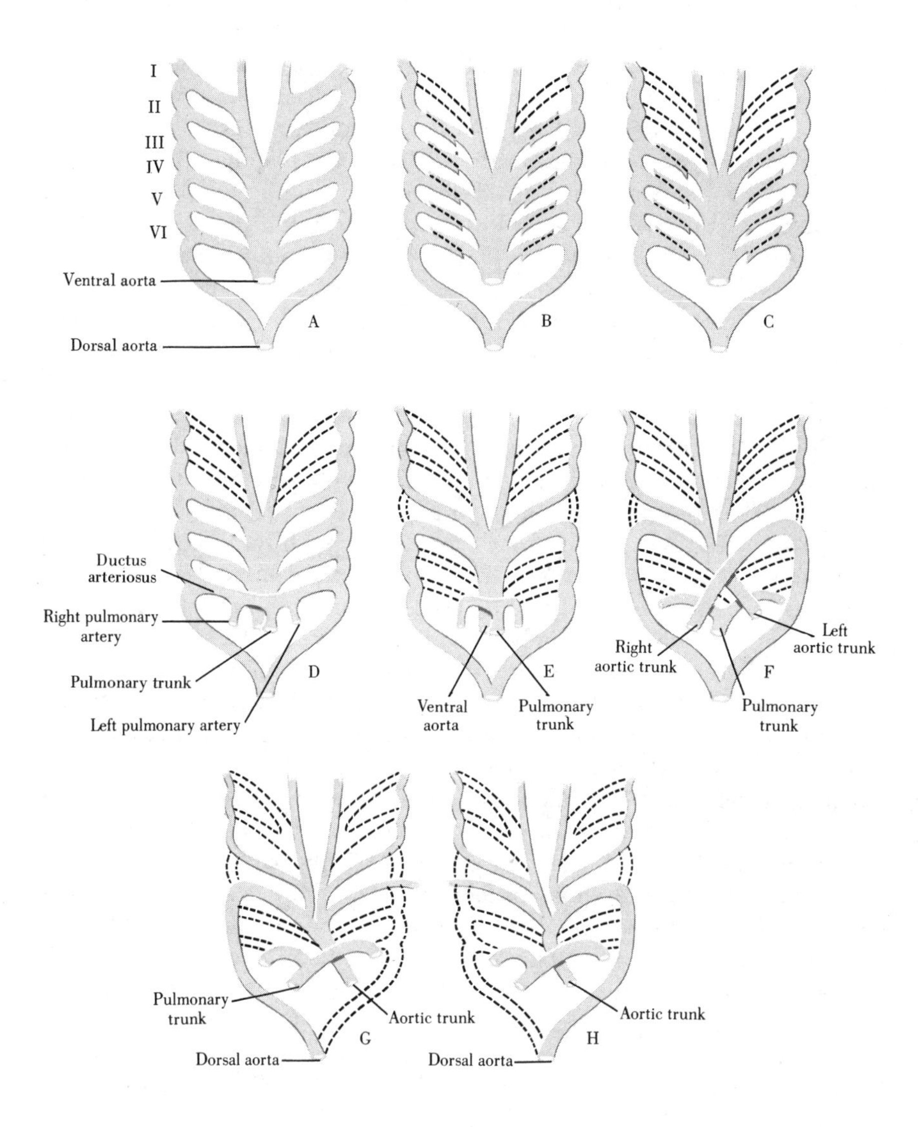

Fig. 12-3 *Modifications of the aortic arches in various jawed vertebrates. (A) Theoretical ancestral jawed vertebrate; (B) elasmobranchiomorph; (C) typical osteichthyan; (D) a tailed amphibian that has retained the fifth pair of aortic arches and persistent connections between the third and fourth aortic arches by parts of the paired dorsal aortae; (E) frog and toad; (F) reptile; (G) bird; (H) mammal. (Modified from Kingsley,* Outlines of Comparative Anatomy of Vertebrates. *Copyright © 1926, McGraw-Hill Book Company. Used by permission.)*

known also as the systemic arches, became the main channels for blood from the heart to the lower part of the body in these animals. Why the fourth arch was selected and not the shorter route through the fifth arch is an enigma. The lungs receive blood through the pulmonary arteries that emerge from the sixth pair of arches, which in frogs and toads lost connection with the dorsal aortae. As a result of these changes, the pulmonary system of frogs and toads became completely separated from the vessels going to the rest of the body. But in tailed amphibians, the dorsal part of each sixth arch or *ductus arteriosus* persists, which is inefficient because all of the blood that enters the pulmonary trunk does not go to the lungs for oxygenation. In frogs, toads, reptiles, birds, and mammals these portions of each of the sixth arches dorsal to the pulmonary arteries are present only until metamorphosis or birth, and then disappear.

Among the reptiles, there occurred the first steps toward eliminating the problem inherited from the amphibians and created by having oxygenated and deoxygenated blood flowing through a single ventricle. The problem was that some of the deoxygenated blood, without having been oxygenated by a passage through the lungs, would be recirculated through the rest of the body. The ventricular cavity has become divided, completely in the crocodilians and incompletely in the other reptiles, thus essentially providing reptiles with a heart that is four-chambered, as is the heart in birds and in mammals. In reptiles, three vessels leave the heart anteriorly (Fig. 12-3F). In addition to the pulmonary trunk from the right ventricle, by further splitting, the remnant of the ventral aorta has become subdivided at its base into two major arteries, aortic trunks; one stems from the right ventricle that connects to the left systemic (that is, fourth aortic) arch, and the second from the left ventricle that connects to the right systemic arch. However, in reptiles, even in crocodilians also, there is a gap at the base of each of these three arterial trunks where they leave the heart, so it is still possible for some mixing of oxygenated and deoxygenated blood to occur there. Furthermore, while the pulmonary trunk primarily receives deoxygenated blood from the right side of the heart, because the systemic arch remains paired, the left systemic arch receives from the right ventricle mostly deoxygenated blood, which is inefficiently recirculated through the body.

In endotherms, all possibility of mixing of the bloodstreams from the left and right sides of the heart has been removed by the evolutionary elimination of the gap at the bases of the arterial trunks. Furthermore, in birds (Fig. 12-3G) the inefficient left systemic arch (present in reptiles) has been eliminated. Mammals (Fig. 12-3H), although they evolved from a group of reptiles different from the one that gave rise to the birds, attained the same type of efficient circulation as birds have, but by a slightly different method. In mammals, the right systemic arch disappears during embryonic development, and a single aortic trunk emerges from the left ventricle to connect with the left systemic arch. Consequently, birds and mammals differ as to which member of the pair of systemic arches they have retained, but in both birds and mammals, the aortic trunk leading to the arch that remains leaves the left ventricle.

Vertebrates from osteichthyans through mammals have a series of vessels, the *lymphatics*, whose main function is to provide for the return of excess *interstitial fluid*, sometimes called tissue fluid, to the circulation. Interstitial fluid is the intercellular fluid in the tissues. An excess is created by the pressure of the circulating blood, forcing fluid out of the capillaries into these tissue spaces. Lymphatics are not connected in any way to arteries. They arise instead from their own capillaries, which have blind tips and empty into a vein. Fluid from the tissue spaces enters the lymphatic system through the walls of the lymphatic capillaries. Osteichthyans, amphibians, and reptiles have *lymph hearts* (contractile regions of the lymph vessels), which aid in forcing the fluid along the vessels. However, in birds and mammals (which are more active animals), movements of the body and its organs provide sufficient external pressure on the lymphatic vessels to move the fluid. The lymphatic system is also involved in fat absorption, and much of the fat absorbed from the intestine enters the circulation by way of this system. This process may be an adaptation to increase the body's efficiency in obtaining these energy-yielding molecules. Fats are relatively large molecules, and the blood pressure in the capillaries may hinder them from entering the main circulation. Lymphatic vessels, however, offer a relatively low-pressure system of drainage into the veins where the blood pressure is lowest.

RESPIRATION Most vertebrates, to satisfy their oxygen requirements, depend only on their gills and lungs. In some fishes (eels, for example), all amphibians, and sea snakes, however, oxygen obtained directly through the integument is a significant supplement to that obtained by gills and lungs. In elasmobranchiomorphs, osteichthyans, and larval lampreys, the respiratory current always enters through the mouth, passes over the gills, and leaves through the pharyngeal (gill) slits. Juvenile hagfishes and adult agnathans also have slits in the pharyngeal wall and these lead into pouchlike chambers (which open to the outside) so that when the animals are not feeding, they can respire in the same way other fishes do. But agnathans face a respiratory problem when the mouth is occupied for long periods of time with feeding, as when an adult lamprey is attached to another fish, because then the water current cannot enter the mouth. Under this circumstance, adult lampreys draw water into the gill pouches through the external openings on the sides of the body. This is accomplished by means of a bellowslike muscular action of the walls of the gill pouches, and water is then forced out the same openings. But hagfishes solved the problem of respiring while feeding in a different way. They have a single nostril at the tip of the snout that connects internally with the pharynx, permitting water to flow into the pharynx even when water cannot enter the mouth. In contrast, the nostril of lampreys is a blind sac; it does not open into the pharynx.

Gas exchange between the blood and the water is enhanced in osteichthyans by a countercurrent flow of oxygenated water and deoxygenated blood. That is, the blood flows in one direction through the gills while the water flows over the gills in the opposite direction. The result is that the blood lowest in oxygen meets the water that has given up much of its oxygen already, whereas the blood that has already taken on some oxygen meets the water which contains the most oxygen. This arrangement allows for highly effective oxygen uptake, some osteichthyans extracting 85 percent of the oxygen from the water passing over their gills.

Most osteichthyans have a *swim bladder* (a gas-filled sac, lying dorsally, above the peritoneal lining of the coelom but ventral to the dorsal aorta), which develops as an outpocketing from the pharyngoesophageal region of the digestive tract. The connection (pneumatic duct) with the digestive tract may persist in the adult, in which case, gas can pass in either direction through it; in most species, however, this duct is lost. When the pneumatic duct is present, it has a muscular valve that prevents accidental loss of the contained gas. The chief role of the swim bladder is that of a hydrostatic organ regulating the buoyancy of the fish; the quantity of gas it contains can be increased or decreased to alter the specific gravity of the fish as it changes depth in the water. The lighter-than-water gas counterbalances the heavier-than-water tissues, thereby bringing the specific gravity of the fish close to that of the ambient water. By adjusting the gas volume of the swim bladder, thereby changing its specific gravity, the fish can maintain a particular depth with the least muscular effort because it becomes neutrally buoyant, expending a minimum of energy maintaining its position in the water. When a fish descends, for example, from sea level to a depth of 10 meters, the increased external hydrostatic pressure causes the volume of the swim bladder to be halved, with a concomitant increase in the specific gravity of the fish, and its buoyancy becomes decreased. Water pressure increases by approximately 1 atmosphere for each 10 meters in depth. The fish must, therefore, add gas to its swim bladder until the pressure in it is 2 atmospheres, thereby returning the swim bladder to the volume it had at sea level and providing neutral buoyancy at the 10-meter depth. If the gas were not added, the fish would tend to sink even further because of its increased specific gravity. A secondary role of the swim bladder in some fishes is a sensory one, aiding these fishes to hear by acting as sound resonators.

Swim bladders are best developed in osteichthyans that are active midwater swimmers. Bottom dwellers, such as flounders, lack a swim bladder. The specific gravity of bottom dwellers is always greater than that of sea water, about five percent greater, which is an advantage because they do not tend to float up from the bottom. If bottom dwellers had a swim bladder, they would obtain no benefit from it. Some primitive fishes that dwell near the surface and have a pneumatic duct fill the swim bladder by gulping air. The rest of the fishes, including those that do not use the pneumatic duct to fill the swim bladder, fill it by secreting gas into the bladder cavity from a network of capillaries in the bladder wall. The secreted gas in fishes living near the surface is much like air, whereas at greater depths, oxygen becomes the increasingly dominant component. The gas in swim bladders of deep-sea fishes can be 65 to 95 percent oxygen. This ability to secrete gas into the swim bladder was an adaptation that eliminated the need among fishes that have this ability to return to the surface to gulp air and may have been important in the colonization of deep water by osteichthyans. Removal of gas from the swim bladder occurs as a consequence of decreased external hydrostatic pressure when a fish swims toward the surface (a) by direct release into the water through the duct (when it is present), or (b) by reabsorption into the blood. Reabsorption occurs in those fishes lacking the duct and is also known to take place even in fishes having the duct that are exposed to only a mild reduction in external pressure. The dorsal location of the swim bladder confers greater stability on fishes than would a ventral location, which would make fishes top-heavy.

The lung of lungfishes, whose lobed fins were described earlier, is, depending upon the species, single or paired, and saclike with few internal folds or subdivided into many pouches. Like the swim bladder, these lungs develop as outgrowths of the pharyngoesophageal region of the digestive tract. There is, of course, a persistent duct running from the site of the original outpocketing to the lungs, through which air can enter and leave them. Unlike most fishes, the nostrils of lungfishes open internally, providing an air passage. Lungfishes do not have a swim bladder. According to A. S. Romer, the swim bladder most likely evolved from a fish lung and not vice versa. Fish lungs occupy the same dorsal position as the swim bladder. A rudimentary swim bladder occurs transitorily in a few elasmobranchiomorph embryos as well as in embryos of some osteichthyans that lack it as adults. It is entirely absent in all agnathans and adult elasmobranchiomorphs.

The evolution of tetrapods that lived on land and breathed air by means of lungs was accompanied by several anatomical alterations and innovations. The nostrils of all tetrapods open internally, providing a passageway for the air. In amphibians, these internal openings are in the anterior part of the primary palate. From the oral cavity of those amphibians that are air-breathing, the air enters the pharynx, from which it passes through the *glottis*, a median ventral opening in the pharynx, into the *larynx,* an enlarged chamber that leads to the *trachea,* a tube that divides ultimately into two *bronchi,* one leading to each lung. In contrast, crocodilians and mammals possess a hard, bony *secondary palate,* which completely separates the nasal passages from the oral cavity. The primary palate is the ancestral roof of the oral cavity, as seen in fishes and amphibians. However, the evolution of the secondary palate resulted in the formation of a new roof for the oral cavity with the air passageway thus shut off from the oral cavity, being located now between the primary and secondary palates. As a result of the development of the secondary palate, air coming into the nostrils of those animals that have a secondary palate has a freer corridor to the pharynx than in other tetrapods, where it must pass through the oral cavity. In mammals, this hard secondary palate is extended backwards even further by a membranous *soft palate.* Birds and reptiles other than the crocodilians show the beginnings of the development of a secondary palate in the form of a pair of longitudinal *palatal folds* that help to form a channel above the tongue for a somewhat freer passage of air through the oral cavity than occurs in amphibians. In frogs and toads, a few lizards, and most mammals, the larynx contains a pair of vocal cords. These cords are ridges containing elastic tissue that can be set in vibration by the passage of expired air between them. The bird larynx has no vocal cords; instead, birds produce sound in an organ, the *syrinx,* comparable to the larynx but located in the region where the trachea divides into the two bronchi.

Tetrapod lungs are always paired. Those of amphibians are saclike, like the simpler ones of lungfishes. Saclike lungs are not as efficient as lungs that are highly subdivided. Multichambered lungs have a much greater internal surface area for gas exchange than do lungs that are simple sacs. While the inner surface of amphibian lungs is often smooth, in some frogs and toads it has a series of ridges that slightly increase the surface area available for gaseous exchange, thus increasing their effectiveness somewhat. The mucus secreted onto the surface of the amphibian integument is an important aid for aerial respiration. This mucus keeps the integument of amphibians moist so that oxygen can diffuse in. When the integument is dry, oxygen ceases to enter. Any animal that does not live in water but whose integument has a significant role in respiration has a means of keeping its integument moist. A respiratory surface must be moist because oxygen can diffuse into a cell only if it first becomes dissolved in moisture on the surface of that cell. This ability of amphibians to obtain a significant amount of oxygen through the integument, up to 40 percent of what they use when not hibernating, helps to compensate for their relatively inefficient lungs. Hibernating amphibians obtain practically all of their oxygen through the integument. The lungs of reptiles are an improvement over those of amphibians. Reptilian lungs have internal partitions subdividing the lungs into blind pockets, called *alveoli,* markedly increasing the surface area that can serve for the exchange of oxygen and carbon dioxide. However, reptilian lungs are not nearly as highly subdivided as are the lungs of endotherms. The lungs of reptiles, birds, and mammals, because of having become internally subdivided, are spongy in contrast to the saclike amphibian lungs. The reptilian and mammalian lungs are built on the same plan; that is, the gas exchange occurs in the alveoli which were formed by the subdividing of their lungs.

Bird lungs, in contrast, are built on a different plan. Birds have an even more complex lung apparatus, which is described below, than is found in reptiles and mammals. This is not surprising in view of their need for large amounts of oxygen during flight. Communicating with the air passages in the lungs of birds are numerous paired *air sacs*, referred to earlier, that invade every major part of the body. The structure of the air sacs suggests that they themselves would take up little oxygen directly; their walls have few blood vessels and no folds to increase the surface area. The air sacs can be divided into two groups, anterior and posterior. Only in birds does each bronchus run to and then through one of the lungs, terminating in a posterior air sac. Anterior air sacs connect to each bronchus at the anterior portion of each lung and posterior air sacs connect to each bronchus at the posterior end of each lung. Furthermore, each bronchus, in passing through a lung, sends branches to the lung tissue, and some of the air sacs also connect directly to the lung tissue as well as to the bronchi. When a bird inhales, the air passes via the bronchi directly to the posterior air sacs. On exhalation, air from the posterior air sacs flows into the lung, not out through the bronchus. On the next inhalation air from the lungs flows into the anterior air sacs. Then, finally, on the second exhalation, air from the anterior air sacs flows into the bronchi and to the outside. As is untrue of other tetrapods, in birds, therefore, two cycles of inhalation and exhalation are required for air to enter and leave the system. Furthermore, in contrast to the blind alveoli in reptiles and mammals where the gas exchange occurs, in birds the gas exchange occurs within the lungs in channels, called *parabronchi*, that are open at both ends; and whereas in the alveoli of reptiles and mammals the air must enter and then reverse direction to leave, in bird lungs, the air flow is unidirectional, always from posterior to anterior.

Amphibians fill their lungs by forcing air into them. Air is sucked into the oral cavity through the nostrils (the mouth being kept tightly closed) by lowering the floor of this cavity. The nostrils then close and the floor of the oral cavity is raised, forcing the air into the lungs. The elastic recoil of the lungs and pressure exerted on them by contraction of the muscles of the body wall help to expel the air from the lungs. The mechanisms animals use to bring oxygen to a respiratory surface and to remove carbon dioxide from it are called *ventilating mechanisms*. The ventilating mechanism that amphibians use, forcing air into the lungs, is the positive-pressure type of ventilation. The rest of the tetrapods fill their lungs by negative-pressure ventilation, that is, the volume of the cavity in which the lungs reside becomes increased. This increase results in a decrease in the air pressure around the lungs to a level below that of the atmosphere and air flows into them. Most reptiles produce this negative pressure in the trunk by moving the ribs outward and forward. Turtles, however, are unable to move their ribs, which are fused with the shell; they depend mainly upon the action of abdominal muscles to provide for inhalation. When these muscles contract, they cause an increase in the volume of the coelom, bringing about a pressure decrease therein which leads to lung inflation. In birds, inhalation depends upon muscles that cause the *sternum* or breastbone to move anteriorly and ventrally and the ribs to move laterally and anteriorly, thereby increasing the size of the coelom. However, because of the unique arrangement of lungs and air sacs, the lungs of birds show little change in volume with the passage of air through them; it is the air sacs that are mainly responsible for air passing through the lungs. The air sacs act as bellows to move air in and out as a result of the pressure changes in the coelom. Mammals, to accomplish inhalation, utilize not only rib action as in birds but also a dome-shaped muscle, the *diaphragm*, which is found only in mammals. It separates the coelom into thoracic and abdominal portions. At inhalation, the diaphragm contracts, flattening itself and thereby contributing, along with the rib action, to the increase in the size of the thoracic cavity. When, in these vertebrates that use the negative-pressure type of ventilation, the muscles that produce inhalation relax, the air is expelled from the lungs. The lungs of reptiles, birds, and mammals and the air sacs of birds have an inherent elasticity, like the lungs of amphibians, that facilitates exhalation.

SENSE ORGANS The more important sense organs of vertebrates are the nose, eyes, and ears. The receptor cells for olfaction (the sense of smell) are in the nasal epithelium. The nose evolved as an olfactory organ and assumed secondarily a role in respiration. Agnathans have a single nasal opening. In lampreys, it is located between the eyes on the upper surface of the head and is a blind pocket, in contrast to the nasal opening of hagfishes, which, as stated earlier, is located at the tip of the snout and opens internally into the pharynx. All other vertebrates have paired nasal openings. Most jawed fishes, unlike the lungfishes, have nasal pockets that do not open internally. In tetrapods, however, the nasal passages always open into either the oral cavity or the pharynx. Although the nasal passages of terrestrial vertebrates are filled with air instead of water, they are lined with a film of liquid in which the airborne odorous substances dissolve. The olfactory cells, as is true of the taste buds, are sensitive only to dissolved chemicals.

Vertebrates have a camera-type eye whose gross structure is similar to that of the octopus and squid. The cornea's index of refraction is so close to that of water, that in aquatic species it is nearly useless as a focusing device, the burden of focusing falling almost completely on the lens. In terrestrial vertebrates, however, the cornea does most of the focusing, with the lens serving only for fine adjustments. However, all vertebrates do not use the same method to alter the focus of the lens for near and distant objects. In reptiles, birds, and mammals, the curvature of the lens is increased in order to focus on near objects. In fishes and amphibians, the general method is the same as that used by the octopus and squid. The lens is moved forward or backward to control focus: forward to adjust for near objects, and backward for distant objects.

Two types of photosensitive cells, rods and cones, have been found in the vertebrate retina. They are named for their usual shapes in mammals. Rods are used for vision at very low intensities of illumination. They can function, in fact, at light intensities that are too dim to stimulate the cones. But in bright light, much of the photosensitive pigment in the rods is bleached, and their sensitivity is greatly reduced. After exposure to bright light, a period of 30 to 40 minutes in darkness is required for the rods to recover their maximal sensitivity. On the other hand, cones function efficiently in bright light and are responsible for color vision. The rods cannot distinguish colors, perceiving only brightness differences. Experiments with man and monkeys have revealed that their eyes have three types of cones, which are differentially sensitive to light of various wave lengths, providing the basis for color vision. One type absorbs blue light maximally; another, green light maximally; and the third, yellow light maximally. The fact that many vertebrates, from osteichthyans to mammals, have color vision has indeed been conclusively demonstrated experimentally.

Eyes of vertebrates exhibit several adaptations that have been correlated with the daily activity pattern of the animal. The visual adaptations of nocturnal (night-active) animals increase the sensitivity of the eye; those of diurnal (day-active) animals increase visual acuity or resolving power. Animals, such as ground squirrels and turtles, that are active only during the daytime (strictly diurnal) have all-cone or virtually all-cone retinas, whereas animals, such as bats and armadillos, that are active only at night (strictly nocturnal) have no cones or relatively few cones. Furthermore, diurnal animals show a decrease in the number of cones connected to a single neuron in the retina, thereby increasing their acuity, whereas nocturnal animals show an increase in the number of rods connected to a single neuron, thereby increasing their sensitivity, but at a sacrifice of some acuity. Another visual adaptation of nocturnal animals is the presence of a reflecting layer (*tapetum lucidum*) on the choroid coat, which lies just behind the retina. If light has passed through the retina once without being absorbed, this reflecting layer directs light back into the retina. Some of this reflected light happens to be reflected out through the cornea and is responsible for the eyeshine so commonly associated at night with nocturnal animals.

The ear structures that are common to all vertebrates are those of the paired *inner ear*, a labyrinth of membranous, fluid-filled canals and sacs concerned with both equilibrium and hearing. There is an inner ear on each side of the skull. The original basic function of the inner ear was probably that of an equilibrium receptor, a statocyst, hearing apparently having been lacking or at best poorly developed in the ancestral vertebrates. The equilibrium-receptor portions of the inner ear are two saclike structures, the *sacculus* and *utriculus*, and the tubular *semicircular canals*. Each of these equilibrium receptors contains patches of sensory hair cells. The sensory patches in the sacculus and utriculus are called *maculae*, whereas the sensory patches in the semicircular canals are called *cristae*.

At the end of each semicircular canal is a bulblike enlargement called an ampulla. In it is a crista. Jawed vertebrates have three semicircular canals on each side of the head. Each of the three lies in a plane at right angles to the other two. Agnathans, however, have only one semicircular canal on each side of the head (hagfishes) or two on each side (lampreys). The sacculus and utriculus convey information concerning the static position of the head and respond to *linear acceleration* (straight-line, nonrotational motion). The semicircular canals, on the other hand, impart information concerning *angular acceleration* (rotational motion of the head). There is evidence from experiments comparing the utriculus and sacculus that the utriculus is the principal organ for sensing linear motion. The fluid in the inner ear is called *endolymph*. Each sacculus and utriculus contains one or more statoliths, whose displacement stimulates their hair cells. On the other hand, the semicircular canals have no statolith. Their hair cells are stimulated by angular acceleration in the following way: The endolymph, because of its inertia, lags behind the canal wall in acceleration, resulting in a relative motion of the endolymph past the hair cells, bending the "hairs."

The hearing, or phonoreceptive portion, of the inner ear is represented in fishes, amphibians, and reptiles other than the crocodilians by a small evagination from the sacculus, the *lagena*. But in crocodilians, birds, and mammals, the lagena has evolved into a highly complex, endolymph-filled organ of hearing called the *cochlea*. The cochlea of crocodilians and birds is essentially a straight tube, but it is coiled in mammals. The cochleas of mammals presumably evolved their compact, coiled configuration because if uncoiled they would be so much longer than those of crocodilians and birds that there would be a space problem fitting them into the skull. The sensory portion of a lagena consists simply of a patch of sensory epithelium containing hair cells and this patch is called the *lagenar macula*, but in a cochlea the sensory structure is elongated, running the length of the cochlea, and is called the *organ of Corti.* It contains the sensory hair cells, and rests on the basilar membrane (Fig. 12-4). Overhanging the organ of Corti is a flap, the *tectorial membrane.* The tips of the hair cells are embedded in the underside of the tectorial membrane. When vibrations produced by the external sound waves are received within the cochlea, a particular portion of the basilar membrane vibrates more than any other in response to a sound having a specific frequency. This vibration induces bending of the "hairs" of the hair cells in the region of maximal vibration against the tectorial membrane, thus stimulating the hair cells which, in turn, initiate impulses in fibers of the cochlear nerve. Frequency discrimination, therefore, depends upon which hair cells are stimulated. Intensity discrimination is based on the fact that higher intensities of sound cause a greater amplitude of oscillation of the basilar membrane, leading to a more intense stimulation of the hair cells.

Conduction of the sound waves from the body surface to the lagena or cochlea is accomplished in various ways. In most fishes, for sound waves in the water to be heard they must produce vibrations in the head of the fish, which cause the endolymph in each lagena to vibrate, thereby stimulating the hair cells of the lagenar maculae. Some fishes, however, as mentioned in the discussion of the swim bladder, make use of the swim bladder as an aid to hearing. In these fishes, either the anterior end of the swim bladder lies alongside a part of the inner ear, thus more directly inducing vibrations in the endolymph, or three or four small bones (*Weberian ossicles*), derived from the most anterior vertebrae, form a chain connecting the swim bladder to the inner ear that transmits sound vibrations from the swim bladder to the inner ear. In those tetrapods that have tympanic membranes (some amphibians lack them, as stated earlier), the sound waves cause them to vibrate. The evolution of the tympanic membrane was accompanied by the evolution of the *middle ear,* an air-filled cavity between the tympanic membrane and inner ear. Within this cavity in those amphibians that have one, as well as in reptiles and birds, is a single bone, the *stapes,* which transmits vibrations from the tympanic membrane to the inner ear. The absence not only of tympanic membranes but also of

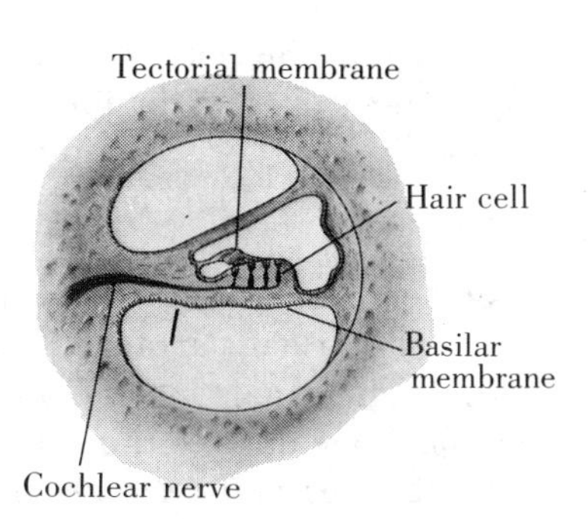

Fig. 12-4 *The cochlea and neighboring structures in the human ear.*

middle ear cavities and stapes in some amphibians, such as the snake-like, legless caecilians, is not thought to be a primitive condition but rather a degenerate one. In mammals, the middle ear contains three bones, *malleus*, *incus*, and *stapes*, that transmit vibrations to the endolymph-filled cochlea. The vibrations in the endolymph, in harmony with the external sound waves, are the immediate cause of the oscillations of the basilar membrane. The tympanic membranes on the body surface of those amphibians that have them have moved internally in some reptiles, in birds, and in mammals. This resulted in the development of an *external ear* on each side of the head, which consists of the external auditory canal, and in mammals the projecting ear pinna, which collects sound waves, guides them into the canal, and also aids in localization of their source.

Each visceral arch contains not only an aortic arch, but also a cartilaginous skeletal element, called a *visceral cartilage*. As stated earlier, the ancestral jawed vertebrate is considered to have had six pairs of visceral arches. However, the jawless ancestor of the jawed vertebrates appears to have had additional visceral arches, possibly three or four more pairs, anterior to the six pairs postulated for the ancestral jawed vertebrate. These additional, anterior visceral arches, with their visceral cartilages, appear to have evolved into the upper and lower jaws of the jawed vertebrates. Furthermore, when tetrapods appeared and they lost their gills, obtaining lungs, the remaining six pairs of visceral cartilages were no longer needed to support the gills, as they do in living fishes, and they appear to have been transformed into the malleus, incus, and stapes and skeletal supports for the tongue, larynx, and trachea. With this apparent evolution of new uses for ancient structures, we have an excellent illustration of how different the functions of homologous structures can become.

Some birds and mammals use *echolocation*, sometimes also called biological sonar or acoustic orientation, in navigation and in locating food. Echolocation by bats is probably the best-known example. While a bat is flying in a completely darkened room, it is able to avoid fine wires suspended from the ceiling. However, if either its mouth or ears are sealed, the ability is lost. Bats avoid such obstacles by emitting high-frequency sounds, higher than can be detected by man, and listening for the reflected waves. Echolocation also enables bats to catch flying insects at night for food. Dolphins and cave-dwelling birds are examples of other vertebrates that employ echolocation.

Lateral line organs are found in fishes, larval amphibians, and permanently aquatic adult amphibians. They consist of canals that run along each side of the body and onto the head in which are embedded sensory hair cells arranged in clusters. These clusters are called *neuromasts*. The canals may be open grooves, or they may be closed with occasional openings to the surface. The lateral line system is sensitive to disturbances and currents in the water, thereby providing the bearer with a "distant touch" capability, enabling the animal to determine the position of another organism that is creating a disturbance in the water. The animal can also determine the position of stationary objects that reflect disturbances in the water produced by the animal itself. Lateral line organs can provide information to be used in orientation, food detection, and schooling. They may also enable the animal to determine its rate of movement through the water. These organs are of evolutionary interest because it appears that the inner ear, which, as noted above, also has hair cells, evolved when an anterior portion of the lateral line system sank into the head.

EXCRETION The excretory systems of the several vertebrate classes show a clear evolutionary relationship to each other, but have a different structural basis from the excretory systems of tunicates and amphioxus that were described in Chapter 11. The basic functional unit of the excretory organs of vertebrates is the *nephron*. It consists of a closed bulbous structure, called *Bowman's capsule*, at its inner end, and a long tubule that connects to ducts leading to the exterior. The excretory organs of vertebrates are paired and lie dorsally, outside the coelom. The nephron does not appear to have been phylogenetically derived from either the protonephridia or metanephridia of invertebrates, including amphioxus. Furthermore, neither chaetognaths, echinoderms, nor hemichordates (the nonchordate deuterostomes) possess any structure from which the vertebrate excretory organs could have been derived, and the tunicate excretory system that utilizes amoebocytes for the deposition of excretory products is not the type that would have given rise to a nephron. The nephron most likely, therefore, evolved independently as an innovation of the vertebrates. Comparative vertebrate anatomy and embryology do suggest, however, that the first vertebrate nephron, instead of having a closed end (Bowman's capsule), had a nephrostome that opened into the coelomic cavity, an anatomical arrangement reminiscent of the metanephridia of invertebrates. Through the nephrostome, each nephron possibly collected excess water that had passed into the coelomic fluid by way of *glomeruli* (tufts of capillaries), which were suspended in the coelomic cavity. Later, the nephrons would lose their direct opening into the coelom and glomeruli would begin to supply a blood filtrate directly to the nephrons through the closed Bowman's capsules. The excretory tubules of invertebrates are generally not massed together to form discrete excretory organs, such as occur in vertebrates. Consequently, the term *kidney* is usually reserved for the excretory organ of vertebrates.

Vertebrate kidneys form from two bands of tissue that extend the length of the trunk, dorsal to the coelom (Fig. 12-5). The anterior ends of both bands soon differentiate into functional kidneys. Each of

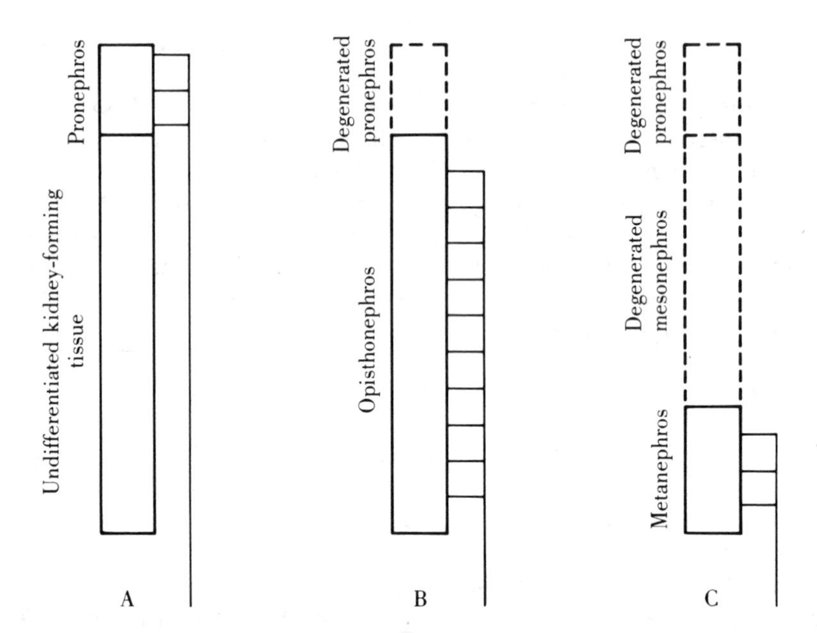

Fig. 12-5 *Fate of the vertebrate kidney-forming tissue. (A) Embryonic state, showing functional pronephros with the rest of one of the two bands of kidney-forming tissue in the still-undifferentiated state; (B) a functional opisthonephros; (C) a functional metanephros.*

these early kidneys is called a *pronephros*. The remaining posterior portions of the two kidney-forming bands are called *opisthonephroi*. The pronephroi are usually short-lived, being replaced by kidneys that develop from the more posterior opisthonephroi. However, in some fishes and amphibians, the pronephroi may persist as functional kidneys until late in development. Nevertheless, in fishes and amphibians each opisthonephros, in its entirety, ultimately forms a functional kidney that takes over the excretory role from the transitory pronephros and persists for the life of these animals.

Pronephroi function as excretory organs only until the nephrons of the opisthonephroi have developed the capability of assuming this role. Whereas in fishes and amphibians, the entire opisthonephros develops into the kidney of the adult, in reptiles, birds, and mammals, each pronephros is replaced at first by still another transitory kidney, called a *mesonephros*. The mesonephros develops from the anterior portion of the opisthonephros and functions (both the pronephroi and mesonephroi of the reptiles, birds, and mammals ultimately degenerate) only until the posterior portion of each opisthonephros differentiates into a *metanephros*. The two metanephroi are the kidneys of the late embryo and the adult among birds, mammals, and most reptiles, but in some reptiles, the mesonephroi continue to function for a short time after the animal hatches, and then the metanephroi take over and the mesonephroi degenerate.

Chordates doubtless originated in the sea from marine invertebrates, tunicates and amphioxus always being strictly marine. The fossil record supports the view that the first vertebrates were also marine organisms. However, study of the salt concentrations in the blood of living fishes has led many investigators to conclude that although the first fishes may have been marine organisms, at some time in the past, an ancestral agnathan stock had to have taken up residence in fresh water where the more modern agnathans and also the jawed fishes (the elasmobranchiomorphs and osteichthyans) then originated. From this freshwater habitat, some of the newly evolved fishes are presumed to have then entered the sea. The concentrations of salts in the body fluids of all living, aquatic vertebrates, except hagfishes, are indeed less than that of seawater, which would support the hypothesis that present-day fishes had a freshwater origin and only later invaded the oceans. Furthermore, it has been suggested, by those who support the freshwater origin of present-day fishes, that the kidney of present-day fishes evolved initially as an organ to eliminate the excess water these fishes in freshwater acquired because of their being hyperosmotic to their environment. In fact, in modern (that is, ray-finned) osteichthyans, nitrogenous wastes are lost by diffusion across the gills, not *via* the kidneys. The primary function of the vertebrate kidney would, therefore, have been the excretion of excess water, and excretion of nitrogen wastes was only a secondarily acquired role. If present-day fishes had a freshwater origin, when some of them later migrated from fresh water to the sea, these migrants were faced with the opposite problem they had in fresh water; because their blood was much less concentrated than sea water, they would have lost water to the environment. Mechanisms to compensate for or prevent water loss were necessary. The blood of marine lampreys and modern marine osteichthyans is only about 10 percent more concentrated than that of their freshwater counterparts. Marine lampreys and modern marine osteichthyans compensate for their water loss by swallowing sea water. The water and some of the salts are then absorbed from the digestive tract. The water is retained, while the excess salts are eliminated *via* both the kidneys and the gills. Elasmobranchiomorphs have a very different solution to the osmotic problem they would face if they were hyposmotic to sea water. Marine elasmobranchiomorphs keep the osmotic pressure of their body fluids equal to, or slightly higher than, that of their environment by retaining a relatively large quantity of urea and trimethylamine oxide, both of which are non-electrolyte nitrogen-containing compounds, in their blood. The total concentration of the urea and trimethylamine oxide in the blood of marine elasmobranchiomorphs is about 2.0–2.1 percent, there being about twice as much urea as trimethylamine oxide. This retained urea and trimethylamine oxide accounts for somewhat more than one-half, near 55 percent, of the total osmotic concentration of the blood in marine elasmobranchiomorphs. The concentration of salts (about 1.5%) in the blood of marine elasmobranchiomorphs is actually about the same as in modern marine osteichthyans; the total concentration of the salts in sea water is 3.5 percent. Thus marine elasmobranchiomorphs face no threat of water loss and have no need to drink sea water. Even freshwater elasmobranchiomorphs retain urea and trimethylamine oxide, the total concentration of these two nitrogen-containing compounds averaging about one-third of that in the blood of marine elasmobranchiomorphs. There is also, as in marine elasmobranchiomorphs, about twice as much urea as trimethylamine oxide present in the blood of these freshwater elasmobranchiomorphs, although their blood would be hyperosmotic to their environment even without the urea and trimethylamine oxide. The concentration of salts in the blood of freshwater elasmobranchiomorphs is about 50 percent of that in marine elasmobranchiomorphs.

Hagfishes are exclusively marine. Their blood is isosmotic with sea water. They are the only vertebrates that are isosmotic with sea water by virtue of the salt concentration in their blood alone, showing no tendency to retain either urea or trimethylamine oxide as do the elasmobranchiomorphs. Surprisingly, the living lobefinned osteichthyan, the coelacanth, which, like the hagfish, is also exclusively marine, has a high blood osmotic pressure because it adopted the elasmobranchiomorph strategy of retaining urea and trimethylamine oxide to raise the osmotic pressure of its blood. The concentration of electrolytes in the blood of a coelacanth is in fact very close to that of other marine osteichthyans. Also, the blood of the coelacanth contains about three times as much urea as trimethylamine oxide, the combined concentration of the urea and trimethylamine oxide being about equal to the concentration of salts in its blood, about 1.5 percent, about the same concentration as in the blood of marine elasmobranchiomorphs and modern marine osteichthyans. Consequently, according to these most recent data, the coelacanth is, in spite of retaining both urea and trimethylamine oxide, still slightly hyposmotic to sea water. All other living marine osteichthyans, as explained above, are, in contrast, markedly hyposmotic to sea water, retaining neither urea nor trimethylamine oxide. The fact that the primitive hagfish is isosmotic with sea water, (not because of the retention of urea and trimethylamine oxide, but because the electrolyte concentration of its blood is the same as that of sea water) is considered by some workers as evidence against a freshwater origin of present-day fishes. They also consider it support for the hypothesis that all present-day fishes originated in the sea, adopting the salt concentration of salts in the sea as their own. According to this theory, the regulatory mechanisms described above only evolved later, some fishes invading freshwater habitats, others remaining in the oceans, and still others, such as the salmon and eel, spending part of their lives in the sea and part in fresh water. However, those who favor a freshwater origin for present-day fishes have proposed that the reason hagfishes have a higher electrolyte level than other fishes is that when they invaded the seas, they failed to evolve a regulatory mechanism that would enable them to keep a low electrolyte level in their blood, such as drinking sea water or retaining urea and trimethylamine oxide, and simply "allowed" their blood to become more concentrated, ultimately equaling the electrolyte concentration of sea water.

Agnathans and osteichthyans are ammonotelic. Elasmobranchiomorphs are ureotelic, presumably because they retain so much urea in their blood for osmoregulation and not because of a restriction in water supply. Because marine elasmobranchiomorphs are either isosmotic or slightly hyperosmotic to their environment, they do not suffer from osmotic water loss.

Amphibians face the dual problem of dehydration on land and dilution when they enter fresh water. Urea is the main nitrogenous waste product of most adult amphibians; but in larval and permanently aquatic adult amphibians ammonia is the main one that is excreted. When amphibians are in fresh water, they take on excess water osmotically. Their kidneys excrete the excess water as dilute urine, and their skin actively absorbs salt from the medium to compensate for that lost in the urine. On land, amphibians conserve water by reducing urine formation to a minimum.

The urine of birds and terrestrial reptiles is a semisolid or solid mass; so much water is reabsorbed along the route of urine formation that the uric acid, their main nitrogenous waste product, which has a low solubility, precipitates. It will be recalled that uric acid is the nitrogenous waste product typically associated with animals from relatively dry environments. Urea, however, is the main nitrogenous waste of aquatic reptiles and all mammals. Birds and mammals are the only vertebrates capable of producing a urine that is hyperosmotic to their blood. This ability is much more pronounced in mammals than in birds. For example, the urine osmotic concentration of some mammals can reach 20 times that of their blood, while the maximum urine/blood concentration ratio that can be attained by birds is only about 2. This ability to produce urine that is hyperosmotic to their plasma is a decided advantage for terrestrial organisms because it minimizes their water loss by way of the kidneys. Reptiles can produce urine that is isosmotic with their blood but cannot produce urine that is more concentrated than their blood. Birds and mammals are able to produce urine that is hyperosmotic to their blood because their nephrons are unique in that a portion of the nephric tubule forms a hairpin-shaped loop, *the loop of Henle*, which in conjunction with a collecting tubule acts as a countercurrent multiplier. The countercurrent flow of fluid through the loop of Henle and a collecting tubule, as will be described, allows for the formation of a more concentrated urine than would be possible otherwise. Those species that can produce the most concentrated urine have the longest loops of Henle. The loop of

Henle is longer in mammals than in birds, which accounts for the ability of mammals to produce urine which is even more hyperosmotic to their blood than can birds. The collecting tubules are the series of ducts into which the nephrons of a metanephros empty, the urine passing through the collecting tubules and eventually being eliminated from the body. Most vertebrates develop a bladder in which the urine may be stored prior to being voided; some elasmobranchiomorphs have no urinary bladder. The anatomical relationship of the loops of Henle and the collecting ducts is such that the ascending branch of each loop of Henle has a collecting tubule running parallel with it, carrying fluid in the opposite direction from that in which the fluid inside the ascending branch is flowing. The fluid that initially enters the loop of Henle is isosmotic with the blood, having passed from the blood into Bowman's capsule and then on into the loop of Henle. As this fluid passes down the descending branch of the loop of Henle it becomes increasingly concentrated because the ascending limb of the loop of Henle pumps out sodium ions into the surrounding fluid, negatively charged chloride ions passively follow these positively charged sodium ions, and these sodium and chloride ions then diffuse into the fluid in the descending limb of the loop of Henle where their concentration is lower than in the surrounding fluid. This fluid in the descending limb then flows around the bend into the ascending limb where the sodium ions are again being pumped out, again the chloride ions following passively, and the fluid in the ascending limb of the loop of Henle becomes less and less concentrated as it passes up the ascending limb. As a result of this recirculation of sodium and chloride ions, the lower section of the loop of Henle and that portion of the collecting tubule which is adjacent to the lower section of the loop of Henle are bathed in a fluid that is hyperosmotic to the blood. The cells of the ascending branch, unlike those of the descending branch, are impermeable to water, so it does not diffuse out when the sodium ions are pumped out. The collecting tubule is, however, permeable to water, so as the fluid passes down this tubule, it passes the zone of high concentration of sodium and chloride ions and water leaves the collecting tubule by osmosis, thus resulting in the formation of urine in the collecting tubule that is isosmotic with the fluid in the zone of high concentration but hyperosmotic to the blood. In this way, birds and mammals are able to form urine, which is more concentrated than the blood from which it was derived.

In addition to the salt-excreting structures already referred to in vertebrates, namely the kidneys and gills, two additional salt-excreting organs have been found in some of these animals. One of the two is the finger-shaped *rectal gland* of elasmobranchiomorphs that removes excess sodium chloride from the blood and eliminates it *via* a duct that empties into the rectum. Second, marine reptiles and birds have *salt glands* in their heads through which they can likewise excrete excess sodium chloride. For example, the salt glands of sea gulls are especially active after these birds have been drinking sea water. These glands most commonly empty into the nasal cavity; but in marine turtles they open into the orbit of the eye, and very salty tears are shed when the turtles excrete their excess sodium chloride.

NERVOUS SYSTEM The nervous system reaches its highest level of development among the vertebrates, culminating in humans, who have the capability of articulate speech for communication. Man's mental development has enabled him to dominate all other animals and to survive in every climate.

The central nervous system of adult vertebrates consists of the brain and spinal cord. The neurons and nerves outside the central nervous system constitute the *peripheral nervous system.* The brain and spinal cord develop from the hollow nerve cord of the embryo, and remain hollow throughout life. Except in agnathans, the spinal cord is enclosed and protected by the vertebrae. Nerves that originate from the spinal cord are called *spinal nerves.* These motor and sensory nerves remain separate from each other in lampreys, as in amphioxus, but in hagfishes and jawed vertebrates they fuse outside the spinal cord to form a mixed nerve, that is, a nerve containing both sensory and motor fibers. Among the vertebrates, as in amphioxus, the fibers that originate dorsally are sensory whereas those that originate ventrally are motor ones. Those nerves that originate in the brain are called *cranial nerves.* The roots of the cranial nerves do not show the dorsal sensory-ventral motor pattern of the spinal nerves.

Brains of adult vertebrates have a common structural plan (Fig. 12-6). Early in the development of the vertebrate brain, three portions can be distinguished; the *prosencephalon* or forebrain, the *mesencephalon* or midbrain, and the *rhombencephalon* or hindbrain. As the embryo develops further its brain becomes five-sectioned. The forebrain subdivides into the *telencephalon* and *diencephalon;* and the hindbrain subdivides into the *metencephalon* and *myelencephalon;* but the midbrain goes on to develop into its adult form without subdividing. The telencephalon thus becomes the most anterior portion of this five-part brain, whereas the myelencephalon is the most posterior portion.

The telencephalon ultimately differentiates into (a) the *olfactory bulbs,* in which the sensory fibers from the olfactory cells terminate, and (b) the *cerebrum,* which consists of a pair of *cerebral hemispheres.* As the vertebrates evolved, the cerebrum underwent a steady increase in size and complexity. In fact, the cerebrum of man makes up about 80 percent of the brain weight. In fishes and amphibians, the cerebrum is concerned with little more than olfaction, but in reptiles, birds, and mammals it is the major coordinating center for all sensory

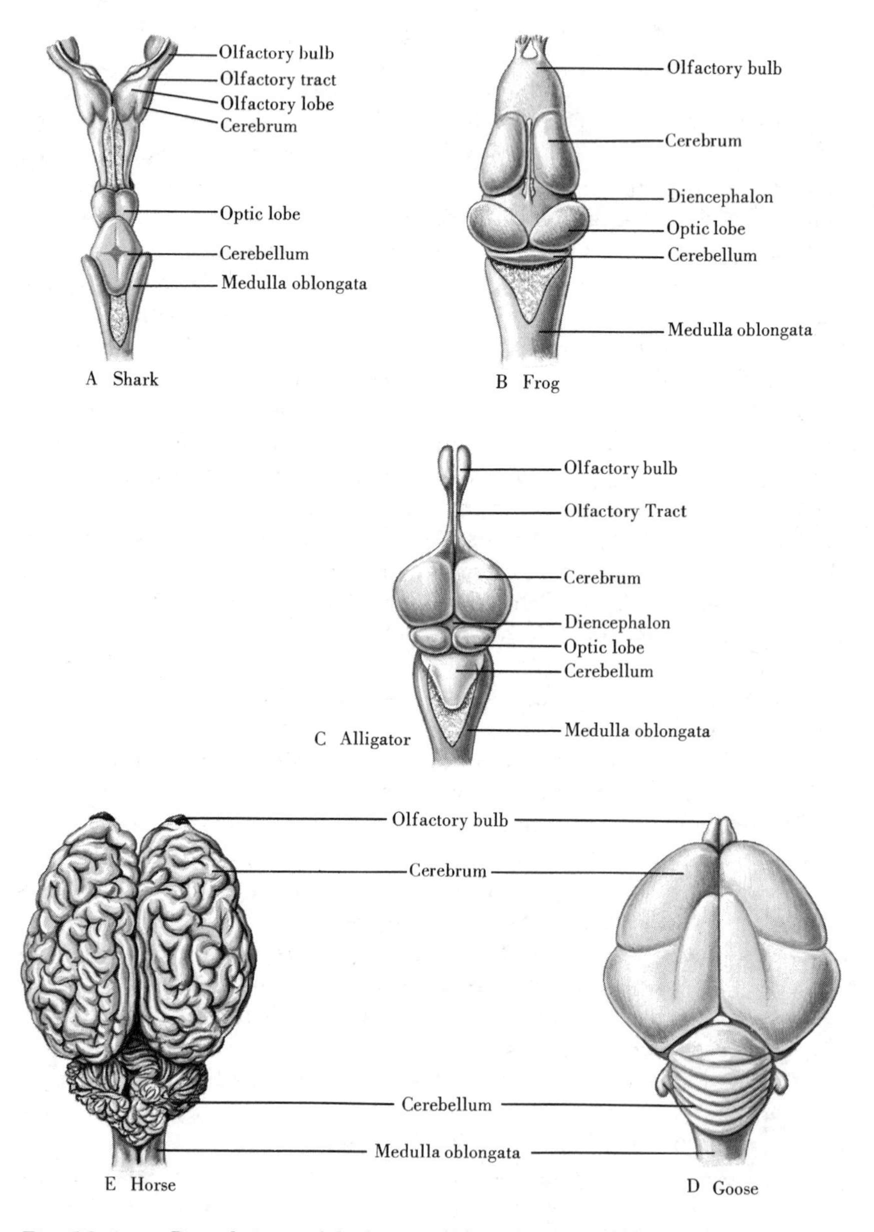

Fig. 12-6 *Dorsal views of the brain of (A) a shark; (B) a frog; (C) an alligator; (D) a goose; (E) a horse. (After Romer after others,* The Vertebrate Body, *2nd ed. W. B. Saunders Company, 1955.)*

and motor functions. Concomitant with the steady increase in the size and importance of the cerebrum has been a steady decrease in the relative size and importance of the midbrain. The midbrain is the region of highest integrative activity in fishes and amphibians; it controls their most complex behavior. In reptiles, birds, and mammals, however, the midbrain now consists only of fiber tracts connecting the forebrain to the hindbrain, and reflex centers for certain auditory and visual reflexes such as the movement of a dog's ears when a sound is heard and the constriction of the pupil when a light is shone into the eye. The changes that produced the mammalian cerebrum may have been the most significant ones that occurred in the course of vertebrate evolution. The activities of the neurons in the mammalian cerebrum provide the physiological basis for the superior memory, reasoning ability, and intelligence of these animals.

The diencephalon differentiates into the *thalamus* and *hypothalamus*. The hypothalamus is ventral to the thalamus. The thalamus has relay centers for sensory impulses going to the cerebrum and for motor impulses leaving the cerebrum. The hypothalamus is the site of origin of many of the nervous mechanisms involved in *homeostasis*, the maintenance of a constant internal environment. It has a variety of control centers for basic functions and drives. Included are centers that participate in the control of body temperature (in endotherms), anger, appetite, drinking, blood pressure, and sexual behavior. In addition, electrical stimulation of various parts of the hypothalamus produces, among other sensations, those of pain or pleasure. The hypothalamus also has neurosecretory cells whose roles will be discussed below.

The metencephalon eventually differentiates into (a) the *cerebellum*, (b) *ventral fiber tracts* that not only cross from one side of the cerebellum to the other but also provide for intercommunication between the cerebellum and other parts of the brain, and (c) part of the *medulla oblongata*. In lower vertebrates, the floor of the metencephalon shows little demarcation from the medulla oblongata (which develops from the myelencephalon), but with the increasing importance of the cerebellum, particularly in birds and mammals, the metencephalon ventrally comes to be occupied by more and more fiber tracts running to and from the cerebellum and from one side of it to the other, producing a swelling in the region. In fact, a mammal has so many fibers here, and they form such a conspicuous swelling on the ventral surface, that this region in mammals has been given a special name, the *pons*. The cerebellum coordinates the activity of the skeletal muscles assuring normal, smooth movements and proper posture. It receives sensory impulses from the inner ear and from *proprioceptors*, or sensory cells, in skeletal muscles and tendons that provide information about the positions of the animal's limbs. The size of an animal's cerebellum is roughly directly proportional to the complexity of its activity. It is best developed in fast-flying birds and in mammals. Removal of the cerebellum does not result in paralysis, but does impair the animal's ability to perform precise movements. For example, a bird without its cerebellum is able to flap its wings but it cannot fly. The medulla oblongata, the most posterior part of the brain, merges with the spinal cord. It develops from the entire myelencephalon. Centers regulating swallowing, the respiratory rate, and the heart rate are present in the medulla oblongata. In addition, nerve tracts connecting the spinal cord and the more anterior parts of the brain pass through the medulla oblongata.

The vertebrate peripheral nervous system has two divisions, the *somatic* and the *autonomic*. The somatic nervous system controls those muscles that are under voluntary control, whereas the autonomic nervous system controls structures such as glands, heart muscle, and the muscle in the walls of arteries and the intestine, structures that are generally not under voluntary control. The autonomic nervous system, sometimes called the involuntary nervous system, has two subdivisions, the *sympathetic* and *parasympathetic*. These subdivisions differ from each other both physically and functionally. One of the physical differences, for example, is that the parasympathetic nerves emerge from the brain and the sacral (most posterior) region of the spinal cord whereas the sympathetic nerves emerge from thoracic (chest) and lumbar (between the thoracic and sacral) portions of the spinal cord. Functionally, the sympathetic and parasympathetic nervous systems oppose each other. The sympathetic nervous system prepares an animal for emergency situations; the parasympathetic nervous system, in contrast, brings on a relaxed state. For example, the sympathetic nervous system accelerates the heartbeat whereas the parasympathetic slows the heartbeat.

ENDOCRINE SYSTEM Vertebrates, just as arthropods, have both epithelial (nonneural) endocrine glands and those derived from nervous tissue (Fig. 12-7). The *pituitary gland*, sometimes called the *hypophysis*, lies below, and is attached to, the diencephalic portion of the brain. This gland has two major portions, the *adenohypophysis* and the *neurohypophysis*, each having a different embryogenesis. The adenohypophysis develops from the epithelium of the roof of the oral cavity while the neurohypophysis grows down from the hypothalamic portion of the diencephalon. The neurohypophysis retains its connection with the brain throughout life and consists of two neurohemal centers, the *pars nervosa* and the *median eminence*, where axons of neurosecretory cells, whose cell bodies lie in (or at least very close to) the hypothalamus, terminate.

There is disagreement over whether the cell bodies of all the neurosecretory cells whose axons terminate in the median eminence lie

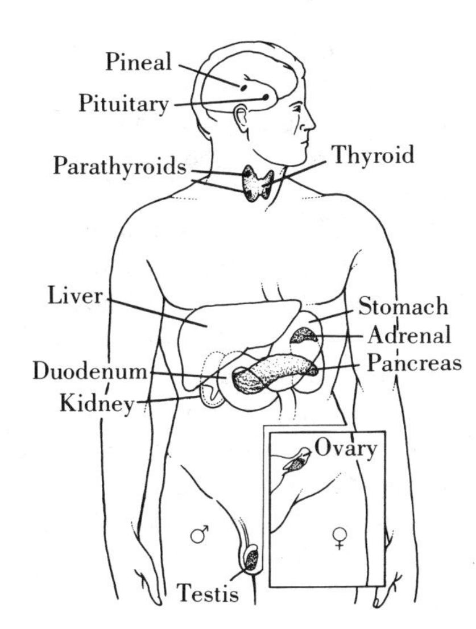

Fig. 12-7 *The locations of the endocrine glands in humans. (Modified from Villee,* Biology, *7th ed. W. B. Saunders Company, 1977.)*

in the hypothalamus. One such group of cell bodies lies in the preoptic area, which is regarded by a good number of investigators as part of the telencephalon, not of the hypothalamus. This division of opinion is due to the fact that the demarcation of the hypothalamus from the telencephalon is indistinct. As stated above, the hypothalamus is derived from the diencephalon. Even if the preoptic area is not anatomically a part of the hypothalamus, the preoptic area is certainly functionally very closely related to the hypothalamus. At least nine hormones that regulate the release of the hormones produced by the adenohypophysis are released from the median eminence. The existence of at least six hormones that stimulate release of adenohypophysial hormones has been demonstrated, the release of growth hormone, thyroid-stimulating hormone, adrenocorticotropic hormone, follicle-stimulating hormone, luteinizing hormone, prolactin, and intermedin being thus stimulated. The hormone that stimulates release of the luteinizing hormone appears also to cause release of the bulk of the follicle-stimulating hormone. But whether a hormone exists that stimulates only the release of follicle-stimulating hormone is not yet known. Furthermore, there are three hormones that inhibit the release of hormones from the adenohypophysis. Release of the growth hormone, thyroid-stimulating hormone, prolactin, and intermedin are thus inhibited, with the same hormone functioning as the release-inhibitor for both the growth hormone and the thyroid-stimulating hormone. Hence, there is a dual system of control by these neurosecretory neurons, stimulating and inhibitory, of the release of growth hormone, thyroid-stimulating hormone, prolactin, and intermedin from the adenohypophysis. All of these release-stimulating and release-inhibiting hormones, with the sole exception of the intermedin-release-inhibiting hormone, are released from the median eminence, entering the blood stream, which will then carry them to the adenohypophysis. However, the intermedin-release-inhibiting hormone appears to be released from axons that terminate in the pars intermedia itself, where, as stated above, the cells that produce the intermedin are located.

The adenohypophysis produces several hormones, some of which directly stimulate other endocrine glands. *Growth hormone* (or *somatotropin*) stimulates body growth. *Thyroid-stimulating hormone* (or *thyrotropin*) stimulates production and release of thyroid hormone. *Adrenocorticotropic hormone* promotes an increase in the size of the adrenal cortex and production and output of adrenal cortical hormones. *Follicle-stimulating hormone* promotes the growth and development of ovarian follicles in females, and the formation of sperm in males. In addition, in a female, *luteinizing hormone* acts synergistically with follicle-stimulating hormone to promote both production of estrogen (one type of female gonadal sex hormone) by maturing follicles, and also to promote ovulation. Also, in mammals alone, luteinizing hormone, without the aid of the follicle-stimulating hormone, stimulates both development of a *corpus luteum*, a glandular structure that develops at the site of a ruptured ovarian follicle, and secretion of another class of female sex hormone (progestogen) by the corpus luteum. In a male, luteinizing hormone causes formation of androgen, or male sex hormone, by the interstitial cells (*Leydig cells*) in the testis. Consequently, luteinizing hormone is also known as *interstitial cell-stimulating hormone*. The pituitary hormones that affect the gonads are called *gonadotropins*. The function of the hormones produced by the gonads will be described below.

Prolactin (or *lactogenic hormone*) is the most versatile of all the adenohypophysial hormones. Its actions among the several classes of vertebrates have been divided into five categories: actions related to reproduction and parental care, actions on the integument and its derivatives, control of salt and water balance, stimulation of growth, and the control of fat and carbohydrate metabolism. More specifically, among its diverse actions are (a) stimulation of feather growth in birds, (b) stimulation of milk production by the mammary glands of mammals, and (c) promotion of fat deposition in birds prior to their migration, the fat providing energy during the migratory flight. *Intermedin*, so named because it is produced by the *pars intermedia* (a subdivision of the adenohypophysis), is a hormone that acts on vertebrate chromatophores (but birds and mammals lack chromatophores). This hormone is also known as *melanophore-stimulating hormone* because its main action is to produce darkening of the integument by causing dispersion of the pigment in vertebrate *melanophores* (chromatophores with brown or black pigment granules). Melanophores are the predominant type of vertebrate chromatophore and the color changes of these animals are due primarily to their melanophores. Vertebrates do, however, also have red, yellow, and white chromatophores. Intermedin also disperses the pigment in these red and yellow chromatophores, but concentrates it in these white ones.

The pars nervosa is the site of storage and release of the *antidiuretic hormone* and *oxytocin*. The antidiuretic hormone acts on the kidneys of terrestrial vertebrates to reduce the volume of urine produced, thus conserving water. In frogs and toads, it also promotes the uptake of water through the skin, thereby enabling a dehydrated individual to rehydrate rapidly after reentering water. In mammals, oxytocin stimulates ejection of milk from the mammary glands of nursing females, oxytocin being released as a result of stimulation of the nipples by the sucking child. Oxytocin also promotes uterine contractibility to assist the ascent of sperm cells in the female reproductive tract after mating, and to expel the fetus at childbirth. The act of mating and uterine distention late in pregnancy appear to provide the stimuli that promote oxytocin release in these last two instances. All the pars nervosa hormones are nonapeptides, composed of nine amino acids. The amino acid composition of the antidiuretic hormone differs somewhat among the several vertebrate classes. The antidiuretic hormone of mammals is known specifically as *vasopressin*, that of birds and fishes is called *vasotocin*, that of reptiles is known as *mesotocin*, and frogs and toads possess both vasotocin and mesotocin.

The *pineal gland*, which develops from the roof of the diencephalon, appears to have roles in color changes and reproduction. This gland contains, among other substances, *melatonin*, which concentrates the pigment in the melanophores of the ammocoete larva of a Tasmanian lamprey, the embryo of an osteichthyan, and amphibian tadpoles. A more complicated situation occurs in the adult pencil fish, an osteichthyan that has a color pattern different at night from its daytime color pattern. Melatonin produces the night coloration, which involves concentration of the pigment in some melanophores and its dispersion in others. With respect to reproduction in mammals, the pineal gland has inhibitory effects on gonadal functions; for example, it inhibits ovulation. The pineal gland may have an antigonadal action in some of the other vertebrates also. The number and nature of the antigonadal hormones actually released from the mammalian pineal gland into the blood is still uncertain. Melatonin, which, as stated, is present in the pineal gland, was at one time the favored candidate as the mammalian pineal hormone, but more recent experiments showed that by itself it is at best not very effective as an antigonadal agent. However, polypeptides from mammalian pineal glands have been found to have antigonadal action. Perhaps melatonin and these polypeptides work synergistically to regulate the reproductive system. The pineal gland through its hormone or hormones appears to exert its effect on the gonads indirectly by inhibiting the release of gonadotropins from the pituitary gland. Interestingly, no release inhibitors of the follicle-stimulating hormone and the luteinizing hormone have been found to be produced by the hypothalamic neurosecretory cells (nor in the preoptic area if it is indeed not part of the hypothalamus)—only the release of these two hormones is known to be hormonally induced by a neurohormone from the hypothalamus (including the preoptic area). Perhaps the pineal gland functions as part of the total releasing hormone system by supplying the release inhibitors necessary for the fine control of the release of these gonadotropins.

The *thyroid gland* appears to be phylogenetically related to the endostyle of tunicates and amphioxus; both the thyroid gland and endostyle develop from the floor of the pharynx. Furthermore, during metamorphosis of an ammocoete larva of a lamprey, part of its endostyle becomes the thyroid gland of the adult. The thyroid gland releases two hormonal variants, *thyroxine* and *triiodothyronine.* They will be referred to simply as the *thyroid hormone.*

The thyroid hormone has a wide variety of actions among vertebrates that fall into two major categories, metabolic and growth-producing developmental. Despite the fact that prolactin actions are divided into five categories and those of the thyroid hormone into only two, the thyroid hormone appears to affect more organs, organ systems, and metabolic processes than does any other hormone. The following are but two examples of the diverse actions of the thyroid hormone. First, it consistently increases the metabolic rate, which is reflected in an increased oxygen consumption, of endotherms, but almost every investigator has failed to find such a stimulatory effect in ectothermic vertebrates. This difference appears to be a consequence of the endotherm adaptation for heat-generation (calorigenesis)—they increase their metabolic rate in order to regulate their body temperature. Excision of the thyroid gland of a rat, for example, produces a decrease in its oxygen consumption, but such surgery has no significant effect on that of the dogfish. However, administration of an extract of the thyroid gland of a dogfish to a rat will increase the rat's oxygen consumption. Second, thyroid hormone is essential for the metamorphosis of amphibian tadpoles, but it does not induce metamorphosis of an ammocoete. Precocious metamorphosis of amphibian tadpoles can be obtained by adding thyroid hormone to aquarium water; thyroidectomized amphibian tadpoles never metamorphose.

Parathyroid glands are found only in tetrapods. These glands lie close to, or are attached to or embedded in, the thyroid gland. *Parathyroid hormone* regulates the quantities of calcium and phosphate ions in the blood. The hormone causes a rise in blood calcium but a fall in blood phosphate. The quantities of these ions must be carefully regulated, because (a) calcium is essential for the maintenance of normal nerve and muscle excitability, the coupling of excitation to contraction in muscle, and normal cell membrane permeability, and (b) phosphate is important in carbohydrate metabolism, in biological energy transformations, as a constituent of nucleic acids, and in the regulation of the acidity of the blood. The *ultimobranchial glands,* which are found in all vertebrates, are the source of another hormone, *calcitonin,* that also regulates the amounts of calcium and phosphate ions in the blood. But calcitonin, unlike the parathyroid hormone produced, decreases in the levels of both calcium and phosphate. The ultimobranchial glands are so named because they are derived from the ultimobranchial bodies (pharyngeal derivatives) of the embryo. In mammals, these glands become incorporated into the thyroid gland, whereas in nonmammalian vertebrates they remain separate.

In all vertebrates except agnathans, microscopic clusters of cells, the *islets of Langerhans,* are the sources of two hormones, *insulin* and *glucagon,* which regulate carbohydrate metabolism. Agnathan islets secrete insulin alone. Insulin lowers the level of blood glucose; glucagon has the reverse effect. In agnathans, the islets are embedded in the wall of the intestine near the point where the bile duct joins the intestine, but remain separate from the clusters of enzyme-secreting, pancreaslike cells that are also embedded in the agnathan intestinal wall. But in jawed vertebrates (except for a few osteichthyans where the islet tissue forms a body completely separate from the enzyme-secreting pancreas), the islets are scattered throughout a compact pancreas. Glucagon may, however, be produced by agnathans also. There is some evidence that glucagon, although it is not found in the pancreas or islets of Langerhans of agnathans, is synthesized by some intestinal cells.

Androgens from the testes and estrogens from the ovaries are responsible for normal development of the *secondary sexual characteristics* and *accessory reproductive structures.* Secondary sexual characteristics are those characteristics that are associated with gonadal hormones but not directly involved with reproduction. These characteristics are most frequently external ones. Some examples of secondary sexual characteristics, applicable to both sexes in various vertebrates, are hair distribution, body shape, voice, plumage, and pigment patterns.

The accessory reproductive structures are the ducts, glands, and other organs, which, together with the gonads, form the reproductive system. The masculine accessory reproductive structures are those ducts involved in sperm transport, the glands whose secretions serve as a vehicle for the sperm, and the structures used to transport the sperm to the female. The copulatory appendages of male crustaceans are an example of such a structure in an invertebrate. The feminine accessory sexual structures include structures, such as the oviducts and uterus, that are necessary for the transport and development of the egg. The brood pouch of female crustaceans is another example of such a structure. The principal androgen of the human male is testosterone. Estradiol-17β is the most potent estrogen released from the human ovary. Progestogens, also produced by the ovary, supplement the action of the estrogens on the mammalian uterus, assuring that the uterine wall is in a condition suitable for implantation and development of the fertilized egg. Progesterone is the principal progestogen in the human female.

The *adrenal glands,* which lie near the kidneys, actually consist of two components that are genetically distinct in embryo, cortical cells and medullary cells. The cortical cells are epithelial endocrine cells, developing from mesoderm, whereas the medullary cells are modified neurons, derived from ectoderm. The name "adrenal gland" is derived from the fact that these two types of cells unite in most vertebrates to form discrete bodies. However, in some fishes, these two cell types remain separate from each other, although in others they are intermingled. In amphibians, these cells are always intermingled. The adrenal glands of reptiles and birds are more compact organs than are those of amphibians, and the two types of cells are again intermingled. But in mammals, although both types of cells form a compound structure, the cells do not intermingle. In mammals, the medullary cells form an inner core of the gland (the *medulla*), and the cortical cells form an outer *cortex* that wholly envelops the medulla. The active substances extracted from the adrenal cortical cells fall into two categories. The *mineralocorticoids,* such as aldosterone, have as their primary function the regulation of body electrolytes and water. Mineralocorticoids act on the kidneys to reduce excretion of sodium and water, but enhance potassium excretion. The main function of those in the second category, the *glucocorticoids,* an example of which is cortisol, is to stimulate the formation of glucose from proteins and fats, a process known as *gluconeogenesis.* Adrenocorticotropic hormone from the pituitary gland is clearly very effective in stimulating the secretion of the glucocorticoids, but has at most only a minor stimulating effect on the production of mineralocorticoids.

The hormones of the medullary cells are *epinephrine* and *norepinephrine.* These two medullary hormones have, in general, qualitatively similar physiological actions, but they show important quantitative differences in the degree of their effects. In general, epinephrine is primarily involved in metabolic adjustments (for example, maintaining a normal blood glucose level), whereas norepinephrine primarily promotes circulatory adjustments (for example, maintaining a normal blood pressure). Epinephrine is about four times as effective as norepinephrine in raising the blood glucose level, but norepinephrine is twice as effective as epinephrine in increasing the systolic blood pressure, that which is measured during ventricular contraction.

A group of *gastrointestinal hormones* is involved in coordinating the activities of the digestive system as ingested food passes along the digestive tract. Almost all of the research done on these hormones has been with mammals. The arrival of food in the stomach causes the release of *gastrin* from the stomach wall into the blood. Gastrin then acts back on the stomach, stimulating the secretion, by the stomach wall, of gastric juice containing the hydrochloric acid and the pepsin. The passage of the partially digested food from the stomach into the intestine then causes the release from the intestinal wall of two additional gastrointestinal hormones. One, *secretin*, causes the flow of a pancreatic juice that contains a large amount of bicarbonate ions, but secretin does not cause the release of any of the pancreatic digestive enzymes into this fluid. The bicarbonate ions neutralize the hydrochloric acid. Release of enzymes by the pancreas into its juice is caused by a second intestinal hormone, *cholecystokinin-pancreozymin*. It not only strongly stimulates release of the digestive enzymes produced by the pancreas, but also causes contraction and the resultant emptying of the gallbladder. Cholecystokinin and pancreozymin had earlier been considered to be distinct substances, but more recent experiments have shown that they are one and the same. Cholecystokinin was the name given when this hormone was thought to affect only the gallbladder, and pancreozymin was the name given to the substance thought to act only on the pancreas.

CHROMATOPHORES AND COLOR CHANGES

The ability to change color by redistribution of the pigment in chromatophores has been observed for numerous fishes, amphibians, and reptiles. Vertebrate chromatophores have fixed cell outlines, as do those of arthropods. At least for frog melanophores, microtubules seem to be the cell structures providing the motive force for aggregation of the pigment granules, whereas microfilaments, which are much narrower than the microtubules, appear somehow to provide the motive force of pigment-granule dispersion in these cells. Microfilaments are minute, solid, fibrous structures, approximately 4–6 nm in diameter, in the cytoplasm. How these microtubules and microfilaments apply a force to the pigment granules to cause them to move is not known, nor is it known how these melanophores control the direction in which the pigment granules are moved.

Among the vertebrates, there is considerable variation in the control of their chromatophores when these animals change color in response to a change in the color of the background. Agnathans, amphibians, some elasmobranchiomorphs, and some reptiles exhibit only a hormonal control of their chromatophores; their chromatophores are not directly innervated. Intermedin is the most important vertebrate color-change hormone. The rest of the elasmobranchiomorphs and reptiles, and all osteichthyans, exhibit either (a) cooperation between direct nervous and hormonal controls, or (b) a dominant, or perhaps even exclusive, nervous control of their chromatophores. For example, if the animals in category b use their intermedin at all it is only for the production of an extremely dark skin coloration during long exposure to a dark background. The chromatophores, if innervated, may be singly innervated (in such a case always a pigment-aggregating fiber) or doubly innervated by pigment-aggregating and pigment-dispersing fibers.

Some northern birds and mammals display seasonal changes in color: brown in summer, white in winter. The times of these changes in such species as the ptarmigan, ermine, and varying hare, are determined not by the environmental temperature but by the seasonal changes of the photoperiod, the number of hours of daylight in the 24-hour day. Experiments with the varying hare showed that it is brown when large amounts of gonadotropins are present in its blood, and white when they are low. The quantities of gonadotropins released from its pituitary gland are directly related to the length of the photoperiod. In winter, the photoperiod is short; consequently, its pituitary gland releases small quantities of the gonadotropins. These color changes can be induced out of season by experimental alteration of the photoperiod. Pituitary extracts containing gonadotropins will also cause white varying hares to shed their white hair and grow brown hair.

REPRODUCTION Vertebrates show a pronounced trend from class to class toward greater parental protection of the developing embryo and care of the newborn. Because of this trend, the chance of an individual's surviving until maturity, especially among the birds and mammals, is markedly enhanced. The number of eggs produced at any breeding period shows a downward trend as the chance of any one surviving increases; at any single breeding period, the average fish produces many more eggs than does the average mammal. As stated in Chapter 11, whereas most tunicates are monoecious, the rest of the chordates are dioecious.

After the evolution of the cleidoic egg, none of the reptiles had to return to water to lay their eggs. The cleidoic egg can be laid on land with minimal danger of desiccation, and the embryo inside its protective shell is safer from predators than eggs and larvae in an aquatic medium. The reptilian egg has a large volume of yolk to nourish the embryo during its development. Within this egg an *amnion* (fluid-filled sac) forms. The amnion encloses the embryo, the amniotic fluid protecting it against mechanical shocks. This type of egg was retained by birds and the lowest mammals, the egg-laying (oviparous) ones, such as the duck-billed platypus. The egg-laying mammals constitute one of the two subclasses of mammals, the Prototheria. The rest of the mammals represent the other subclass, the Theria. The therians eliminated the eggshell, becoming live-bearing (viviparous) instead of egg-laying. There are two infraclasses of therians, the marsupials and the placental mammals. An *infraclass* is the category of classification below the subclass, and above the superorder. Infraclass is one of the additional categories, beyond the major ones (see Chapter 1), that was created to assist in classifying large phyla. Marsupials, also called pouched mammals (kangaroos, for example), retain these shell-less eggs in a portion of the reproductive tract, the uterus, only a short period of time, the young being born in a very immature condition. These young then enter the mother's *marsupium* (abdominal pouch, hence the names "marsupials" and "pouched mammals" for these animals), attach firmly to a mammary nipple, and complete their development. Most species of mammals (humans, for example), however, have not only eliminated the shell, but in addition, retain the young in the uterus until development is complete, nourishing them through the umbilical cord and placenta, the placenta being attached to the uterine wall. Consequently, these mammals are called placental mammals. The embryo of a marsupial goes through early development enclosed within an amnion, whereas a placental mammal goes through its entire development enclosed in one. Because they form an amnion, reptiles, birds, and mammals are collectively known as *amniotes*.

PHYLOGENY The classes of vertebrates do not appear to have evolved one from the other along a single straight line (Fig. 12-8). Agnathans, the most primitive vertebrates, apparently gave rise directly not only to the elasmobranchiomorphs but also directly to the osteichthyans. Consequently, elasmobranchiomorphs would not be directly ancestral to the osteichthyans. The osteichthyans then appear to have given rise to the amphibians, the amphibians to the reptiles, and the reptiles to both the birds and the mammals. The agnathan ancestors of the elasmobranchiomorphs and osteichthyans were the now-extinct ostracoderms first mentioned in Chapter 1. Although neither present-day agnathans nor elasmobranchiomorphs have any bones, the fossil record reveals that the ostracoderms, and at least some of the ancient elasmobranchiomorphs, were partially covered by an armor of bony plates. (In some ostracoderms, the internal skeleton of the head was also bony.) The members of an extinct subclass of elasmobranchiomorphs, the placoderms, seem to have been not only the first jawed fishes but also the first elasmobranchiomorphs. Placoderms were fishes that had bony armor plates also (although their internal skeleton was apparently completely cartilaginous), and placoderms, with the loss of their bony armor, apparently gave rise to the present-day boneless elasmobranchiomorphs. The absence of bone in present-day agnathans and elasmobranchiomorphs appears, therefore, to be due to a process of reduction from an original bony condition and not a primitive characteristic. At one time, it was thought that osteichthyans had evolved from elasmobranchiomorphs which were completely devoid of

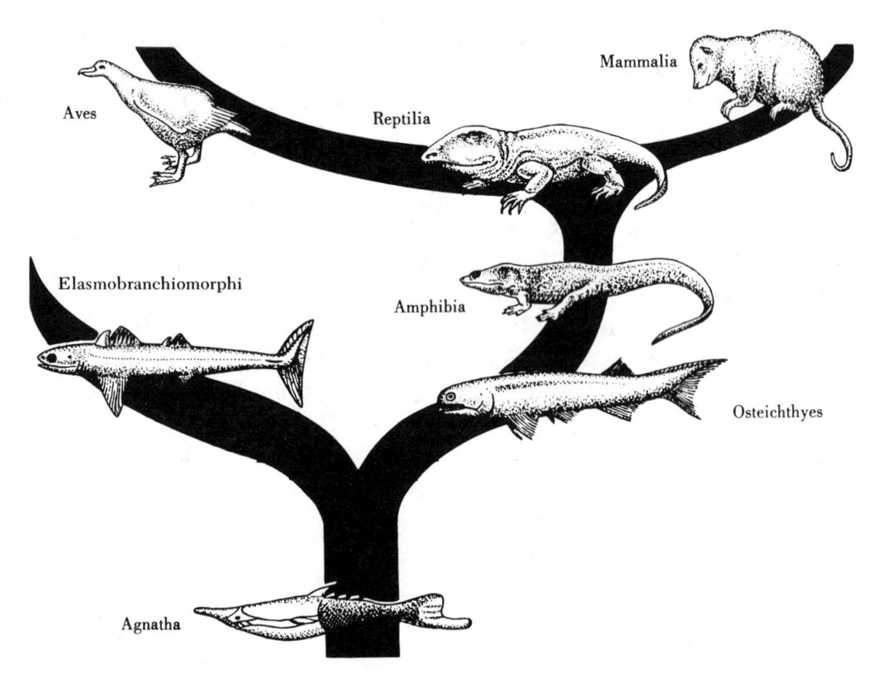

Fig. 12-8 *The phylogeny of the vertebrate classes.*

bone, and that bone was a new acquisition made by the early osteichthyans. It now appears, however, that osteichthyans merely retained and improved upon a bony skeleton they inherited from their ancient agnathan ancestors. According to the best interpretation of the fossil record, birds (recall *Archaeopteryx* from Chapter 1) evolved from the subclass of reptiles, the Archosauria. The dinosaurs were members of this subclass. Present-day crocodilians are the only living members of this subclass. Mammals, on the other hand, appear to have evolved from a different, and now extinct, reptilian subclass, the Synapsida.

The next chapter deals specifically with the primates, the order to which the human species, *Homo sapiens,* belongs. Much of the chapter will be devoted to a discussion of what appears to be the evolutionary pathway that led to modern man. The origin of the human species has long been a subject of great interest to humans.

FURTHER READING

Andrews, S. M., R. S. Miles, and A. D. Walker, eds., *Problems in Vertebrate Evolution.* New York: Academic Press, 1977.

Bakker, R. T., "Dinosaur Renaissance," *Scientific American,* vol. 232:4 (1975), p. 58.

Bellairs, A., d'A., and C. B. Cox, eds., *Morphology and Biology of Reptiles.* New York: Academic Press, 1977.

Buffetaut, E., "The Evolution of the Crocodilians," *Scientific American,* vol. 241:4 (1979), p. 130.

Davis, D. D., "Origin of the Mammalian Feeding Mechanism," *American Zoologist,* vol. 1 (1961), p. 229.

Fenn, W. A., "The Mechanism of Breathing," *Scientific American,* vol. 202:1 (1960), p. 138.

Gans, C., and K. A. Gans, eds. *Biology of the Reptilia, Vol. 8: Physiology B.* New York: Academic Press, 1978.

Hoar, W. S., and D. J. Randall, eds., *Fish Physiology, Vol. IV, The Nervous System, Circulation, and Respiration.* New York: Academic Press, 1970.

Hodgson, E. S., and R. F. Mathewson, eds., *Sensory Biology of Sharks, Skates, and Rays.* Arlington, Virginia: Office of Naval Research, 1978.

Marx, J. L., "Warm-Blooded Dinosaurs: Evidence Pro and Con," *Science,* vol. 199 (1978), p. 1424.

Osborn, J. W., "The Evolution of Dentitions," *American Scientist,* vol. 61 (1973), p. 548.

Ostrom, J. H., "Bird Flight: How Did It Begin?" *American Scientist,* vol. 67 (1979), p. 46.

Romer, A. S., "Major Steps in Vertebrate Evolution," *Science,* vol. 158 (1967), p. 1629.

Romer, A. S., and T. S. Parsons, *The Vertebrate Body,* 5th ed. Philadelphia: W. B. Saunders, 1977.

Schmalhausen, I. I., *The Origin of Terrestrial Vertebrates.* New York: Academic Press, 1968.

Schmidt-Nielsen, K., "How Birds Breathe," *Scientific American,* vol. 225:6 (1971), p. 72.

Tesch, F.-W., *The Eel: Biology and Management of Anguillid Eels.* New York: Halsted Press, 1977.

Young, J. Z., *The Life of Vertebrates,* 2nd ed. New York: Oxford University Press, 1962.

chapter **13**

Chordata Continued—Primates

As related in Chapter 1, mammals first appeared during the Triassic Period, but they did not become the dominant group of land animals until the Cenozoic Era—an elapsed time of more than 100 million years. The fossil record also reveals that long before the end of the Mesozoic Era, the ancestral mammals gave rise to two subclasses that we still have today, the egg-laying mammals (Prototheria) and the rest of the mammals (Theria). Later, but before the end of the Mesozoic Era, probably in the Cretaceous Period, the primitive therians gave rise to the two infraclasses still living today, the marsupials (Metatheria) and the placental mammals (Eutheria). Then, very early in the Cenozoic Era, according to the fossil record, out of the ancestral placental mammals came a line that

Fig. 13-1 *Selected primates. (A)* Hylobates lar *(gibbon)*; (B) Macaca mulatta *(rhesus monkey)*; *(C)* Pan troglodytes *(chimpanzee). (Courtesy of A. J. Riopelle, Louisiana State University, Baton Rouge.)*

ultimately gave rise to the order Primates. This order has great significance to us because we (*Homo sapiens*) are included in it. In addition to humans, this order also includes the apes, monkeys, and less familiar animals, such as the lemurs and tarsiers (Fig. 13-1). The nonprimate mammals that existed along the line that led from the earliest eutherians to the primates, and which were the immediate ancestors of the earliest primates, were most likely insect-eating, arboreal, shrewlike animals, with long tails.

THE EVOLUTION OF THE PRIMATES The very earliest (now extinct) primates, called early prosimians ("premonkeys"), were, like their apparent immediate ancestors, shrewlike in appearance, with a fairly long snout and a long tail. The early prosimians eventually gave rise to at least five suborders, four of which survive today.

The most primitive suborder consists of the modern prosimians, a group to which belong animals such as the lemurs and tarsiers. Modern prosimians are present on the island of Madagascar and in Africa and southeastern Asia. The snouts and tails of modern prosimians are still long. Also, as in their early prosimian ancestors, most of their digits have flattened nails instead of claws. Their eyes have moved well onto the face from the sides of the head, but are still directed outward more than in higher primates, whereas the eyes of a typical nonprimate mammal are on the sides of its head. In the rest of the primates, known collectively as the anthropoids, all the claws have been replaced by nails, and the snout of anthropoids, compared with that of the prosimians, has become reduced. A nail is simply a broadened, flattened claw. Only primates have nails.

New World monkeys (ceboids), found in Central and South America, constitute another suborder; Old World monkeys (cercopithecoids) form a third one. While present-day Old World monkeys are native only to Africa and Asia, in the past they were also spread widely through southern Europe. The monkeys and higher primates have forwardly directed eyes, an adaptation that imparts binocular, stereoscopic (three-dimensional) vision. New World monkeys descended from a group of primates that became isolated in South America during the Tertiary Period and evolved independently of the primates elsewhere in the world. The tail in most, though not all, New World monkeys is prehensile, being used like a fifth limb. In addition, New World monkeys have widely separated, forwardly directed nostrils that give the nose a flat, wide appearance, whereas the nostrils of Old World monkeys are closer together and point downward, and the nose is narrower. Old World monkeys have a tail also, but it is not prehensile.

The fourth suborder, the hominoids ("manlike" animals), consists of the apes (gibbon, orangutan, gorilla, and chimpanzee) and the family Hominidae or the hominids. The family Hominidae consists only of modern man, fossil man, and their humanlike ancestors. The hominoids lack an external tail and have larger brains than do the monkeys.

Exploitation of the arboreal environment was a very important factor in the shaping of primates. If the insect-eating ancestral stock that gave rise to the primates had not been arboreal, it is unlikely that the modern human body form would have evolved. Compared with the rest of the mammals, primates are a relatively unspecialized group of animals. Specializations of nonprimate mammals include the prehensile trunk of elephants, which was formed by elongation of the nose, allowing this heavy-bodied animal to reach food easily in trees and on the ground, and the elongated hoofed limbs of horses, which are a specialization for rapid running. In horses, the number of functional digits in each limb has become reduced to one, the third, which is enlarged, and its claw is thickened to form a hoof.

The early prosimians kept on with the arboreal way of life of their nonprimate, shrewlike ancestor and were, as a result, isolated from the terrestrial mammals. Because this isolation removed the primates from competition with ground-dwelling animals, the primates avoided the specializations they would have needed to compete successfully if they had not stayed in the trees. The evolutionary flexibility (that is, lack of specialization) among the early primates ultimately allowed this order of mammals to become the important order it is today. Life in the trees presented a set of problems entirely different from those of a grazing animal. Nails replaced claws in the new environment through the process of natural selection. Claws do not allow as firm a grasp of a thin branch as do flattened nails. The primate skeletal and muscular systems became modified for jumping, swinging, and grasping.

A notable change occurred in the hands, where the pentadactyl (five digits) plan so typical of tetrapods, though retained, became modified among the primates so that the thumbs were opposable. The significance of this modification cannot be overestimated, as it allowed the early primates to grasp and handle objects more effectively than can any other animals. Primates were then able to get a firmer grip on tree branches. A shift of the eyes from the sides to the front of the head also occurred. Eyes on the side of the head are satisfactory for a grazer on the lookout for a potential predator while feeding, but for a primate that leaps from branch to branch as a way of life, eyes directed forward offer a great advantage by the enhancement of depth perception, allowing better judgment of distances, which is so crucial for leaping about in trees. Excellent coordination between the eyes and the hands is a necessity for such leapers. Forward-directed eyes are also important in merely crawling hand over hand along a narrow branch and maintaining balance while doing so.

Through natural selection, vision ultimately became the dominant sense in primates, and as olfaction was reduced to a sense of lesser importance, the snout decreased in size. For most nonprimate mammals, smell is still the most important sense. The vision of a plains animal, for example, is often restricted by shrubs and tall grasses. In order to survive, it needs a well-developed sense of smell to detect predators. However, as primates evolved, they became less dependent on a sense of smell and more dependent on vision for survival. Concomitant with the evolution of an opposable thumb and the increased dependence on the eyes, there occurred a general enlargement of the brain, particularly of the areas involved in vision, muscle coordination, memory, and learning. However, the olfactory portions of the brain became reduced as the dependence on the sense of smell decreased.

Once an opposable-thumbed hand and forwardly directed eyes

evolved, natural selection would favor mutations that further increased brain size (and intelligence). That is, increased dependence on hands and eyes had an evolutionary effect on the brain—it got bigger. As the brain enlarged, the skull bulged upward to accommodate the brain. The increase in brain size produced more intelligent primates. Apparently, not only primate body shape, but also primate intelligence, was originally an adaptation to arboreal life. Life in the three-dimensional realm of the trees is more complex than on the two-dimensional ground. The need for precision timing, judging distances with little error, excellent coordination of muscular movements, and mental adroitness in order to survive in this three-dimensional world would have favored the evolution of a larger brain, thus the development of a higher level of intelligence. Precise coordination between the eyes and the limb movements was particularly necessary. Support for this concept that increased use and dependence on the hands and eyes fostered the evolution of more intelligent primates is the fact that a comparison of various species of primates has revealed a definite positive correlation between caliber of hand-eye coordination and level of intelligence. Although brain size is not always an accurate measure of intelligence, it is generally true that among species of the same size, those with larger brains are more intelligent than those with smaller brains.

THE EVOLUTION OF MAN

About 15 million years ago, a hominoid stock became divided into the apes and the hominids. After the hominids branched from the common hominoid stock, they descended to the ground from the trees, dwelt for some time on the floor of the forest, and ultimately (about 14 million years ago) left the forest to become plains dwellers. This descent from the trees and subsequent migration to the plains may have been stimulated by the thinning out of tropical rain forests which occurred at that time. The thinning of the tropical forests was due to a worldwide change in weather patterns, caused, in large measure, by the emergence of the Himalayas, which produced a barrier to global wind circulation. This new barrier helped produce cooler climates, some tropical regions became temperate, and tropical vegetation dried out. Furthermore, the fruits and nuts that had been available at all times of the year became seasonal. The change in the food supply provided a stimulus for the hominids to venture out of the forests and feed on the plains. The hominids, taking advantage of this opportunity, did indeed move out of the forests to feed on roots and meat, and became plains dwellers.

The apes remained ecologically tied to the forests. Although the gibbon and orangutan rarely descend to the ground, the chimpanzee often descends to the ground, and the gorilla is now essentially a ground-dwelling animal, limiting its arboreal life to the large lower limbs of trees. Apparently, it was not until after the hominids had descended from the trees and also become plains dwellers that they evolved into bipedal organisms, walking erect on two legs. They quite likely went through a "knuckle-walking" stage, walking the way chimpanzees do today, which is intermediate between quadrupedalism and bipedalism. The shift from forest-dwelling to life on the plains was probably the most significant factor in the evolution of bipedalism. Bipedalism would have been advantageous, enabling these hominids to move rapidly across the open plains. Certainly, also, the ability to stand erect and peer over the tall grass would have provided a selective advantage. Natural selection would have favored the evolution of bones and musculature that provided for bipedalism and erect posture. With the advent of bipedalism, the forelimbs were then free to be used in a variety of different ways, for carrying food, throwing a rock, and fashioning a tool from a stone, for example. Natural selection would have favored individuals that were more adept in performing these acts, thereby leading to further increases in the size and complexity of the hominid brain. Why these early hominids responded to the change in the food supply by moving out into the plains, whereas the apes remained ecologically tied to the forest, is not clear. Perhaps the apes were able to compete successfully against the hominids for what food there still was in the forests, forcing the hominids to look elsewhere for their nutrients.

When the hominids began to hunt prey on the plains, the hominid tooth pattern changed from that of apes. Compared with the teeth of apes, those of a hominid have reduced canines (the canines of apes are distinctly longer than the rest of their teeth), but the molars of hominids are larger in proportion to their incisors than are the molars of apes. The result of these changes is that hominid teeth are all about the same length. This transition from the tooth pattern of apes to that of hominids was probably also related to (a) the newly acquired bipedalism and the concomitant freeing of the hands from use in locomotion, allowing the hands to be used increasingly in such new activities as hunting and tool-making, and (b) the consequent decreased dependence on large canine teeth, in particular for capturing food and for defense. With the advent of bipedalism and tool-making, there would presumably no longer have been an adaptive advantage in retaining apelike teeth, and natural selection would not have favored their retention. The hominid tooth pattern is a better one for chewing the food of an omnivorous animal than is the tooth pattern of apes.

The transition from quadrupedalism to an efficient way of walking erect for long periods of time while maintaining proper balance and posture necessitated evolutionary changes in the shape and proportion of the hominid feet, legs, and pelvic bones, and also changes in the musculature of the legs and around the hip joints. The feet evolved into flat platforms, the legs became straighter and closer together, and a twisting and flattening of the two large pelvic flanges occurred, which not only resulted in the trunk being set more vertically on the legs, but also provided better leverage for the muscles of the buttocks that are used in walking. Modern man differs anatomically from the apes in several other ways also. The following are some of man's distinguishing characteristics: man's brain is absolutely larger, being approximately two and a half to three times bigger than a gorilla's; man is relatively hairless; the points of attachment of the skull to the vertebral column are moved forward, so the skull is better centered, hence better balanced on the vertebral column; the nose became more prominent, with an obvious bridge and tip; and the arms became shorter. It seems clear that the process of evolution to modern man was not only dependent upon our ancestors having had an arboreal existence, but it also required a subsequent descent to the ground.

THE FOSSIL RECORD Fossils that many investigators consider to be of the earliest hominid have been found in India, Kenya, Greece, Turkey, Hungary, and Pakistan. The material from Kenya is 12.5 to 14 million years old, whereas these fossils from the other sites are more recent, 8 to 12 million years old. It is the consensus of the investigators that these ancient fossil hominid specimens from all these countries are the remains of members of a single species, *Ramapithecus punjabicus*, but the Kenya material is somewhat more primitive than that from the other countries. As compared with apes and the modern human, *Ramapithecus* exhibits only an early stage of reduction of the canines; nevertheless, its teeth are clearly not those of an ape. The premolar and molar teeth of apes always form two parallel rows, whereas in hominids, they form a curve with the widest part at the rear. The teeth of *Ramapithecus* show this hominid pattern. *Ramapithecus* was most likely the hominid that first ventured from the forest out into the plains and became a plains dweller. All that has been found of *Ramapithecus* are jaws and teeth. Consequently, we have no information on brain size or stance. *Ramapithecus* apparently was on the line that culminated in modern man. *Ramapithecus* appears to have descended from the large ape *Dryopithecus* that first appeared about 20 million years ago. *Dryopithecus* seems actually—by 14 million years ago—to have given rise to three lines, (a) one that led to *Ramapithecus*, (b) one that led to the modern chimpanzee and gorilla, and (c) one that led to the ape *Gigantopithecus*, which became extinct about one million years ago.

According to the best estimates we have, about six million years ago *Ramapithecus* gave rise to at least one new species of hominid. According to some authorities, the *Ramapithecus* line actually split into two, or even three, hominid lines at that time, one of which led to modern *Homo sapiens*. Regardless of the number, a new hominid did evolve at that time and it was either *Australopithecus africanus* (Fig. 13-2) or *Australopithecus afarensis*. Africanus was first discovered in 1924, in South Africa. The oldest Africanus fossils that have been found are about 2.5 million years old. However, in 1973, the remains of even older bipedal hominids, 2.9 to 3.8 million years old, were found in East Africa. These specimens are clearly not *Ramapithecus*, but instead are the remains of an australopithecine. Some investigators have decided that these older specimens represent a new species, and it has been given the name *Australopithecus afarensis* after the Afar region of Ethiopia where most of these fossils were found; they have also hypothesized that Africanus was a descendant of Afarensis. On the other hand, many investigators hold the view that Afarensis is simply an early, more primitive Africanus. Although Africanus was first discovered in South Africa, the oldest Africanus fossils have been found in East Africa; no Africanus fossil has been found outside Africa.

Africanus had a wide, flat chin; the forehead was low and sloping, with prominent brow ridges. The jaws protruded more than those of the modern human, and Africanus had, at most, a very small chin. These characteristics gave the face a decidedly apelike look. Like that of an ape, the braincase was quite small compared with the protruding face. But, in spite of the apelike facial characteristics of Africanus, the skull and other bones do show obvious hominid characteristics. The teeth are even more like those of a modern human than are those of *Ramapithecus;* they are not those of an ape. The canines of Africanus are even further reduced in size than those of *Ramapithecus*, but the Africanus canines are still larger than those found in modern populations of *Homo sapiens*. Furthermore, while the premolars and molars of *Ramapithecus* do exhibit the hominid characteristic of diverging toward the rear (in contrast to the parallel condition in apes), these teeth diverge even more in Africanus. The structure of the pelvis and legs reveal that Africanus was bipedal and capable of standing but with a slightly bowed stance, being incapable of standing fully erect. Furthermore, the bone structure suggests that Africanus was a better runner than walker. Africanus probably walked with a shuffling gait, not being capable of the modern human's striding gait. Africanus males, being about 137 centimeters tall and weighing about 35 kilograms, were larger than the the females.

Before some very important fossils were discovered recently, the idea that Africanus was on the direct evolutionary line to the modern human had become generally accepted by investigators of human evolution. Furthermore, the earlier finds seemed to indicate that Africanus directly gave rise to another hominid, *Homo habilis*, about two million years ago, and that Habilis was also on the direct line to the modern human. The fossil remains of Habilis appeared to be essentially those of a larger, advanced (more humanlike) form of Africanus.

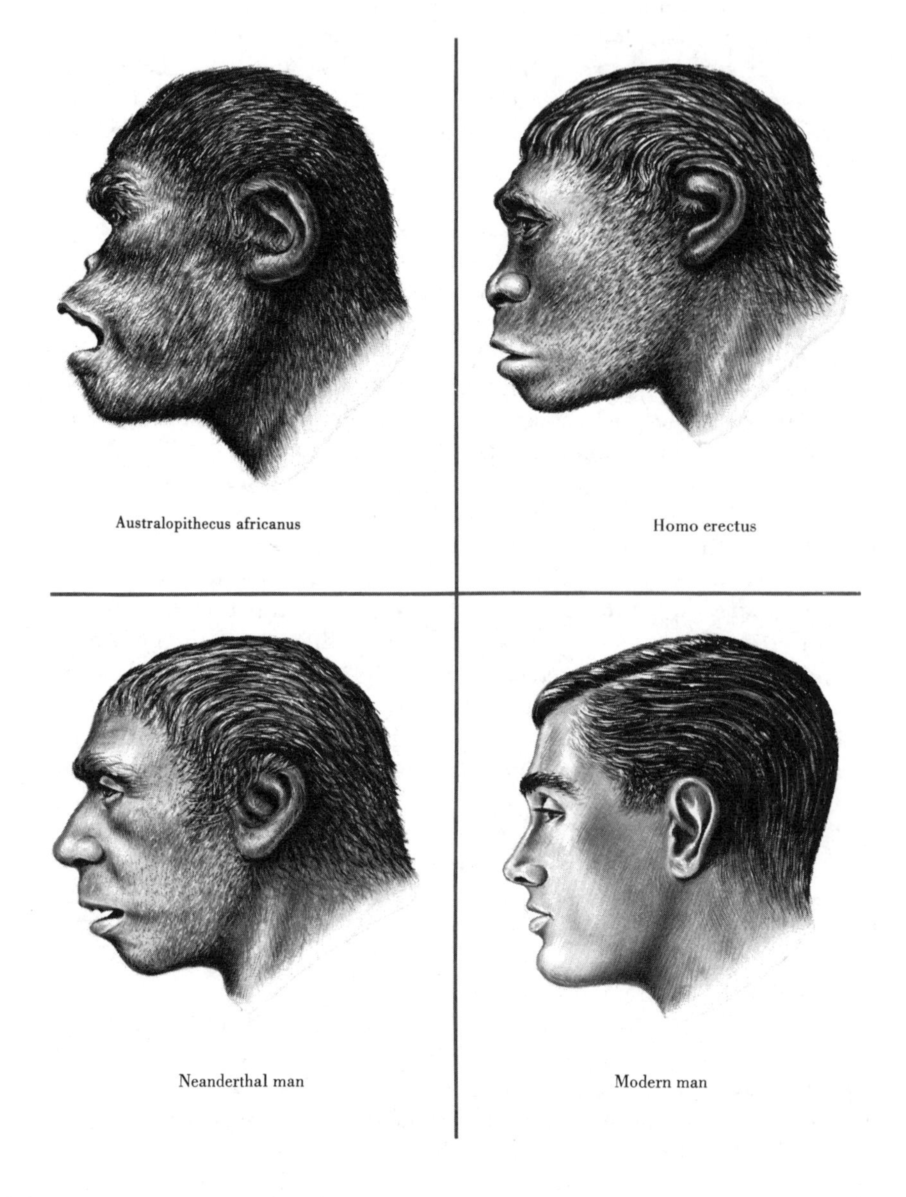

Fig. 13-2 Australopithecus africanus, Homo erectus, *Neanderthal man* (Homo sapiens neanderthalensis), *and modern man* (Homo sapiens sapiens).

Whether it merited being called a new species created some debate. Some investigators simply called it another *Australopithecus africanus*, some called it *Australopithecus habilis*, but it is now generally considered to belong truly in the genus *Homo*. Habilis teeth are more modern and the brain appears to have been larger than those of any australopithecine. No specimens of the less advanced and smaller Africanus (as compared with Habilis) younger than about two million years have been found, Africanus as such ceasing to exist about two million years ago.

The use of the term *Homo* reflects the fact that Habilis now is considered by most investigators to have been truly human. However, more recent finds of Habilis material from Kenya and Ethiopia show that he had been living in Africa longer than had previously been thought to be the case. These fossils, which were at least 2.6 million years old, surprisingly revealed that Habilis was contemporaneous with Africanus. In fact, the two lived in the same areas for some time, but no intermediate forms have been found. Consequently, it seems that they were not only genetically isolated from one another, but that Africanus was not on the line to the modern human, whereas it is generally agreed that Habilis was ancestral to *Homo sapiens*. In order to have both Africanus and Habilis living at the same time, with no interbreeding between them, either *Ramapithecus* stock would have had to split and give rise directly to Habilis on the one hand and Afarensis-Africanus on the other, or *Ramapithecus* could have given rise to Afarensis (if Afarensis was indeed not simply an early Africanus) but not to Habilis, and Afarensis would then have given rise directly to both Africanus and Habilis, with only Habilis leading directly to the modern human.

Habilis males were about 150 centimeters tall and weighed about 45 kilograms. The females were slightly smaller than the males. While definitely still apelike in appearance, Habilis does look less apelike and more like a modern human than does Africanus. Habilis had a larger chin, less prominent bony ridges over the eyes, smaller jaws with less forward thrust to them, longer legs, and a larger brain than Africanus. Africanus had a brain size of about 450 cubic centimeters, whereas the brain of Habilis averaged about 650 cubic centimeters, one having been found as large as 684 cubic centimeters. In comparison, the value is 410 cubic centimeters for the average chimpanzee, 510 for the average gorilla, and 1,400 for the average human.

Deciding whether a fossil represents the remains of a hominid that had not yet become truly human, and whether it should be placed in the genus *Homo*, often depends upon which criteria are applied. One authority set 700 cubic centimeters as the minimum brain volume for a true human, another set 750. Other authorities use tool-making as the criterion, a true human being one who not only used tools but made them, as opposed to a "man-ape" that simply used tools, that is, used suitably shaped natural objects as tools but did not modify them in any way. The decision is based upon the nature of the objects found at the site where the fossil was discovered. On the basis of brain size, then, Habilis would not meet the true-human criterion; but Habilis does meet the tool-making criterion. Habilis was definitely a toolmaker, fashioning tools from rocks by chipping off pieces to produce a cutting edge. Africanus was undoubtedly a tool-user, and may have made them also.

The remains of at least one additional species of australopithecine, named *Australopithecus robustus*, have been found in South Africa. Robustus was, however, an evolutionary dead end, becoming extinct about one million years ago. The oldest fossils that have been identified as Robustus are about two million years old. Robustus was heavier than Africanus. The bone structure suggests that Robustus walked fairly well, but not as well as did Africanus. The male Robustus weighed up to 68 kilograms and was 152 centimeters tall; the female Robustus weighed 36 kilograms and was 137 centimeters tall. The Robustus brain was about 500 cubic centimeters in volume. The jaws and teeth of Robustus were very large, less like those of a modern human than are those of Africanus. Robustus was apparently herbivorous. The molars resemble the grinding molars of herbivores more than they do the molars of the omnivorous modern human, which, as mentioned in Chapter 12, are intermediate between those of strict herbivores and strict carnivores. In addition, Robustus has a ridge of bone (skull crest) running from front to back on top of the skull similar to that in a gorilla. This ridge was presumably the site of attachment of large muscles needed for chewing vegetation. Herbivores require large chewing muscles because they have to break up the cell walls of plant cells in order to expose the cell contents to the action of digestive enzymes. (Such large chewing muscles are not necessary in meat-eaters because animal cells lack cell walls.) Africanus does not have the skull crest. Robustus was apparently too specialized to evolve further, and became extinct.

What is possibly still another species of australopithecine, which also appears to have been an evolutionary dead end, was originally given the name *Australopithecus boisei*. However, it now seems quite likely that Robustus and Boisei are actually representatives of one and the same species. The similarities between them are numerous, and the differences between them may simply be local population ones. Boisei was the largest australopithecine of all; it also had a skull crest and large grinding molars, and was, therefore, presumably also herbivorous as Robustus was. Boisei, like Robustus, also died out about a million years ago. It lived in East Africa. The knees of Boisei were proportionally farther apart than in other australopithecines, indicating that its walking ability was the poorest of all the australopithecines. Furthermore, the arm bones were proportionally larger, which suggests that Boisei also stood the least erect. Boisei males were up to 168 centimeters tall and 91 kilograms in weight; the females were about 15 to 25 percent shorter and weighed only about half as much. The brain of

Boisei had a volume of about 530 cubic centimeters. Boisei teeth 3.7 million years old have been found, showing that Boisei first evolved at least that long ago.

Those who hold the view that Robustus and Boisei are a single species have also suggested that Boisei-Robustus evolved directly from *Ramapithecus*, about six million years ago, the *Ramapithecus* line having split at that time to form this herbivore, Africanus (or Afarensis), and perhaps also Habilis. Another possibility that has been suggested is that Robustus and Boisei are indeed separate species, with only Boisei evolving from *Ramapithecus*, Robustus arising from Africanus. Regardless of the origins of Robustus and Boisei, both were evolutionary dead ends. There is an obvious need for more fossil evidence to fill in the gaps in our knowledge about the correct relationships of these early hominids to each other.

It is interesting that Habilis coexisted in Africa for many years with at least one species of *Australopithecus*. One wonders what transpired when a roving band of Habilis, the hominid from which the modern human was derived, met a band of these australopithecines. There is disagreement over whether these species of *Australopithecus* and Habilis were capable of any speech at all. Some experts hold the view that these early hominids were all capable of speech, while other authorities have concluded that speech was beyond the capacity of these hominids because of the small sizes of their brains. However, there seems to be general agreement that about 1.6 million years ago Habilis evolved into *Homo erectus*, and that Erectus was the immediate ancestor of *Homo sapiens*. There is no question that Erectus is deserving of being called *Homo*, a true human. Erectus made good tools and even built shelters. The brain volume of Erectus was about 1,000 cubic centimeters. About as tall as Boisei, the posture of Erectus was very similar to that of a modern human, hence the specific name. Erectus apparently walked erect in the same manner as modern man. Although Erectus still had a low, sloping forehead, a large jaw with only a slight chin, and prominent brow ridges, these characteristics were not as pronounced as in the immediate ancestor of Erectus. As the brain increased in size, the head became higher-domed and the forehead receded less than in Habilis. In addition, the jaw and teeth became even smaller as Erectus began to make use of fire (the first known user of fire) and cook food, and no longer needed a massive jaw to chew and tear raw meat. Erectus was probably capable of rudimentary speech. One authority has stated that from the neck down, the difference between Erectus and a modern human could be detected only by an experienced anatomist.

When the Java man, Peking man, and Heidelberg man were first discovered, each one was given its own name, *Pithecanthropus erectus*, *Sinanthropus pekinensis*, and *Homo heidelbergensis*, respectively. But now we know that they are really just different representatives of *Homo erectus*. Erectus has also now been found in Africa, Spain, France, and Hungary, having apparently spread widely; what differences there are among these fossils represent merely local variations in a single species of the sort one finds today among modern humans.

About 250,000 years ago, *Homo erectus* evolved into *Homo sapiens*. Skulls of this age have been found in Steinheim, Germany, and Swanscombe, England, and indeed appear to be those of very early *Homo sapiens*. These skulls are intermediate in form between those of *Homo erectus* and a full-scale Neanderthal man who evolved later, about 100,000 years ago, but the brain of the Steinheim-Swanscombe man was fully as large as that of a modern man today, showing that the Steinheim-Swanscombe man deserved being called *Homo sapiens*. Neanderthal man is now considered to have been a representative of *Homo sapiens* that evolved from the Steinheim-Swanscombe man. From Erectus through Steinheim-Swanscombe man to Neanderthal man, there was a steady progression with the brain case getting larger and the brow ridges smaller.

The question of where hominids first evolved is a tantalizing one, both India and Africa having been suggested. In addition to *Homo erectus* from Java, even older fossils (1.9 million years old) have been found there. It is not certain yet to what species these fossils belong, but at least one investigator considers them to be Habilis. In view of the presence of *Ramapithecus* and Erectus in Africa, Europe, and Asia, and Habilis possibly in Java, it has been suggested that hominids did not originate in Africa, but in India. However, in view of the facts that (a) the *Ramapithecus* fossils from Africa are older and appear to be somewhat more primitive than those from India, and (b) Africanus, Afarensis, Robustus, and Boisei have been found only in Africa, it seems more sensible to look upon Africa as the site of origin of the hominids, with *Ramapithecus* migrating from there.

The last of the archaic men was the Neanderthal man. His skeletal remains were first discovered in the Neander Valley located near Düsseldorf, Germany; they have since, however, been found elsewhere in Europe, and in Africa and Asia. The classic Neanderthal man first roamed the earth about 100,000 years ago, as stated above, and survived, according to current data, until about 40,000 years ago. Unfortunately, the carbon 14 dating system can only be used to measure time up to about 40,000 years ago, and the techniques that utilize other radioactive elements provide dating only from about 500,000 years ago back to the earth's beginning. Consequently, between 40,000 and 500,000 years ago, there is a dating gap, and in that gap is the time of the Steinheim-Swanscombe and Neanderthal men. There is, however, a new dating technique that gives promise of providing accurate dates for the fossils from those important years of hominid evolution. This new technique, which involves determination of the extent of amino acid *racemization* (realignment of the atoms in a molecule to produce the mirror-image form of the original molecule) in fossil bones, is based upon (a) the fact that all the amino acids constituting the proteins of all living organisms are "left-handed," and (b) our knowing that after organisms die, their amino acids slowly undergo racemization into "right-handed" amino acids, the rates of racemization having been determined for at least some of these molecules. According to at least one authority, this technique may well show that Neanderthal man did not last as long as it is now generally considered that he did. He may have actually disappeared about 10,000 years sooner; that is, about 50,000 years ago. Neanderthal man was similar enough to the modern human to justify being classified as a member of the species *Homo sapiens*. But he was given the subspecies name "*neanderthalensis*" to indicate that while he was a *Homo sapiens* he did show some differences from the modern human, who has been given the subspecies name "sapiens"; hence the modern human is *Homo sapiens sapiens*. The Neanderthal skull has a lower, flatter crown and bulges more at the back and sides than does that of the modern human. Neanderthal man also had a receding chin, and, compared with those of modern humans, still had prominent brow ridges. Neanderthal man was just over 1.5 meters tall. Neanderthal man has been badly maligned; he has been depicted as a dull-witted brute, when in reality his brain was fully modern in size. Neanderthal man made a greater variety of tools than did his forefathers and initiated the practice of burying his dead—sometimes along with weapons, food, and tools—as if believing in an afterlife. Flowers have also been found in Neanderthal graves. Perhaps the Neanderthals thought the flowers had medicinal properties, or the flowers may have been put there as an expression of grief, as we do. Neanderthal man was capable of speech; but measurements of Neanderthal skulls revealed that the shape of the air passages in these skulls was not yet modern, and the speech would have been consequently defective by modern standards. These air passages help generate speech by altering the simple tones produced by the larynx. Neanderthal man was apparently incapable of saying the consonants *g* and *k* or the vowel sounds in words such as *car*, *moo*, *deep*, and *bought*. Neanderthal man is the oldest man for which there is a complete skeleton.

Modern man (*Homo sapiens sapiens*) replaced the Neanderthal man. The first modern human is known as the Cro-Magnon man. He deserved to be called *Homo sapiens sapiens* as much as any person living today. "Cro-Magnon" is the name of a rock shelter in southwestern France where skeletons of this first modern human were found initially, but the name is now generally used to refer to all modern humans everywhere who lived from the time of modern man's emergence, about 35,000 to 40,000 years ago, until around 10,000 years ago, when humans turned to agriculture and metal-working (ending the Cro-Magnon period). Cro-Magnons were hunters who fashioned stone tools. They were anatomically completely modern. Their skeletons are well within the range of those found among Europeans today. Males averaged 173 centimeters in height, and were taller than the females. These modern humans invented pottery-making, built better homes, and made finer tools and clothing than did their predecessors. They had artistic skills, making statuettes of bone, ivory, and clay, but are most famous for their magnificent cave paintings, depicting not only a variety of animals such as the bison, horse, deer, and boar, but also what appear to be magicians or sorcerers. Cro-Magnon was the first ancestor to possess both the mental and physical capacities to talk as we do.

The fate of the Neanderthal man is not agreed upon. Clearly, he vanished about 40,000 years ago and was replaced by the fully modern Cro-Magnon. But the reason for this disappearance is disputed. There are a number of hypotheses. It has been proposed that Neanderthal man was an intermediate stage in the evolution of Cro-Magnon man from *Homo erectus*, Neanderthal man having evolved directly into Cro-Magnon man; that Neanderthal man hybridized with another stock of men, resulting ultimately in the disappearance of the full-scale Neanderthal man and the appearance of Cro-Magnon man; that another stock of men (either Cro-Magnon or a stock that would evolve into Cro-Magnon) exterminated Neanderthal man in warfare; that Neanderthal man became extinct because of some disease, with another stock of men evolving into Cro-Magnon man; and that Neanderthal man became extinct because he had diverged too far from the main evolutionary line running from Erectus to Steinheim-Swanscombe man to Cro-Magnon man to have evolved into Cro-Magnon man, having developed too many specialized structures and consequently having lost evolutionary flexibility, while some other stock of archaic men evolved into Cro-Magnon man.

The hypotheses dealing with the fate of Neanderthal man that require the existence of some other group of men either to hybridize with Neanderthal or to evolve on their own into Cro-Magnon (with Neanderthal man having been an evolutionary dead end) all suffer from the same basic weakness; namely, there is no evidence that any men other than Neanderthal man himself existed during Neanderthal times. The hypotheses that Neanderthal man was eradicated in warfare, that he became extinct because of a disease, and that he became extinct because he was too specialized imply that Neanderthal man was an evolutionary dead end that had split off from the main evolutionary line leading to Cro-Magnon man and was not in any way ancestral to modern man; that is, Neanderthal man neither evolved by himself into Cro-Magnon man nor hybridized with another stock of men to produce Cro-Magnon man. On the other hand, those who support the view that Neanderthal man alone was the immediate ancestor of Cro-Magnon man can point to (a) the fact that no fully Neanderthal fossil has been found that is more recent than 40,000 years old, (b) the fact that the oldest reliably dated (by carbon 14) Cro-Magnon man, from Czechoslovakia, is about 26,000 years old, although carbon 14-dated artifacts, such as bone spear-points, found in sites occupied by Cro-Magnon man and presumably made by him, indicate that Cro-Magnon man first appeared 35,000 to 40,000 years ago, and (c) the existence of fossils less than 40,000 years old (about 36,000 years old), which were found in a cave (Mugharet es-Skhūl) on Mount Carmel in Israel, that showed a blending of Neanderthal and modern traits. The Skhūl fossils are intermediate in time and form between the classic Neanderthal man and the earliest securely dated Cro-Magnon man; this is to be expected if Neanderthal man evolved without hybridizing with another stock of men into Cro-Magnon man by a steady progression of small changes. The term "Neanderthaloid" was coined for these intermediate forms, acknowledging that they were not full-scale Neanderthal men. Apparently, most authorities now hold the view that Neanderthal man was in the direct line of ancestry to *Homo sapiens sapiens* (Erectus to Steinheim-Swanscombe man to Neanderthal man to *Homo sapiens sapiens*).

We have seen that the modern human descended from the early prosimians, but nevertheless his brain now sets him apart from all other animals. Man alone, through use of his brain, is able to modify his environment for his advantage instead of being at its mercy, as are all other species. *Homo sapiens*, furthermore, is the only species that has the ability to transmit accumulated knowledge by means of articulate speech and writing.

FURTHER READING

Birdsell, J. B., *Human Evolution*, 2nd ed. Chicago: Rand McNally, 1975.

Campbell, B. G., *Human Evolution*, 2nd ed. Chicago: Aldine, 1974.

Constable, G., and the Editors of Time-Life Books, *The Neanderthals*. New York: Time-Life Books, 1973.

Edey, M. A., and the Editors of Time-Life Books, *The Missing Link*. New York: Time-Life Books, 1972.

Editors of Time-Life Books, *The First Men*. New York: Time-Life Books, 1973.

Howells, W. W., "*Homo erectus*," *Scientific American*, vol. 215:5 (1966), p. 46.

Johanson, D. C., and T. D. White, "A Systematic Assessment of Early African Hominids," *Science*, vol. 203 (1979), p. 321.

Leakey, M. D., "Footprints in the Ashes of Time," *National Geographic*, vol. 155 (1979), p. 446.

Leakey, R. E., "Hominids in Africa," *American Scientist*, vol. 64 (1976), p. 174.

Pfeiffer, J. E., *The Emergence of Man*, 2nd ed. New York: Harper and Row, 1972.

Pilbeam, D. R., *The Ascent of Man*. New York: Macmillan, 1972.

Poirier, F. E., *Fossil Man, An Evolutionary Journey*. St. Louis: C. V. Mosby, 1973.

Prideaux, T., and the Editors of Time-Life Books, *Cro-Magnon Man*. New York: Time-Life Books, 1973.

Robinson, J. T., *Early Hominid Posture and Locomotion*. Chicago: University of Chicago Press, 1972.

Simons, E. L., *Primate Evolution*. New York: Macmillan, 1972.

Simons, E. L., "*Ramapithecus*," *Scientific American*, vol. 236:5 (1977), p. 28.

Simpson, G. G., *This View of Life: The World of an Evolutionist*. New York: Harcourt, 1964.

Solecki, R. E., "Schanidar IV, a Neanderthal Flower Burial in Northern Iraq," *Science*, vol. 190 (1975), p. 880.

Tauxe, L., "A New Date for *Ramapithecus*," *Nature*, vol. 282 (1979), p. 399.

Washburn, S. L., "The Evolution of Man," *Scientific American*, vol. 239:3 (1978), p. 194.

White, T. D., "Evolutionary Implications of Pliocene Hominid Footprints," *Science*, vol. 208 (1980), p. 175.

Wood, B. A., *Human Evolution*. Somerset, New Jersey: Halsted, 1978.

Index